AF540236

WHEN GODAVARI COMES: PEOPLE'S HISTORY OF A RIVER

Journeys in the Zone of the Dispossessed

R. UmaMaheshwari

WHEN GODAVARI COMES: PEOPLE'S HISTORY OF A RIVER
Journeys in the Zone of the Dispossessed
R. Umamaheshwari

Published 2014

ISBN 978-93-5002-308-2 (Hb)

Published by
AAKAR BOOKS
28 E Pocket IV, Mayur Vihar Phase I, Delhi 110 091
Phone : 011 2279 5505 Telefax : 011 2279 5641
info@aakarbooks.com; www.aakarbooks.com

Printed at
Sapra Brothers, Delhi 110 092

Dedicated to the Memory of Ramani-Rajamani
(Amma-Appaji)

&

To the Love and Inextinguishable Spirit of Godavari's Marginalised People, Alienated from Democracy

Contents

Acknowledging

Strangely, I did little else than think, or write, about the issue this book deals with, from years 2006 (apart from, of course, complete a PhD, *not* on this subject). Unintentionally it has come to be a chronicle of my life's journeys in these many years.

And though the act of writing or circumstances for it may be individual or personal, this book happens due to the direct and indirect support of friends, acquaintances and institutions whose names I cannot leave out. But there may be people whose names aren't mentioned here or missed out but I acknowledge their contribution none the same.

Firstly, I thank the people of Godavari, some of whom I wish to name – Muttu Rama Rao and family, Illa Rami Reddi, Kondla Gangaraju, Madi Muttem, Kumari, Veerappa Reddi and family, Sitaramachandramurthy and family, Saikrishna and family, Mutchika Suramma (late), Posi Babu, boatman Punnam Mutyam, Mr. Radhakrishna, Raju, Varada, Saroja, Malladi Gangadharam and family, Karri Bhagyam, Malladi Posi, John Babu, Suri Babu, Raju, Gangireddi, Gangaraju, Padala Veerabhadra Rao and family, Ekka Rajanna Dora and family, Kosi Ramalakshmi (and others from M.Ravilanka, Chinabhimpally, Pedabhimpally), Pydipaka Peddaraju, Sitamma, Valairpadu Sarojinamma, and several others from East and West Godavari districts, Warangal and Khammam districts (who cannot all be named for want of space), for giving me their precious time, and many a times food, water and shelter, and their love and trusting. Without the always available space at Kondamodalu, Kathanapally, Devipatnam, Manturu, V.R. Puram, Polavaram or Rajahmundry my Godavari journeys would never have been what they became, an extended self-searching journey. I hope for them all to hold this book one day as their own and see in it their immense contribution in every page.

Thank You, V. Sai Shashank (Golu) for your wise words (from the time you were just ten, till date) that have kept me walking. Thank you, Siddharth (Mrithyunjaya) and Karthik (Dhananjaya) for the always cheerful conversations and other help. Thank you, Viji-Bala *athimber* and Balu-Lopa for moral support from a distance. For emotional bond and many kinds of help, not letting me give up nor give in, through the darkest of times, I thank my old friends (from the 'Godavari' hostel vintage in JNU), Susan George (and family: Josey John, Ruth Annamma, Kabir Jonathan, Geeta

Das, Mischa, Fidel and Anne and aunty), Tanuja Kothiyal (and Keshav, Arunima), Yasmin, Manjari (and family). And thanks, Kavita Datla and Rasna. I acknowledge with utmost sincerity the support of P. Sainath.

I have benefitted immensely from feedback of some people who read earlier drafts of this manuscript, those who gave constant moral encouragement and active support and few who tried to help in their own ways. These include Prof. Padmanabh Jaini (through his belief in this work as much as mine on the Tamil Jainas), Mr. Ramaswamy Iyer, Dr. Rohan D'Souza, Ramachandra Guha and Felix Padel and my old teacher, Prof. Shereen Ratnagar; thanks to Noam Chomsky for kind words on a small part of the manuscript during a particularly bleak moment of my almost giving it up.

From Hyderabad, I thank friends M.L. Narasimha Reddy, Radhika Desai-Tushar Vaidya, Devi, Jyothy, Udayalaxmi-Bhaskar, Anu-Sagar, Ramesh and Ramgopal-Prabha. Many thanks to Kasturi, Sagar (Zaheerabad), Usha and Shailaja (Machnur), C. Jayasri (DDS) and Kashinatha Rao (the senior-most city-guide of Hyderabad). Dr. Jagadish Kumar helped me hold my pen again, without his fees for months, an invaluable help. And I thank his family, Pooja, Siddharth and Agnivesh for their love. I thank Sangeetha for moral support when most needed. I thank Padmini Krishnamurthy and family and Mrs. Leela Natarajan and Prerna, and neighbours – Ambujavalli Seshadri, Tulja, Mr. Chaman Singh, Jaya-Tirupati Rao and P. Vatsala and Ms. Leela.

I thank Dr. Vidyasagar Reddy, Mr. Radhakrishna and staff of the Rajahmundry Head Post Office, advocates Mr. Nagaraju, Mr. Trinadha Rao, Mr. Rambabu (Laya), State Silviculturist Office, BSNL-Rajahmundry, Mr. Babji (APVVU), Andhra Pradesh Rythu Sangham, Prajashakti Book House, Agency Girijana Sangham, AP Rythu Coolie Sangham, Mr. Gandhi Babu and his staff at ASDS, Mr. Haribabu (Bhadrachalam), CPI (M) office and cadre (Bhadrachalam and Polavaram), Mr. Krishna Reddy (CPI-ML), Mr. Bose, and Vijayalakshmi at Sivananda Asramam, Bhadrachalam, staff of the Panchayat Office, Polavaram, Forest Officials of Kunavaram, V.R. Puram and Rampachodavaram, among others, and staff of M.S. Swaminathan Research Foundation (Kakinada) for all their help. Thanks also to the staff at Gautami Grandhalayam and IIC library, Rajahmundry.

The following scholarship and fellowships made these journeys possible for an independent journalist: 2006 - Prem Bhatia Memorial Trust (New Delhi), 2008 - National Foundation of India (NFI, New Delhi), 2010 - FES-Infochange (CCDS, Pune). The following academic fellowships allowed an independent academic to make sense of it from a distance: Nehru Trust (Delhi) and the Victoria & Albert Museum (London) for giving me the Jain Art Fund UK Visiting Fellowship (which also took me to Godavari's colonial past, by chance); School of Women's Studies, Jadavpur University, Kolkata for inviting me as SRTT National Visiting Fellow (besides offering space and care for both Malli and me) and Devika (Tiruvananthapuram) for suggesting it; IIT-Madras for giving me the short-term Post Doctoral Fellowship, which bound the journeys together. I also thank organisations and Trusts, PEACE (Delhi, late 2007), Prakriti Foundation (Chennai, 2011). I thank Mr. Anil Chowdhary and Mr. Ranvir Shah, respectively. Without Ranvir's help, I wouldn't have been able to stay in Chennai. I also thank V. R. Devika for the suggestion.

I thank also these media institutions and individuals who supported an independent journalist with either space for words or help in translation, etc: Prajashakti (Telugu daily, their staff in Hyderabad, Polavaram, Kovvur, Rajahmundry), Prajashakti Publications – Mr. P.S. Vijaya Rao, Ms. Sharavani, Ms. Satya, Balu, et. al.; Andhra Jyothy – Mr. Ramachandramurthy (former Editor) and Mr. Srinivas (Editor); The Hindu and Frontline Mr. N. Ram, Mr. Vijay Shankar; The Hindu Sunday Magazine – Ms. Nirmala Lakshman, Shalini Arun; The New Indian Express – Mr. Aditya Sinha (former Editor), and Mr. Sreevalsan (back in 2009-10); Mr. Rammanohar Reddy (EPW); Ashvin and Subbu Vincent (of indiatogether.org); Hutokshi Doctor (infochangeindia.org and Agenda). I thank also A.T. Jayanti (Deccan Chronicle, Hyderabad) who gave me my first break as a freelance journalist (and later as a reporter) more than a decade ago, helping me both fund my PhD and giving me a great learning experience at the reporting desk.

Among those who are no more (and would have been happy to see this) but deserve my sincere acknowledgment were the old Hyderabadis Kadambi Acharya, founder-proprietor of the erstwhile Kadambi Booksellers (with postcard invitation for tea at his bookshop often), and Ashvin photo studio V.C. Shanmuga Mudaliyar (marking envelopes of photographs with handwritten "very beautiful work"). Donating books or selling books on credit and printing photographs on credit happened often with these two generous souls.

I thank the following organisations for their indirect help: the InDG (i.e. India Development Gateway) team at C-DAC, Hyderabad, and Dr. Saratchandra Babu. I thank DDS, Medak district, for assignments and their trust, which helped in their own way to fund my journeys in 2006, 2007, and 2011.

I wish to specially mention some people in some places whose help in their own ways helped me, and thereby my work:

England (2008-09) – Peter Flugel (SOAS), Beth Mckillop and her family (including Alfy and Poppy) for their home and love; Rosemary Crill and colleagues at the V&A Museum, London for an enabling environment, Peter Bryant-Claire Mason and family, British Library staff for using the India Office collections and the manuscripts, Peggy (for opening a home and heart on a bleak winter), Sameera, and Soumya (a fellow resident at the Indian YMCA, London).

Karnataka, Bengaluru (2009) – Ananthu, Gopalan and everyone at Navadarsanam, for the space to eat and write in peace for free - some words happened here; Viji-Adarsh.

Kerala (2010-11) – Sheeba, Sudhakaran, Annie, Saju-Srija.

Kolkata (2011) - Director and staff at School of Women's Studies, Jadavpur University, Dr. Samita Sen, Joyanti Sen, Dr. Jasodhara-Amiya Bagchi, Jayeeta Bagchi, Suchetana Chattopadhyaya, residents of Srishti apartments and Manju – Surojit Dasgupta, Swapan Majumdar, Sudip Ganguly.

Chennai (2011-12) - Prof. Sudhir Chella Rajan, and staff and faculty of School of Humanities and Social Sciences, IIT-Madras, Mr. Mani, Suguna, Shashi, Bala, Jayakumar, Dr. Arti Kawlra, Dr. Rajesh, Anjana Raghavan and administration staff of IIT-M; Nivedita, Sudheesh, Banti, Anu, Teena, Gajendran, Bharathikannan and students of MA (Integrated) Development Studies, BTech, others for asking questions, helping me articulate. A special thanks to the Cholamandal Artist Association and the staff and workers at the Cholamandal Artist Village (Nandagopal-Kala, Gopinath and family, Haridasan-Padmini, Douglas, Nandan, Shanti, Jhansi, Smyla, Chandra, Ponnusamy, Narayanasamy), for not only welcoming but accepting in their midst a non-'artist' with her dog. I also thank Chandran (craftsman, wood-worker), Augustine-Stella, and Nasreen. Some very intense thinking about belonging, displacement, dislocation and alienation happened in this ambience. I also thank the staff at MIDS, Chennai. Thank you – Mary Stone (for a dramatic entry, many good wishes and connecting me to well-wishers in Hawaii), Dr. Brigitte Sebastia-Christian. Thanks to Michael Mccaughan for having listened to part of my draft.

Guwahati (2012-13) – Subeno; Dr. Bharati Baruah, Rimi Tadu (also in Hyderabad); Phulmani-Juti-Dipika-kakaideo; Balli uncle-Anita; Ranjita Das-Mrs. Das.

Others who supported with funds include Sajji Mani, Amman Madan (Kanpur), Gurumurthy Kasinathan (Bangalore), Narasimhan K (Hyderabad), Madhav Rao (Pune), Srinivas Mandalika (USA), Kiran Mutukuru (USA), Amar-Sujana, Udaya (Miryalguda), Viswesh, Vamsi, Vishnu and Arvind-Anita Singh.

It had to be at Indian Institute of Advanced Study, Shimla, for it to see light of the day, though my Fellowship project here does not pertain to the subject of this book. I gratefully acknowledge the support and cooperation of the IIAS institutional family under the stewardship of the Director, Prof. Chetan Singh, the entire Staff and the extended circle of Fellow- & Visiting Scholar-friends. I must, however, thank specially one or two people here whose everyday help meant a lot towards the last lap of this book – Uma Dutt ji, Ms. Chandrakala and Mr. Ramesh (at the Stores), Ram Singh ji (Carpenter) and Padam ji. Let me also express my gratitude to Albeena, Rajeev, Sunera-Richard. And to Nandita Gupta, each of whose help contributed in the final lap.

I wish to also thank Profs. Uma and Anand Chakravarthy for their deep interest in the nature of this work, support, encouragement and invaluable suggestions.

And thanks to Mr. K.K. Saxena and Aakar team for keeping their word. I also thank Mr. Manoj Kumar and Mr. Sudhir Vats at Arpit Printographers for handling this work with extreme patience and dedication.

And I am thankful, all the richer, and freer, for help and support of institutions one approached which did *not* come by!

Last but not the least – though scriptural language is not her forte – I want to place on record my deepest gratitude towards Malli's truly quiet and reassuring companionship through good and bad, all travels and circumstances, helping me/my words. And thanks to Ms. Satya Butt's team and their support system in Hyderabad (in an otherwise largely animal-unfriendly city) which made my many travels possible and I also thank some of the staff/workers (across categories/designations) of the Indian Railways, in Hyderabad, Kolkata, Guwahati and Chennai for help with making travel with Malli, when utterly called for, always possible at the end of it all.

Foreword

Felix Padel

This book represents 'the real story' of Polavaram, showing how, over at least ten years of agonised struggle, what started out as one of India's strongest social movements against a destructive dam got fragmented into a story of subverted, undermined resistance - a process that can only be properly understood through the myriad of convoluted, telling details presented here, and through an author allowing herself to become a medium for hundreds of diverse voices of affected people, the land and its forest, and the river herself - one of the world's greatest, soon to be stopped in her tracks. The resistance is not over yet though, and has seen a new lease of life in 2014. Thousands of lives have already been ruined in the scams and pressures surrounding land acquisition. What can be learnt?

This book requires a deep submergence into the ethnography of displacement. It seems that the battle to write it and see it through publication has paralleled the battle to stop Polavaram dam. Often, many of the 'reading classes' seem deaf to village people's voices. Having researched and written a lot in the area of displacement and 'R & R' (Resettlement and Rehabilitation), this book speaks to me in a way that is very rare. Most policy statements and literature are haunted by an appalling gap between written constructs and reality. Very little R & R literature even begins to evoke the full horror experienced by tribal, farming or fishing communities when they are displaced from their land and livelihood. Most end up adding insult to injury, and quantitative analysis often compounds this, since every counting process is subject to complex distortions. By contrast, this book is full of the real 'hard facts' - verbatim statements from people as they stare displacement in the face, and go through the dehumanising maze that constitutes the displacement process, where nothing is as it seems, and promises are systematically betrayed.

In 1984 I started a long bicycle journey on the banks of Godavari, crossing it on a remote ferry, seeing for myself the slow-paced tribal world that embraced the author, before it started to unravel before her perceptive eyes, through the series of journeys presented here in the form of a diary. The human interest moves through many types of households and characters that welcome her, with the warmest hospitality in the remotest tribal villages.

Estimates vary of course, but Polavaram is likely to exceed Sardar Sarovar dam on the Narmada, and every other dam project in India, in the sheer scale of displacement, approaching 300 villages submerged and 300,000 people displaced. A majority are Adivasis of the Konda Reddi and Koya communities, and their stories and voices are evocatively captured here: '...here are some of the few communities who practise agriculture in synchrony with the changing rhythms of the forest; and these may well be the last few pockets left of this kind of forest-agrarian economy that will soon be wiped out.' This Adivasi economy is dismissed by supporters of the dam and most government officials as 'primitive', when it is based in highly developed skills that are being devalued and destroyed. 'The tribal communities are being delinked from lived histories'.

In the view of a revenue official promoting resettlement, 'Tribal people are mostly encroaching on government land. We are giving them pattas... In a way the Polavaram is a great opportunity for tribal families affected by flood on a permanent basis.' This is not how any of the tribal people interviewed perceive their situation! 'The officials here are telling lies. Godavari comes once a year and goes. We've learnt to live with it. It is normal for us to climb to higher reaches of our hills when that happens. But our homes and our lives were never destroyed totally.'

During 2006 floods, 'Government was using Godavari in spate as a bait to convince or coerce people to give their lands for the project while resistance was relatively strong.' In fact, forms were printed out classing people as 'Godavari-affected', rather than Polavaram-affected, and people were filling these in for compensation without realising this was officially construed as consent for displacement! One Sub-Deputy Collector admitted to the author that 'We are not using "Polavaram". We have to acquire the lands by any means....' In other words, floods were used to force people to sign away their lands.

As with other dams and rivers, evidence on Godavari floods suggests that – far from 'flood control' – big dams built upstream have made the impact of floods much more chaotic and destructive than Godavari's natural floods, which communities knew how to deal with. Moreover, the author finds 'microcredit sharks' swarming to flood-affected villages even at their worst moments, on debt-enforcement missions!

This text presents, almost as an ethnography, the extraordinary range of techniques through which people's resistance – in areas outstanding for a history of resistance to unjust laws, from the Rampa rebellion of 1879, and the *tirugubatu* (or *'fituri'* of the colonial records) of 1920s, to the Telangana Armed Struggle of the 1940s-50s, the Naxalite uprising of 1969, and since - has been undermined, and people forced from a position of complete opposition to parting with their lands for the dam, into dejected acquiescence in a situation where they know their lives and communities are facing complete annihilation.

Arrests, 'false cases', and physical oppression form one range of techniques. Many kinds of pressures and funding inducements have also converted political parties, and NGOs, from a position of opposition into grudging defeat and even, sometimes, collusion. As one young KondaReddi woman, who had managed to reach the position of *sarpanch*, said: 'What's the point of me as a sarpanch if I'm not consulted by the officials? In our gram sabha last month (January 2007) we had resolved we do not want Polavaram... The Mandala Praja Parishad members too passed a resolution in five mandals that people here do not want Polavaram.' Yet, at the same time, District Magistrates claim they have convinced people in these areas! The lack of proper gram sabha consent is one of the most notorious features of Polavaram.

Another technique – reported from Singur and many other places to cow down farming communities' spirit of resistance - involves dumping waste material onto fertile lands, forcing farmers to move by making their land unfit for cultivation. 'Packages' of compensation involve an extraordinary range of corrupt practices – 'illegalities within illegalities in the Polavaram compensation scam'. 'Land for land' has been promised yet rarely realised. By far the largest monetary compensation inevitably end up coming to big landlords, some of whom have made fortunes by parting with lands that were theirs on paper, even when they were known to be alienated tribal lands, and the actual agriculturalists – Adivasi or Dalit – receive no proper compensation at all.

In the words of a Dalit labourer: 'You landlords will get your land. What about us? Where will we go? Drown in Godavari? What do we eat there? These packages will kill people like us, who have no land anyway. Will your government be happy then? Maybe Polavaram project will be happy if small workers like us die. We just have to cry holding onto these trees maybe.'

While Adivasis are losing their communities and whole way of life, Dalits in some ways are even worse off, since they have few rights in Scheduled Areas, and without land, don't get compensation. The same is true of fishing communities, both in the mangrove forests along the mouth of the Godavari and along its length, and their coverage in this book is significant, since officially most are not even categorised as 'project-affected'. These fishermen have already witnessed a massive depletion of fish stocks, due to industrial pollution and the proliferation of tourist boats as well as mismanagement of dams upstream, which have caused the disappearance of *hilsa* and several other species already.

The tourist industry, and its unthinking pollution of the river, works alongside the Telugu film industry, for which – in contrast to the sensitive 1960s film *Muga Manasulu* (Silent Hearts) – the river and mountains have become just a backdrop offering pretty scenery for characters delinked in every meaningful way from the river's ecology and ancient cultural history.

The role of media in portraying the Polavaram issue – often failing to do this in a way that reflects rural people's voices – is brought out vividly in this book; as also the role of politicians from all the main parties, including tribal politicians, some of whom appear as traitors to their communities.

For many of these, Polavaram is an obvious benefit – a 'money-spinner', and symbol of 'development', since the Godavari waters are 'wasting into the sea' instead of being properly 'utilised'. For local people by contrast, whose communities have always lived along the river, it appears obvious that interfering with the river's flow to the sea is like playing God, destroying a life-giving force that 'educated', city people simply do not understand, for whom 'talk of nature is disregarded as romantic fantasy'.

Tribal people are appalled that large tracts of land they have always lived from and safeguarded are being taken over by the Papikondalu National Park, including large areas due for submergence. 'Every tree, every stone here is akin to our gods… We don't eat a mango or other fruits without first offering worship to them.'

As senior members of the Forest Department talk about the coming deluge, 'Animals will just flee when the dam waters come down, and the birds will fly away' - as if they do not understand the enormity of what will die. 'Are foresters trained so as not to feel pain for these losses?' the author asks. A tribal political activist is quoted attacking the environmentalist discourse that ignores *girijan* symbiosis with wildlife: 'The government speaks of protecting elephants and tigers but not of tribal people in whose habitat these animals survived for centuries…'

In dams, as in mining projects, destruction of ecosystems and of ancient communities happens as one, interlinked at every level. This is clear when you look at deforestation – whatever forest is drowned directly by a big dam such as Polavaram, or Upper Indravati, due north of the Polavaram submergence zone in southwest Odisha, where I have seen this myself, is compounded by the inevitable deforestation by communities whose lands and livelihoods were drowned by the reservoir. With no proper compensation or resettlement, in the words of an old man I met at dawn, about to carry a bundle of wood to the market town a dozen kilometres away, 'We have to cut the trees to survive. If I cut trees and sell the wood, I can feed my family today, otherwise, they starve.'

The ideology of development is the real fantasy, when it is driven by unfulfillable dreams of a better life, and fuelled by money. 'Development-Induced Displacement' is a misnomer, since for thousands of people – several hundred thousand people in Polavaram's case - what is displacing them cannot be called development, since for them, it represents the antithesis of real development, devastating their highly evolved ways of life along with the complex ecosystem they adapted to over countless generations. What fuels the process is money – financial investment aimed at increasing profits for certain industries and those running them. So the proper term for what is happening to thousands of families is 'Investment-Induced Displacement'.

And why is so much money flowing in for the Polavaram project? Obviously, in the interests of construction companies, electricity companies, and a variety of industries. The Special Economic Zones in coastal Andhra near Kakinada are seeing a vast array of new factories, power stations, oil extraction, planned aluminium factories and much more. These offer jobs – but for how long? On what long-term basis and security? What is the arithmetic saying that hundreds of thousands of traditional livelihoods being sacrificed in the submergence zone are of no value compared to the new jobs coming up in the SEZs? Are those jobs likely to continue for the next thousand years? The traditional lifestyles being lost along a free-flowing Godavari offered a completely different quality of security.

The faults of big dams have been demonstrated in any number of studies – the hydro-power comes at far too heavy a cost in terms of biodiversity as well as human lives, sedimentation invariably depletes capacity faster than planned, while financial costs are far higher, flood control often counterproductive, and irrigation benefits often a lot less than the fertile lands drowned. But the erstwhile state of Andhra Pradesh (prior to creation of Telangana State) was selling an image of itself as 'Good for Business', offering major financial benefits to numerous companies, politicians, financial institutions and their associates, with the fantasy of future benefits floated for thousands of others. This fantasy and lure of profits drive a process termed 'development' that involves devastation for the ecosystems that life depends on, jeopardizing the well-being of future generations. But the people who see this most clearly are those with least access and subjection to the media.

As with Narmada Bachao Andolan and many other movements, the failure to prevent the devastation is immensely dispiriting to witness. Yet the author confronts this head-on, finding silver linings in the spirit of resistance, human warmth and detail, memories, and analysis - not least of the continuity in the colonial role of controlling a river to bestow its benefits on the people, from near deification of Arthur Cotton with his mandate that 'The river must be restrained from wandering', to eulogies of YSR and other politicians – forgetting thousands of lives ruined by uprooting them from their lands, and thousands of scams and frauds in the race for compensation.

This book will make an enduring contribution to debates about land acquisition, displacement, and people's movements; to analysis of the diverse power structure of politicians, revenue officials, political parties,

NGOs and media involved in promoting or opposing 'development projects'; and in its moving ethnography of Godavari and her people, including the livelihood and fate of fishing communities who have a long symbiosis with India's rivers, raising important questions about who has what rights to which commons. Also, the book traces a continuity between past and present. The ideology promoted by the British colonial regime of 'taming the waters' to put them to profitable use has been reproduced repeatedly by a nationalist ideology of 'growth' based on hydropower and extractive industries. In Godavari's case, this continuity is particularly clear, as is the shockingly callous neglect in populist pro-dam discourse of the river's indigenous farmers, hunters and fishing and ferry people. The voices of ordinary people bring Godavari to life in these pages, giving a view from below that breathes inspiration for a multitude of movements.

Felix Padel is sociologist and anthropologist trained in Oxford University and Delhi School of Economics. Teaching in various institutions in India since 2010, and a Research Associate at the Centre for Environmental History, Sussex University. He is author of *Sacrificing People: Invasions of a Tribal Landscape* (1995/2010), *Out of This Earth: East India Adivasis and the Aluminium Cartel* (with Samarendra Das, 2010), and *Ecology, Economy: Quest for a Socially Informed Connection* (with Ajay Dandekar and Jeemol Unni, 2013).

Gift of a Metaphor and a Context for the Godavari Journeys

Where I met Godavari (Guttala Gangaram)

Muttu Rama Rao and Surakka

I met Godavari, for the first time in year 2004, at Guttala Gangaram village of Warangal district.[1] It was by the riverside thatched-roof hut of a lone fisherman Muttu Rama Rao and his wife Surakka that I met Godavari, the person – the river, the metaphor of life. In their statement, "*Godavari occhinappudu*"; "*Godavari oste, ooru vellipotaamu*"[2], I was gifted a metaphor for a lifetime: she comes.[3] And ever since, I began to see the river that does not 'flood'. I stopped seeing the river as a geophysical entity. In my subsequent meeting (twice thereafter) with Rama Rao's family, on another hot afternoon (15th June it was, year 2005)[4] I got more intimate with Godavari as I tasted her for the first time. She virtually filled my senses in the sugared river water Surakka had offered, to quench my thirst. Godavari was the person, living amidst the most marginalised, on-the-fringes people. I wasn't to realise then that I would travel along her course, two years thereafter, and these many years. I felt this Godavari was safe and secure in the midst of those who perceived her as one of them; a lifelong presence.

I had known of a physio-geographic Godavari[5] from my school text books, flowing past five Indian states, until she reached the sea at the south eastern coast of India. Later I learnt of the political Godavari and the dispute over river-sharing and that she had a new Indian State to cruise through, created well after independence – Chhattisgarh, making the political entities she passed through, six in number. And now, there will be two more new States of Telangana and Andhra Pradesh to cruise through. Gradually, through a long period of stay in Hyderabad, I learnt of an ideological and politico-regional Godavari, the problematic of inter-regional distribution of Godavari waters. All these fragments and fractured senses of one long river, in her upstream and downstream journeys, had dissolved in one glass of Godavari water at the Rama Rao home. The metaphor of 'coming' was to recur, subsequently, over several of my journeys that were to follow (unknown to me at that point, as to how life would pan out) with, over, along, beside, into Godavari which was otherwise a river which, in techno-political or even economic jargon, has been "wasting into the sea", and must be "utilised" to the core. People gave me a new set of lens (apart from my other, old, Konica T24 Reflex SLR camera[6]) and heart to understand 'flooding' and a river, outside of Geography lessons and political debates. So to Godavari, a life source, Muttu Rama Rao and his family and the entire communities of the former East, West Godavari and Khammam districts (of my travels), this book owes its being. It is my attempt to record the people's lives and ideas of the river and its fate that will always coincide with theirs. People's understanding is simple yet profound, like the depth of the river Godavari; a truthful evoking of a larger life-system principle of permanent flows. It reflects the meaning of life and its interconnectedness. The state's perception is as shallow as surface water, with convoluted mathematics created over it to generate uses that will make the river a paying proposition. It is the idea of ownership and possession, and like all things in life, the moment an idea of possession comes in, it brings with it conflict and power – like the ideas of 'wasting into sea' of that which could have been possessed, controlled and manipulated to stop on command and give dividends on expenditure when released as a ration. The language of policy makers, state and technocrats has been one signifying death (for me), in terms such as "dead storage", of cusecs and tmc, of shares and portions. It looks at consumption in commodity terms. For people there is no "dead storage", the river flows its course; Godavari must meet the sea. There is no consumption, but *use* of a river, not as a one-term mammoth exploitation, but over centuries of interface and for sustaining lives and basic livelihoods. In popular discourse, Godavari does not flood. The Telugu language has the term '*varadalu*' for it, but these communities call it '*ostundi*' - 'she comes'. They treat it as a natural event or even a non-event, in the sense that for them it is a natural phenomenon, with its relevance to life. References to past coming ('flooding') are typically prefixed the year of that coming, thus, '*enabhai-aaru Godavari*' for the 1986 coming in spate. In mainstream history, the river has had massive 'flooding' ever so often. In the people's history of the river, she 'comes' and 'leaves'. *Yet,* these categories are not half as simplistic or romantic in a simplistic sense, as life itself is not, for most of the tribal, dalit, and other marginalised groups of East and West Godavari and Khammam districts, which is the territory I have focussed on in this book. And, in the present context, many of the people I refer to are the ones who have lost, or are losing or will lose lands, forests, livelihoods in the wake of the Polavaram dam, which may turn out to be one large event in the course of their many historic life-negotiations in the past. For people like Appayamma, Subbanna, Peddaraju, Gangadharam and others (and Muttu Rama Rao, too, in other ways[7]) fluctuations in the fortunes of Godavari means fluctuations in their lives and livelihoods (and Godavari's fortunes can fluctuate depending on the course of politics of Andhra Pradesh) but it is in serious question as to how far have their aspirations, and their relationship with Godavari as a life-source seen reflection in political discourse in Andhra Pradesh, since its formation in 1956 with a conjoined contentious battle over Telangana statehood, ever since, now culminated in creation of Telangana and residual Andhra Pradesh. And it is on this river that a major global game - of exchanging people's lived histories for a single-settlement *packaji* (Telugu for R&R 'package' - compensation) is being played in the format of a National Project, called Indira Sagar Polavaram Dam.

People's history of a river sounds like an ambitious title for a work of this nature. An explanation for the choice of this title is called for. It is not a history in the sense of a linear, chronological sequence of events with a beginning, a 'middle' period gone by, to be analysed. History is not seen in terms of a linear archival past, but as a present fast moving into the past or a 'past' in the process of being constructed, each day, as their Godavari is set to be 'tamed' yet again and imprisoned by those who have never seen nor will ever see water or river as anything more than an economic 'resource'. People's histories are how they live and fight their battles to protect their homes by the river that both connects and disconnects them from the larger politics of gigantism and monopoly. Crossing over happens almost like an everyday sequence sometimes, and there are also lines drawn of community histories in the ripples of water every time a fisherman's boat is suddenly shaken up by the rushing past of a tourist launch, unmindful of how each clan of the fisher communities, for instance, has its historic fishing zone on a single stretch of Godavari, which they respect and never trespass. Global economy, tourism, industrial greed trespass many of these lines several times over as a divinely ordained right they have assumed. Godavari's history is made up of histories of these people. It takes several crossing-overs to fathom the depth of these lines and ripples. Meanwhile, I am also one of the people there, a part of the developments, even if it was in the traditional sense of a journalist, historian, or simply a woman, recording certain 'truths' of the moment that needed to be told, of people who were to lose their all. I am not removed from the people in that I do not see them from a higher vantage point, unaffected by what is going on; and was unable to be that thoroughly 'professional' media person. People's history, in a larger sense cannot help but be my history, too, and my sense of historically locating the events I witnessed as much as locating myself. History here refers to lives lived in collective contexts of caste, tribe or any other social form under different kinds of political forms with Godavari as the perennial witness. These stories tell of a life closely linked to Godavari, where every metaphor, analogy, idiom has Godavari and her ebb and flow in it. Where memories are about the last time Godavari came. Or, of the before and after of it; being the chronology of a life-altering event, or an event even not *that* life altering, really. Hence, people's history of a river: in its control and dominance, their histories altered, recreated, dislocated or constructed anew. Both the river and the marginal people are, or seem to be, merely 'incidental' to the larger construct of a state or some would say 'nation'. In making policies marginal people are told it is for the larger good ("public good", unknown to them who figures in this vague 'public' category) benefits of which will trickle down to them at some point or the other. The river, too, awaits construction of different forms as a source of power for industry, or agriculture, depending on which lobby has more power at a certain political time. Politics is played out on the river and the marginal people in similar manner. In that sense, too, people's history[8]: of a river that flows past. It is in the everyday language of people that you see the true meaning of Godavari and the river's role in people's histories. It is also in people's everyday language that is closest to the idea of a river as it should be, in nature's larger scheme (alone: hence, flooding) - the ecological flow of a river is visible and audible in the language and dialect, in the region the river touches in its flow. At least one way out of, or into their villages is by the river route almost entirely except in few pockets connected by roads built centuries ago, or in exceptional cases, a decade or two ago through the benevolence of a local MLA or MP, and in rare cases, when the Panchayat president got some funds to correspond with his or her inclination.

That Godavari, the river, is statistics, is one perception. Then there is the Godavari, the metaphor of constant flow of nature, and perhaps one of the fewest left of perennial seasonally flooding rivers. She comes, and stays. And leaves: not a trace of permanent disruption of life. Instead, as she leaves, new signs of life emerge – there is the alluvial soil waiting to be cultivated with post-monsoon (post her coming) crops. The dalits of Bhairavapalem in I. Polavaram (i.e. Island Polavaram, not Polavaram of the dam site) by the estuary of East Godavari district (not in the submergence zone of the dam) within the Coringa Wildlife Sanctuary area and Mallada Narasimhamurthy, and other fishermen on the creek – where Godavari ends as Gauthami before she will meet the sea – await her coming. In the years of her coming unrestrained and in total exuberance of a freedom to be, the fish are aplenty, and the soil more fertile for those downstream. And for others cropping on the *lanka*s (the alluvial deposits by the river when she recedes) it means a quick and good crop guaranteed. The homes remain as they were when water recedes and people are back to their normal routine. Not that they do not face the hardships in the period of her coming and going. They do. But they know that once the water recedes, and it will, they will go back to the same homes, and resume life. Their home addresses do not change; their roofs at times need several repairs, at others, even the thatched roofs are intact, and their walls show the tell-tale signs of Godavari's visit (if made of mud-brick, even more clearly so), and people are left with the memory of how high she rose the last time, which they will show you with both a sense of having witnessed a great natural event – 'last time she came up to this point'. Or they will say, 'she took (herself) back within a day or two' – *Teesindi*. Or, *'ninna ratri koncham teesindi'*, 'she took herself back a bit yesterday night', is another term that conveys the ease with which the river and its flooding patterns have been internalised through centuries of 'socialising' with a natural phenomenon and accepting its presence. *Teesindi* should literally translate as 'took' and in the context of Godavari it means 'she took herself back' (she retreated). After this coming, letters may perhaps arrive later than usual, because the postman's home, too, has been paid a visit (and not to forget the money lender's as well – as we shall see), but arrive they will at the same address. It is in this meantime that it will become a matter of intervention of the administration and the usefulness or otherwise of this vast apparatus will be tested; not the faith of the people in Godavari. As one of them once said to me - "for those of us who live on her, she never let us down; she never destroyed us; it is just a matter of inconvenience for a few days; this dam is throwing all of us out of here forever! Godavari never did that! Since my grandfather's times we have been hearing of her coming and we are still living here, where we heard those stories; but will we be able to tell these stories to our grandchildren in this same village, after the dam is built?"

A headman in a fishing community's hamlet in Kobbarichettupeta (Tallarevu mandal in East Godavari district), Mydu Satya Rao, had brought home the point of the flow succinctly – "Godavari must go to the sea; that is the law of nature (*'adi prakruti dharmam'*). Cotton *dora* (Sir Arthur Cotton) built the barrage at Dowlaishwaram, yet Godavari flows; how much water will you hold in a barrage? What has to flow into the sea will do so..."

One *cannot*, at the same time, romanticise the river and its flow and the havoc caused due to regular flooding of the river. The flooding has been addressed in the language by the colonial powers by building a dam, to store the so-called surplus flows. At the other end, the flooding patterns are fairly regular and systematic enough for the itinerant and seasonal fishing communities settled along its banks to know as to when Godavari will rise, how much, and what will be its impact of that particular season. There are others who observe Godavari's temperaments and time of rising levels to – the dalits and the tribal communities. Yet, there is another reality – that despite having the river flowing by their villages, many of these communities do not have the wherewithal to access the same for their agriculture and drinking water purposes. Some with resources manage to route the river water through crude pumps and motors, while others depend solely on rain-fed agriculture; average rainfall in these areas not as bad as in other parts of AP, by and large.

Every community living by the river observes and respects the movement of the river, and has negotiated its survival. Yet, at times the pathos of these negotiations is highlighted by the state's lack of engagement with these communities, today threatened by displacement. At one end is the positive economics of living in a more sustainable, pollution-free environment than what most of us live in today. At another, there is talk of 'civilising' the tribal people[9] and settling them amidst these devices of modern living. The debate was never sharper.

The DPR[10] of the Indira Sagar Polavaram Multi-purpose Project and Some Thoughts

Before moving further, let me elaborate here what the Polavaram project entails (which was declared a National Project only this year), since my journeys from 2006 began in this context. I retain verbatim the 'multi-purpose' project here[11], in government's own language in a Detailed Project Report that was shared with media in the years 2005-06.

"The Polavaram project is conceived as a multipurpose project conferring irrigation benefits to an extent of 7.20 lakh acres in the upland areas of West Godavari, Krishna, East Godavari and Vishakhapatnam districts, water supply of 0.664 Tmucum (23.44 TMC) for industries in Vishakhapatnam Township and Steel Plant, besides domestic water supply to 28 lakh people in 540 villages enroute and generation of Hydel power with an installed capacity of 960 MW, development of pisciculture and providing recreation benefits and diversion of 2.226 TM cum (80 TMC) of Godavari waters to Krishna river."

"The components of the project are as follows:-

Head Works - Earth-cum-rock fill dams...across the main river with spillway on the right flank, and power house on left flank on downstream slopes of 'd' hill. The Earth-cum-rockfill dam in the main river course is 2310 meters long with top level of 53.32 meters and top width of 12 meters. The Power House...with necessary tunnels penstocks and tail race channel with installed capacity of 12 units of 80 MW, i.e., 960 MW finalised by the APSE Board in consultation with the Central Electricity Authority.

The Spillway is 897.50 meters long is provided with 44 nos. radial gates of size 16mx20 m to pass a maximum flood discharge of 1.02 lakh cumecs (36 lakhs cusecs). The sill level of spillway is +25.72 meters. The full reservoir level is +45.72 meters.

Saddle Dams 'E' and 'F' and KL – Suitable earthen bunds are proposed at E&F saddles to form subsidiary reservoirs inter connected with tunnels.

Left Main Canal – takes off from the Nelakota subsidiary reservoir and runs to a length of 181.50 km crossing major streams like Sarada, Mamidivakagedda, Varaha, Thandava, etc. The Left Main Canal is proposed as a lined canal and designed to provide irrigation facilities to an extent of 1.62 lakhs hectares (4.00 lakhs acres - 400,310 acres) of ayacut in East Godavari and Vishakhapatnam, besides industrial and water supply of 23.44 TMC of Godavari waters to industries in and around Vishakhapatnam city and drinking water to villages enroute. The Left Main Canal is contemplated to run for a length of 110 kms in East Godavari district benefiting 16 mandals, namely – Rajahmundry, Korukonda, Sithanagaram, Rajanagaram, Rayavaram, Biccavole, Gandepalli, Z. Ragampeta, Jagampeta, Kirlampudi, Peddapuram, Pithapuram, Gollaprolu, Thondangi, Sankhavaram, Tuni for a length of 71.50 kms in Vishakhapatnam district benefiting 10 mandals, viz. Payakaraopeta, Nakkapalle, S.Rayavaram, Yellamanchili, Rambilli, Atchutapuram, Kasimkota, Munagapaka, Anakapalle, Paravada...The total demand of Left Main Canal (including water supply canal) calculated as per modified penman method works out to 127.14 TMC as per the crop water requirements.

The Right Main Canal is proposed as a lined canal and designed to provide irrigation facilities to an extent of 1.29 lakhs ha. (3.194 lakhs acres - 319,405 acres) of ayacut in upland areas of West Godavari and Krishna districts besides diversion of 2.266 TMCumecs (80 TMC) of Godavari waters into Krishna river near Vijayawada as per GWDT award. The total demand of right main canal (including diversion to Krishna river)...works out to 174.24 TMC as per the crop requirements. It provides for drinking water to the thirsty enroute villages. The right main canal is contemplated to run for a length of 109 kms in West Godavari District benefitting 17 mandals, viz. Polavaram, Gopalapuram, Tadepalligudem, Kovvur, Chagallu, Devarpalle, Nidadvole, Nallajerla, Unguturu, Dwaraka Tirumala, Denduluru, Eluru, Pedavegi, Pedapadu, Pentapadu and Bhimadolu and further to run for a length of 65 kms in Krishna district benefiting 4 mandals, viz, Bapulapadu, Ungutur, Gannavaram and Vijayawada rural. The total land required for excavation of canal is 14061 acres.

Submergence
No. of villages under submersion – 276 Nos
Population to be rehabilitated – 1,17,034
Area under submergence
Agriculture unirrigated – 22882 ha
Poramboke – 12081 ha
Forest – 3223 ha
Total = 38186 hectares

Cost of the project – Rs. 8198 crores[12]
BC Ratio (Benefit Cost Ratio) – 2.54: 1

According to Dereservation proposals the extent of **forest area coming under Submersion in various Forest Divisions** is as follows:
Eluru Division – 248.60 ha
Kakinada Division – 112.31ha
Vijayawada Division – 40.83 ha
Bhadrachalam South – 2271.00 ha
Bhadrachalam North – 16.00 ha
Paloncha Division – 533.61ha
Total – 3222.35 or say 3223 ha

Extents proposed for compensatory afforestation is as follows:
East Godavari district (Kakinada division) – 2275 ha
Vishakhapatnam (Paderu Division) – 1052 ha
Total - 3327 ha[13]

Resettlement and Rehabilitation in Respect of Indira Sagar (Polavaram) Project – "The State policy provides for allotment of equal land that is lost by the tribals duly acquiring in the command area. This will be a boon to the development of the otherwise deprived tribal population to come into the mainstream. The project will bring new light into the people of four Districts who are otherwise in poverty due to lack of irrigation facilities and frequent drought conditions...Among the total population of 1,17,034 , 47 % belong to the Scheduled Tribes, 14.96 % belong to Scheduled Castes and 22.8% belong to Backward Caste...Remaining 15.26% belong to the forward caste."[14]

Archaeology and Heritage

Between 1985 and 1991 two reports (in two Parts) were submitted by the Archaeology and Museum Department (Government of Andhra Pradesh) which is mentioned in the DPR thus - "While exploring sites of Archaeological and Historical importance the Archaeology and Museum Department had submitted reports...suggesting 12 sites on either banks to be excavated in the submergible area...and shifting of sculptures numbering 30 found on the left bank area to Rajahmundry and 50 Nos on R/S to be shifted to Khammam which requires dismantling, transportation and reconstruction of (1) Siva temple at Vaddigudem, (2) Sri Santhana Gopalaswamy temple at Virivendi (V), (3) Sri Veerabhadraswamy Temple of Mottigadda all of 13th century AD (4) Koundinyamukteshwara Swamy Temple Koundinyamukhi (V) of 15th century AD...All the temples are under the protection of State Archaeology and Museums Department, Hyderabad....It made a mention of five sites of Archaeological importance on Left Bank of Sabari. Finally (3) Part-III report (1992-1994) discussed on survey of Archaeological and Historical sites on Sabari basin and Sileru in Khammam District. The State Archaeology and Museums Department has furnished no objection certificate for clearance of the project..."[15]

What I think of it: Only a problematic understanding of history could lead towards a 'no-objection certificate' in this case. Obviously local communities would never be considered to be part of the process of identifying the historical value of the place or antiquities in their own terms at any stage. Only some of the above mentioned 'monuments' are understood as having 'historical value'. Note that only temples built from the 13th century have been recorded, which in itself is problematic; what was the State Archaeology department going there to 'see', and what are the parameters of validating something as being of importance 'historically' in exclusion of something else? They do not mention the history of tribal societies and they do not take into account historical value of the tribal societies and their gods and goddesses, and the trees propitiated (for instance a tree with an anthropomorphic figure carved on it, which I mention later). Whose historical or archaeological value are we looking at, to "dismantle, transport and reconstruct"?

Though in the DPR the government accepts the truth that the tribal population has been 'deprived' all these years, they conjecture that only this project will bring them into the mainstream. The assurance of giving land to them in the command area is mere empty promise because the alternate land shown to them, wherever, whenever, has no connection whatsoever to the proposed command area of the project. "Frequent drought conditions" is a falsehood perpetuated by the government, because the people for whom the water has been assured will in fact get water for a third crop (as has been stated by several analysts) and there has been no reported drought in these parts in the delta districts in many years now. It may be noted that the maximum submergence of the Polavaram project happens to be in Khammam District, followed by East and West Godavari Districts.[16] It is extremely significant to note as to who are to benefit from the project submerging the lands of maximum number of tribal communities, as is mentioned above – these are areas already irrigated and focussed on the coastal region, including Vishakhapatnam. The intention is built into the project from its very conception stage. This has been a crucial political point raised by several people, especially important to the Telangana statehood discourse. As for the lands to be submerged, almost all of them are among the most fertile and productive agricultural lands in the Godavari region (irrespective of which district they fall under). Incidentally, agricultural loss itself is phenomenal taking, merely, agricultural produce in the submergence area. According to one estimate[17], cotton is grown in over 10,000 acres and gives an average of 150 person days of work. Paddy is grown in 10,000 acres and each acre gives 75 person days of work. Tobacco is grown in 6,000 acres and gives 250 person days of work per acre. And so forth. And one is not even mentioning losses of other livelihoods.

A Word or Two about the Structure of the Book

I call the larger sections of the book 'journey', but within each 'journey' there were several short or extended visits to the villages in the 'submergence zone'. The journey, then, becomes more of a record of some pertinent issues of the time, which each individual visit highlighted. When I returned home, or when I could catch a moment of solitude, in the midst of all that activity, I recorded some points that I distinguished in my dairies, etc as 'reflecting'. Many a times these were my points of anchor. I feel the need to share all of these, too, as moments of my own truth, in this record of these many years of going back and forth along Godavari. Godavari and people of, around, by the Godavari are central to this book. The making of the Polavaram dam begins to shape these lives in newer ways. There are two thematic threads to the book: people's opinions, voices, in the current context of the ongoing dam (though not always about the dam) interspersed with some historical facts (seeing the past from the present) during the course of five journeys (with short and long 'visits' within each journey) and; one's distance from the scene (circumstances and some choices in order for the book to happen) which allows me the space and context to look at larger issues that surround this significant political event of AP. My journeys are presented in the form of a daily journal I maintained in the order of the years I undertook them and what they signified – people, state, democracy, contradictions, and simply, life, at times, and my reflections on these. This book follows exactly in the form that my journeys happened, and the forms in which I recorded the events that I witnessed, in either a single visit or several sequences, at one time or at different intervals of time. Just as there were moments of my recording my state of mind, or the voices of people I met – in a cassette tapes on a personal stereo, on notebooks, diaries, images captured on lens of a camera outmoded by the digital technology, the immensely useful mobile phone[18] allowing me to videograph and type 'notes' when writing on a notebook was just impossible (either on a boat ride, or a rickety bus or auto ride).[19] Till date the 'outdated' camera remains the truest record of things as they were, with no possibility of a recall or auto-correction, just as in life. Development in my own thinking, etc, follows my journey; each of the stories and thoughts here were written in that moment in time as mentioned, and I have been true to their date and time, though I have merely 'copy-edited' and added to relevant parts. Truth in documenting even my own truths seemed important to me in this book. The act of recording two kinds of truths – a truth outside of me, as narrated by the people I met – and the circumstances I saw, and a truth within, of my feelings and reflections, was important for me. As far as possible, I truthfully share both – of seeing outside, and within. So I intersperse my records of the people and the places, with my 'diary' or 'cell phone notes', wherever possible. At times there was silence within and my personal diary was not as much as touched (and these silences need not be expressed. I let them be).

Essence of the Journeys, Explained

This book has written itself as a reliving of those, my Godavari, journeys. Each Journey can even be read as a single episode, or all the journeys as several episodes highlighting one basic reality or multiple realities within the Polavaram Dam Development Story (as to how each year one finds a new aspect to the construction of it). I could not take away from the many moments, many narratives of each year. I could not 'trim down to size' what may not be chronicled again for over six years (past) of many moments and seasons of an overwhelming cataclysmic event for most of those people who speak in this book.

2006[20] , or *enabhai-aaru* (in Telugu), was a journey all about gathering all that I could about the dam – information, sights, sounds, voices – just about everything there was that I felt was important to be collected. It was an almost frenzied, mad phase – for me –where information was important. It was also the point of the 'event' of the Polavaram dam, which is reflected in the nature of reports I had written, developments I had recorded, my own thoughts on the same. I was anxiously engaging with the possibility of an impending displacement. Hence, taking in, recording, whatever I could lay hands on. Later, as one went along, in moments of relative peace, one realised, as to how information, *per se,* becomes the source of power, for those who have all that information, especially in the cities, even among self-styled activists, or intellectuals or organisations, or the departments of governments in power. But at the same time, how information becomes a matter of last hope that people cling to, to save themselves, especially in the villages one visited. In the villages, they would push everything into your hands, without you even asking for it, even as you spoke to them – pamphlets, handwritten letters, petitions with thumb impressions, *patta* titles, if they had them, newspaper clippings they would religiously collect. It was my journey towards understanding (at the end of it; as I returned home and reflected) these subtle realities, the philosophy behind information and concentration of knowledge. After all, isn't it information that the governments in power and many others use in their favour and abuse at the same time? Isn't this the thing that networks are formed around? Gradually, the madness to collect information had subsided by the beginning of the second Godavari journey (2007) and I had calmed down to getting to listen to people at their own pace, in terms of what they felt was important to 'codify' into a newspaper article that they felt might bring justice to their cause (whether it did, or would, was another story altogether). People do make you write, and want to know what you are putting down in your notebook. It was also to listen to long silences in between their talking, when I, too could just sit with them, without the need to put everything on paper anxiously, together watching the river flow past, or the Sun hide into the hills by the river. And in those precious silent moments we both together saw our respective, relative helplessness against a massive construct that this democracy has come to mean, at both ends – mine and theirs with our own respective marginalities, differing in degrees, but pervasive, at the same time. And in that time, someone would say, 'come eat', or give me a cup of tea, and others would leave to attend to their work on the fields, or at homes, and life would go on. And it was time for us both to move on.

In 2008, I discovered some colonial records at the capital city of an erstwhile 'empire', and its well-preserved monument, in a sense, to their own history of colonialism, namely, London, at the India Office Collections of the British Library. I seemed to have 'sought out' Godavari there, and found her! Some of the chapters of colonial history are very important to the understanding of Godavari. Year 2009 was crucial for Andhra Pradesh (and India too) electorally speaking. We saw the State and General elections. Though my visits this time were few I covered an important aspect to the Polavaram dam: tribal candidates contesting on both pro- and anti-Polavaram dam platforms; as well as the clever evoking of Godavari by the late YSR (who died within months of his re-election as Chief Minister and who almost single-handedly watched over one of the largest irrigation scams in the State, though media glorified him as the 'charismatic pro-farmer' CM).

In year 2010, it was like returning to the source, as I met the fisher communities of Godavari to learn about how histories have also been written on a river, by itinerant fishermen, and I tried to retrace steps (unsuccessfully) to Muttu Rama Rao's home in Rajahmundry to learn that this man actually worked against the grain in charting a new course.

The book has a section titled '*Godavari Journeys*' which records each of my journeys to the field from 2006 until 2010 (as also brief visits in 2012-13). At the end of this section I also included some of my personal experiences, adventures, etc which to me were intrinsic to the entire exercise. A second section titled '*Reflections*' deals with my seeking to understand the situation within larger contexts – some of these thoughts happened post-2010 in the course of bringing the book to shape (which has taken longer than was intended). And then an 'epilogue'-moment happened in years 2012-13, as if to complete the journey-phase thus far, which I record in the book. I end the book with significant information around the Polavaram dam, which I share as a duty. At the end of each year's Journey, I provide information about the progress on land acquisitions. This part is to be seen as a signifier of state power. The state's dogged pursuit of land acquisition becomes the *meta-narrative*, in spite of petitions, misgivings, and protests. Several farmers were losing land each year. And these are figures on papers that are given as public information to portray the 'transparency' of the system but what is not transparent is what happens *behind* that makes farmers sign those consent papers. In fact, in these many years, the government succeeded in creating a new mass of mediators and 'negotiators' among MLAs, local scribes, petty contractors and also some NGOs to acquire land. What you will not see in the official records are posts of resistance where tribal villages have not yet signed consent letters. Their numbers are fewer. Yet, important.

I write this book with the hope that the characters within the two flaps will remain at the addresses and locales where I first met them, at home (in their homes, fields, their own region) and through them several others – not in the sense of their present socio-economically limiting contexts but more in the sense of being rightful owners of a shared historical, cultural landscape they helped shape through centuries; or even if these were destinies shaped by others for them, yet there are pages of their resistance to it, in this region they call home. The state and other agents are seeking to uproot forcibly and impose a different kind of history, or landscape, or economy on them which they shall no longer own or define as their own; for they did not build it themselves. This is not 'my' book, as I have come to realise, though the words may be mine; many a times in these journeys I felt like the instrument, the medium of expression of people of Godavari and river Godavari. In many ways all the physical journeys opened my eyes to the dichotomies of people's lives. My journeys are the ones taken by people, exactly in the way they undertake their journeys through the means available to them – buses, ferries, catamarans, auto rickshaws. In sharing these spaces one has immensely benefitted from insights shared by people about the course of developments in their villages. In one or two cases, thanks to a friend or benefactor, I could take a hired cab with fellow scribes from the region. In some cases I rode pillion on a motorbike owned by either an activist or reference of a local NGO (the last only in the case of Khammam, otherwise, I largely kept away from NGOs in these journeys). Many a times these journeys have led me on to paths I had not planned earlier and issues I saw for the first time or linkages where I saw none before.

What happens between journeys? Life catches up in many ways...

The limited resources at my disposal always forced long intervals in between two journeys. But it was a strange irony that, in spite of these intervals, things did not change drastically, so one could pick up the thread from where one last left and resume conversations. At the same time, things *did* change: the AP government secured clearances in quick succession (environment, forests, etc, or so it claimed once every few months through media announcements); and acquired land. And yes, in between these intervals in my journeys I would see the Sarpanch I met last, loud in his protest against the dam, a proud farmer owning three acres of land, become a meek, landless tribal beneficiary sitting outside the MRO (Mandal Revenue Officer)'s cabin by my next visit. One couldn't ask questions from them (as I initially used to, quite insensitively) as to why they changed their positions. One understood that life is a larger game which most cannot fight uncompromisingly. In between my journeys, several proud farmers gave up their rights for a piece of 'I consent' letter, which promised them cash and an unofficial assurance to cultivate their original lands until the mighty dam was built. Several illegal landowners became legal beneficiaries, for their illegal occupation in the so-called Agency (Schedule V) Areas. In between two visits I found newer 'crorepatis' (millionaires) in the village and several landless, unable to harvest their last crop.[21]

Between two Godavari journeys, there is something to say about life. For the marginalised, it is life in between a past that was a matter of subsistence and a future unknown. In between, is a life of several negotiations after years of open confrontation which may have found urban middle class support and later, even that din fades away, usually to metamorphose into an off-ground, cyberspace resistance and petitions online. Between policy formulation and policy implementation, it is that life, in-between; where both the former and the latter come out of high corridors of power and of literate elite. Lack of money, loans, uncertainty – a lot here makes you go through at least part of the experiences, in terms of your responses and reactions to your situation, which Godavari's people may be going through. What connected us? Perhaps vulnerabilities in the face of a volatile nature of economy? And a structure that thrives on these very vulnerabilities? Even I was not immune to it. I used to feel I could never 'get there' on time. Something always happened to our lives in between which showed up as a new development between two journeys or two visits, at times. For me, life in between was also about garnering support to make that next journey, while for them (which was more important), whom I would meet, it was that one step or series of steps between seeing something of their own sweat and blood dissipate into nothingness. For the tribal communities, it was like getting a Forest Rights Act in place to recognise their *podu*[22] lands but see it come to naught with a National Park declaration, fencing them out of their own historical spaces. The futility, at times, of my writing these articles after intervals and the grim reality of people would resound like a loud slap across my face at these moments. And it still does. Yet, in these long intervals and during every visit I also found those who refused to be afraid; those who resisted, till their last breath. It was the resilience of these people that helped one carry on, in spite of shrinking spaces and resources. Then there were people fighting their battles for compensation, running back and forth to the MRO or the RDO (Revenue Divisional Officer), or the Joint Collector, with numerous requisitions put together by the literate among them, with thumb impressions and few signatures. Every moment would be an effort to get the best deal yet. Be it from accepting the Polavaram package or refusing it. They never gave up. Sense of humour intact, same laughter that I had seen the last time, while one could not say so for oneself. For instance, in the year when the Forest Rights Bill was being discussed, Boragam Rama Rao of Mamidigondi pointed out to me the irony. By then he had already given up his cultivated lands to the Polavaram project. But as a Sarpanch he was asked to set up a committee to identify community *podu* lands in the forest. He said. 'Look at the forest guys (*forest vaallu*); so many years they treated us as coolie (wage labourers) to work on our own forests; then came the Polavaram project and they sold the forest to start digging (canals); they stopped us from collecting forest produce. Today they are pleading with me to identify our old *podu* patches! Some of our old *podu* lands have already been dug up!' He then laughed aloud. His laughter was infectious. The pathos biting.

In initial phases, I used to seek out the 'popular movement' format and wonder why there was no 'movement' (against the dam). I realised, over the years, that there has always been a popular movement, in these very villages, but not in spectacle format which has become popular thanks to images on TV. The movements here are not spectacles but daily struggles and fights and resistance of a different nature. They happen not in one grand format on a huge ground or stadium or a park, but in government administration offices, outside of them, within them, inside jails, with tribal farmers and landless tribal people and dalits demanding answers to their questions, filling up offices with their petitions, many of them hand-written (written with the help of some educated youth among them or a local advocate, or activist). They happen when they contact local scribes, stringers, or someone far away to come and see what is happening and gather together to discuss the issue. There are no cheer-leaders, nor icons, nor is there sloganeering; not even OB vans of TV channels lined up here. On the other hand, the resistance I saw in the villages by Godavari has history, depth, pain and the spirit. In my journeys, recorded in these pages, Godavari always introduces me to people and their stories – of people who suffer, and those who make money from others' suffering. It takes all kinds to make a village. And I end up seeing all kinds – brokers, contractors, lobbyists, sufferers, protestors, radicals and even the neutrals. Even within tribal villages, while I met those who have not given up their lands, I also met others who became passive, afraid, or simply desired that one-time settlement of a lakh and fifty thousand rupees they had seldom seen or would ever see again. This amount would help them settle old loans or solemnise marriages. So long as the government gave them house sites. Even if it meant that the government gave only the sites but the houses would have to be built from the scratch by the people out of their own money, released in installments, based on conditions, and through bribery at every level. Yet, they found these steps far simpler than the idea of fighting or resisting the state with its police force and threatening tactics. After all, if the police could open fire on a crowd of protesters (against the dam) accompanied by even a MP or MLA (as in Khammam in 2007 January), what to say of 'lesser mortals'?

Is it a Forced Construct?

Some thoughts kept recurring through my journeys relevant to the context of the dam and beyond. And these go beyond the immediacy of the displacement and come from insights gained from people's thought processes and questions. For instance, in the unassuming terminology of Godavari's coming, can floods be seen as a natural phenomenon with its own relevance for the natural scheme of things? Floods, in their occurrence carry the inherent metaphor of a kind of churning and renewal needed every once a while to revive, to throw up, and bring up to the surface life forms; re-energising the wetlands in their wake with fresh vigour. And without these, what would happen to our seas and oceans? Without rivers meeting their seas? Is just about everything on earth (and now include the skies and space and Mars) meant for the human species' immediate consumption (alone) like there was no tomorrow? Can we, then, with our technologies, alter every natural phenomenon to suit our greed? But can we exist in a vacuum without other life forms that are part of a larger cycle? Finally, how much water does one human being need for daily needs? There are statistics to answer this question. You can also find a random figure for how much electricity (where available, in India) a human being needs for running a household or managing agricultural needs. How much food do we need? There are calculations for this, too. But these calculations always stop at the level of our rural communities, in rural areas, where the government administration knows exactly how much to release of each necessity – water, power, food. The difference between our villages and the cities is the same difference/distance between the so-called first world and so-called third world in terms of how consumption needs are defined and as to why there can be no end to the 'need' constructed for power generation and no end to the idea of meeting those needs through big dams. Sleeping in one of the villages, perhaps Kathanapally, I watched as Saroja (a Kondareddi) tended that young sapling in her home (which has a separately built kitchen outside the house within a compound made of bamboo and thatch), wondering what, about allowing for a sapling to grow into a tree, should seem so inconsequential to the 'larger public good' which, the global state convinces us, is the Metro rail, reaching employees to far-off multinational offices at break-neck speed and to work 24 hours without concepts of night or day? What in the Kondareddis' act of living as peasants seems so unimportant to the larger scheme of things, which include homes like mine, in the city? Are we, in/of the cities, the 'public good' that Saroja and others in her hamlet must pave way for? For Saroja and others like her, there is no 'luck' of a reversal of the fortune written out for them by the state. Endless conversations with Godavari's people reveal much more than the immediacy of displacement on their minds. They reveal minds that have self-reflected deeply on their existence and survival questions which are reflected in the terms they use. In people's conversations, the term '*akkada*' (there) would occur in a context of contrasts. *Akkada*, 'there', is a nameless, disempowering, vague, abstract notion of a place they will be 'thrown into' ("*padestaaru*") following which these people will lose their identities, livelihoods. Even if they happened to know the name of the R&R Colony (in two cases, in Polavaram mandal) it is always "there" "that" (vague, nameless, abstract) place which will never be good enough as this, or here ("*ikkada*"); no matter how difficult lives have been here – it is still their home. The "here" and "there" are two poles – between their own ("my own") and "that" which does not belong to them or where they do not belong but will be dumped into. The same language of government machinery ("*ikkada meeru em chestaaru*"? "*Adavilo untu*"? "*akkada anni sadupaayalu untayi*" (what will you do 'here' living in these forests; 'there' you will get all kinds of facilities'). What facilities? Will bank loans be waived 'there' was a question several SC, ST communities asked in their demands for answers from the government authorities who used to try convince them of the '*packagi*'.

There is no count of how many Telugu films have been made in these environs, yet they hills, fields and Godavari and the beauty remain a passing fancy, like the dream sequence in film songs, a fantasy from which you have to get back to the real world, which cannot be these fields and these hills and forests. Not anymore. In fact, not a single tribal woman or man will be spotted even by a freak chance in these sequences, or films made against this backdrop. The lush fields, the forests, the river, remain backdrops to the central characters of the film, usually urbane, usually from Hyderabad (films made in the last few years). Similarly, Hyderabad and its fast urban pace will be the central protagonist or central plot to the fast fading natural environs of Godavari. Quickly fading picture postcard backdrops cannot influence nor inspire any policy change, surely? They just remain picture postcards. More than displacement, per se, the complexity of the region that has written itself into this book comes from the question of the utilisation of Godavari, intervention on its flows, claimed by the state and polity in contradistinction to the use and flows of a natural river system and access by people for seemingly smaller needs than that of the state and its select enterprises. There is a caste angle, a definite caste-tribe conflict, and economic angle to the issue. The immediate displacement (of nearly 300,000 people), worrying as it definitely is, fails to go beyond the statistics and implications in the present circumstances. Never before was the river seen as a commodity serving larger interests – as Sir Arthur Cotton did – than mere irrigation for smaller fields and subsistence agriculture. In contemporary times, where several experiments have shown successes with eco-friendly means, where these are invoked time and again in the list of possibilities, one must understand that neither the colonial power nor the Indian state (particularly post-'90s) have been interested in the 'small'. Large dividends from large investments has been their mainstay. What is happening at present follows from the single experiment in the past (at least in the case of Godavari) seeking to consolidate the interest of the rich peasantry as it does of industry, with some of the largest ones owned by the same rich, upper-caste farmers' lobby that availed benefits of irrigation access. Most exploitation of water and land has been done by members of this landed class, availing benefits of both agrarian reforms, where they happened (in form of tax reforms), and the industrial, global-looking process. Somewhere, the rush of settlements of non-tribal upper castes on tribal lands within close proximity to the Godavari may be traced back to the idea of land and land-use which the colonial administration put in place. I will just cite two examples here. Hemingway, compiling the *Gazetteer of Madras for the Godavari Districts* (1915) notes that, "Previous to the building of the Dowlaishwaram anicut the cane grown in Godavari was a thin, reed-like variety, similar to, if not identical with, the canes of Ganjam, South Arcot, Trichinopoly and other districts, which was called the *desaváli* or country cane. Its hard rind enabled it to resist the attacks of jackals, so that it was possible to grow it at a distance from the villages; it did not require much water; and the jaggery it gave was small in quantity, though very sweet and white. When the anicut was made, softer, larger and juicier canes were introduced. The *síma* variety, a stout dark kind sometimes called the Mauritius cane, was introduced about 1870 by Messrs. Cotton and Rundall for their factory near Rajavolu (Razole), but the history of the other species is obscure."[23]

And, similarly, on the nature of intervention on sugarcane – "Attempts are being made to improve the quality of the tobacco grown in the district. Messrs. T.H. Barry & Co. of Cocanada have established a tobacco factory in that town and foreign seed has been imported by Government for experimental cultivation in the lankas leased to Mr. T. H. Barry. The chief defect of the existing tobacco is the excessive thickness and dark colour of the leaf. It is sold in other parts of India and Burma and, to a limited extent, in Mauritius, Bourbon and London."[24]

The discourse on globalisation necessarily having started developments in this region – dam, and all the rest – needs to be reviewed in the context of another historical truth of this region. The delta region's language of power, economy, from water, it seems, was the dominating language for all time; aspiring to that level became important for political control and domination, as well, post-Independence. And that politics has always defined the nature of discourse over water and agri-industry combine. Godavari today, or Polavaram today, and even the Telangana discourse today, would be much easier to locate in this canvas, for the *language* of politics has hardly changed, though the *terminology* of it may have, but even that is loosening itself from the 'burden' of 'socialism' around its neck. World market capitalism in Godavari region has hardly been a new phenomenon for its upper class. It is a game they have played since centuries. People's history of a river is closely defined and made by these developments. There were certain agrarian rhythms intrinsic to the agrarian communities before they got engulfed in the mainstream ideas of production and profit. Even the tribal communities of Kondareddis, Koyas (among the tribal communities which can be credited with introduction of agriculture, including the earliest varieties of paddy, when compared to the others who relied on forests and the river) had their own cycles of farming. The Godavari Gazetteer makes mention of this aspect. "The Godavari ryots divide the six months from June to December into twelve *kártis* of about a fortnight each, called by the names of various stars. To each of these periods some agricultural operation or other is considered particularly appropriate. Even the Koyas and hill Reddis, for example, believe that the best time for sowing paddy is the *mrigasíra, kárti,* which begins about the end of the first week in June; the *anurádha kárti* (the latter part of December) is a name of happy augury, suggesting the harvest and the fulfillment of ryot's hopes; thunder on the first day of the *mágha kárti* is the happiest possible omen for the future and 'will make even a pole on a fort wall grow'; and so on. On the day before harvest the ryots run round their fields thrice repeating the name of the village goddess and crying out that she has given them a good crop. They then cut three handfuls of ears to represent the goddess and sacrifice fowls to them. When measuring the first heap of paddy of the first harvest of the year, they pour boiled rice-flour over it to propitiate the belly-god."[25]

Is it the history of a people around a river or is it the river's history that shapes people's histories? If you removed Godavari from the people here, do these people have an identity then? What is that identity like? In every conversation the number of times Godavari occurs is amazing. Can there be a history without this metaphor, this name for a people with lived histories that span several centuries (perhaps a millennium in some cases)? Are human histories devoid of/irrespective of other life-forms or are they shaped by the intricate web of each and every organism with its own life-cycle – the trees, forests, the insects, birds, animals, the minutest algae, fish? Life is a delicate balance between all these. But corporate commerce views the human species as an independent or non-dependent, self-made entity, living in a vacuum, earning money. The rich have 'professions' while the poor have 'livelihoods': mostly, physical labour, in the name of 'employment guarantee'. It will be a life decided by a monetary paradigm. Life, as an evening by the setting sun, under a massive tamarind tree, looking at a river flow past, will have no meaning; has no meaning. At the same time, such ideas – of nature, quiet, ecology – have been assumed to be preserves of the elite; as if the economically/monetarily poor in the villages by the Godavari do not have moments of that deep connection with forests and rivers and moon and the sun. Nature and its 'pristine-ness' has become another commodity to be sold to the urbane rich tourists who understand nature. Ecologists, environmentalists are assumed to be ones form the literate classes in the city, fighting on behalf of the tribal societies. The tribal societies on their own, in their language which none of us understand (speaking of their not understanding English, the language of policy) are assumed to have no other sense of connect with the ideas of nature than just being born into that context, with rituals surrounding nature, and hence (*hence*) having symbiotic relationship with their forests and trees or rivers and streams. They are not allowed, in mainstream perception, that quiet one-to-one moment with the sunrise, or a conversation with a tree, which seems to have gone into ideas of romanticism reserved for the idle rich; or at the worst, the unfeeling eco-tourist. And it is this assumption that also makes the governments believe that 150 square feet of 'house site' and a small patch of land anywhere they deem fit, is all that is needed as compensation for the tribal or other marginalised, displaced from spaces where a certain kind of nature-human connect exists, which is reflected in their drum beats, songs and rituals. But this is not confined to the tribal society alone. There is need to understand that the language of investment, dividends, has stopped relating to any idea of seeking a world where indeed the sun and the moon and everyday living (or livelihood) could in fact be interlinked in ways none can imagine. One might just mention that in terms of geology one was amazed to find that the submergence zone, including the Papikonda Wildlife Sanctuary-turned National Park, have among the oldest Achaean rock formations. In regular talk, we do not see these connections in a paradigm that looks at human existence as being defined by a singular purpose in life – to commercially exploit and earn the GDP to show 'economic growth'.

Why is the freedom to live in a context (they could have left for cities long ago giving up their lands had they so desired; they did not) not available to the people today called (to remain) marginalised (though they are the centre, and their worlds are the center if we looked at other worlds as being on the fringes of these people)? From where do we draw the circles or squares or lines where the centre and the margins are decided? And the fringes are decided? Isn't this the problem, that somewhere a human life seems to exist (as defined by the ones who have assumed power) only as if to feed this assumed 'centre'; a centre within the centre of a few who 'own' the earth's resources? My mind is filled with questions such as these each and every time I enter into and exit from the Godavari 'zone'. More and more questions; few or no answers.

Seasons and Stories

In every season, there was a story, reflecting, or resonating with, the mood of that season, almost, and the time and pace of that season. The journey on a luggage carrier steam boat, under the rain and clouds of the approaching south-west monsoon, has me sitting beside the story of Raju, the Koya boy, sometimes seeming to be quite an upstart; he is the future of this place though his future is uncertain, as he completes his education at Rajahmundry. The clouds, pregnant with rain, seem to correspond with the dense silence surrounding Raju on that boat-ride as he looks on at the river, breaking the silence only now and then when, if, I ask him some questions; just as the promising-to-come rains stop at the manifestation of an overcast sky, I stop too, from getting to know his entire story. But one can understand a bit about the young Koya boy, about to complete his schooling, a teenager, landless, from a village on the submergence map, and the train of thoughts in his mind. He was my first guide for nearly two days in my first trip to Godavari to meet the Polavaram-affected people. And anyway, not all these journeys were about gathering information with a mike in my hand thrust at the faces of people asking them – "so, what do you feel about losing your lands to the project?" Your question already loaded with the answer. Not all these journeys were about filling up reams with thousand and one stories from the submergence zone. Many a times, one just did not wish to ask another question, especially about the dam, the project, the compensation '*packagi*'. Sometimes, in silences, in between all the conversations, a lot more was said and felt that no words can describe in truthful detail except that the feeling was usually one of being witnesses, as against being the actors, in charge of their (our) destinies. In the peak Summer months, with water receding and giving way to a vast stretch of burning sand, you walked in pain, almost burying your feet in the sand (in spite of the shoes), even as you waded through it, the river bed now expanded, from the river bank to the first step of the stairs up to Kondamodalu village. Yes, it is not enough to have waded through the burning sands, climbing the seeming-to-never-end steps uphill and to have 'reached' Kondamodalu. Panting, you climb stone steps, gathering strength, bearing with the heat and the disturbed quiet, again, to walk a few kilometers further. '*Ido, ikkada!*' they will tell you – it is right here, but you know, as in the hills, here too, there is a vast gap between the real 'here' and the perception of it – yet, just three kilometers becomes a 'good' distance if you are walking on burning sand, or up and down hilly stretches, or through the vast barren fields with mid-day sun over your head; wishing you could just dump your heavy (always heavy) bag and rest. But then you have only so many hours and so many days in which to 'do' this trip, and meet as many people as humanly possible, for you do not have the luxury of resources to stay for extended periods. So, another kilometer or so, and I am usually led to a home to gather my strength, perhaps take a nap (*koncham restu teeskondi*, 'rest' meaning a small nap), may be have lunch (if there is time) and tea (the latter, a necessity), take in the flowers, the peace and quiet, spend time with Kumari, always having just about enough time to hear her story, when it is time to move on, further into the neighbouring villages to meet other people. So, in the heat of the summer, Kumari, and her story: a mother and her distance (physical and perhaps otherwise) from her two educated sons who never seem to want to return to their village. Her husband cultivates his three acre field that was once his own, but now taken on lease form the tribal (Girijana) sangham (committee) here. Kumari, though from the Kapu non-tribal community, has had cordial relations with the Kondareddis here, and her husband tills land that was once reclaimed under a tribal peasant agitation in 1969. In the beautiful in-between season of pre-monsoons, let us assume 'Spring', one met the school boy Premchand (indeed!) in a shared auto rickshaw ride from the *laancihila revu* (where the boats anchor) to Rudramakota village. This Premchand loves English literature and wishes to be a teacher: his desire or ambition in life, leaving all the traces of helplessness and despair behind, urging you to believe in a future time, just as he did.

The journeys led me from simple displacement to daily lives, deeper issues of caste, land ownership, alienation, exploitation and criminal proceedings against the tribal people, democratic institutions turned upside down, perspectives of a river. Over these visits, I also learnt about organisations changing stance on the subject or political parties changing stance once in a while, depending on the mood of the time or their use. And ultimately, the people of Godavari stand alone in their battle, either holding on to their lands, or resigning to their fates, as their patience to see a huge uprising against the dam wears itself thin. Of course post-2009 Telangana agitation, 2010 and 2011 saw some scattered, again fragmented, attempts to resume resistance to the Polavaram dam. There is something to be said about the legal process as well. Though petitions were filed in the Supreme Court on various grounds against some of the aspects of the dam, some of these were Greek and Latin to the tribal communities on whose behalf they were filed. Yet ironically, Kondareddis and other tribal communities of Godavari have had some of the longest battles in court on the land issue and access to *podu*, some even going back to the British times.[26]

Both these truths exist in juxtaposition to each other. As does another truth – among all those who protested, the police slapped cases only against the tribal people, leaving out the activists from the city. Barring a single case, where police clamped down on CPI(M) MLA, MP and activists protesting against the dam in January 2007 in Khammam district. Tribal activists and protestors from the submergence zone are still fighting to clear their cases. In one case, though, a local NGO director was constantly summoned by the police for his involvement in the anti-dam movement threatening to create difficulties in his FCRA clearance.

The Godavari region has played an important part in Andhra Pradesh political economic history. But there was one whole chapter of history that might have remained a closed book had it not been for the insights I gained from many conversations with many of the tribal communities in the region as also from conversations with the dalits, including the fishworkers of Godavari who I would include among the dalits considering the relative silence about them that renders them most marginal to the entire river debate. The lived experience of many communities by the river brings forth knowledge of greater significance than can be gleaned from narrow circles of those who seem to fight a different cause in different regions in different times. Or different causes at the same time, from knowledge already gained. One has avoided 'expert' opinion and expert 'quotes' here. The people whose voices are quoted in the book are ones on the ground, facing the situation head-on, long after the last group of flag waving supporters and the last advocates left. I also recorded religiously the scattered meetings, the street protests in the region, non-visible deliberations between groups hoping to get together on a single platform but never quite making it (in one or two contexts I have had to record statements of political leaders, if relevant). All this has led me to understand the construct of movements, as well. Of what makes a popular movement in India today, and why some movements do not happen. Or happen but not quite in the same forms as others. I have witnessed some of the street protests gradually slipping into funded 'projects' as I have seen organisations threatened into submission to the project mode or change their terminology of protest against the dam and joining the ever growing tribe of civil society groups with Polavaram as one of the many projects. By 2011, *Padayatras* (long march), meetings and 'round table' deliberations started all over again within the Telangana movement canvas. In general (not necessarily in case of Polavaram) political resistance takes a back-seat whenever funded protests assume centre stage.

Sounds, Sights, Images

Every moment I could sneak out on my own – early mornings, late evenings, night – was to capture, in a bid to take all that I could for posterity of what would all disappear – these sounds, which are only heard in silences of the mind and soul which cities have managed to crush out of our existence. Though with my Konica autoreflex SLR I could not always capture the darkness of the night, yet I did try and did manage the sights that poetry is made of – moonlight over Godavari, sitting with a group of fishermen on the banks, without use of flash (which I did not possess), and my cell phone became a proud possession at times of need, supplementing or making up for, the camera or notebook, sometimes the only tool I could use, whilst walking with people, or alone, or on a ferry, or on a crowded boat, moving along the backwaters during the Godavari in spate, etc. I could record as many bird calls as I could on this mobile phone (which remains my only phone, because it stores all those memories), call of insects, and sounds of the raindrops which sometimes I would, when I could, in a network zone, call a friend to listen to. Recording was important, with a sense of agony, that all this too will be submerged, killed, destroyed, if the dam comes up. The idea was to share all these – videos, sounds, images – from the camera, a cassette recorder (a personal stereo recorder which was used in initial two Godavari journeys until it stopped functioning), later a voice recorder and a digital camera handed down to me by a friend. From 2006 until 2010, my last physical journey, I recorded everything, capturing the developments, along with all these other elements on a sustained basis. I had sought to record it all and share it all with as large an audience as possible, which did not happen as envisaged. At times, the film roll in the SLR camera got over just when the most important scenario emerged. There is always that perfect or important shot that happens just when you are exhausted of your supply of film. But that is the truest record of the time, and closest to life, where you do not have second chances. It is true of the situation and hence a record of your situation or circumstance at that moment in that particular time in a certain village. So what you come back with happens to be the truest exposition of the reality that stared at you in your face. Digital cameras and endless memory cards take away from that truth and reality. Nevertheless I am glad by the end of the second year I had alternatives to use for my work. But till date the best moments happen to be those that I could take into my SLR camera and the best comments of people recorded happened to be on my old personal stereo. Something must be said about the technology of speed (as I think of intrusion of speed in every sphere of life, including that of a journalist's) or speed of technology which is used to kill the slowness and diversity of multiple, millions of species that one perhaps walks past in those forests, which are not even listed in the forest records, and even these silences of the night and the dawn which one recorded, to someday say to the world, hoping I would, that even all this is going to be washed with the force of the water that comes down the reservoir someday. And lives of people of Godavari is incomplete without all of these. You cannot see them in isolation. 'Social' impact is not merely about loss of 'workdays' in money terms, but also a life lived in a certain ecosystem. People lived for centuries, imbibing from these sounds and silences and darkness and shades of light, as well. The market only understands the sounds of FM radio channels, TV channels playing 24 hours a day, psychedelic lights and neon street lights at night; tried and tested and failed methods of producing, harming the parts that are irretrievable. Even in the forest species lists of the Forest department, there is a gradation of important and unimportant trees, animals, birds, trees (insects and reptiles do not even figure in these lists, nor even spiders nor amphibians); I am sure they do not even consider the micro-organisms present by the thousands in these dry deciduous and in part, semi-evergreen forests in the submergence zone. Talk of nature is disregarded as a romantic fantasy, even in times of global warming and climate change. Not a moment of our lives is dedicated to silences so we may be aware of all the lives around us. Does the R&R concept take into account life as an inter-connected wholesome unit? A forest official told me – "oh, the animals, they will run away when the dam waters gush out. Of course the weaker ones will die, we cannot help, can we? The birds will of course find other forests." The 'weak ones', among humans, include the old and infirm and the physically challenged, for whom there is no special provision. As it happens, they also have to give way to the strong, abled, and the young. The attitude towards the tribal and other marginalised communities in India today is almost similar to the attitude towards species that have come to depend on our grace and benevolence.

Media

Today Polavaram is back in discussion in mainstream media. But speaking of the years between 2006 and 2010, initially it was covered extensively in regional newspapers (Telugu) in the main editions, but reports gradually moved into the district editions and finally ended as event-based reports once in a few months and some of the stringers who regularly followed up on the issue from their field area in the villages (some of them who still remained ethical and felt it their duty to report on the avalanche of corruption developing over the R&R package, and the middlemen among the officials) found their reports getting little or no space even in the district editions. I mention one of the stringers from Polavaram in this book. I had watched the gradual disinterest and dejection that had enveloped this man, over the years in my journeys, to an extent that though he knew more than most about the scam, he chose to not write about it. Many stringers in rural areas are actually shopkeepers, or influential men in some cases (almost always men) who work part-time, gathering news and sending it. In fact, in Devipatnam, the stringer I met in 2006 was an owner of a small café. He was an upper-caste landlord and a beneficiary of the Polavaram compensation package. The struggle to get space for reports can at times completely disempower and make you feel nearly disembodied. But in my next journey, whenever that happened, I would know that somewhere those I met regularly in the villages understood. And the ball would roll again – the latest developments, how much land has been taken, who has taken the *package,* survey of officials, the new Act, the new RDO, the new ITDA P.O, and so forth. The problem is not an individual one, though individual circumstances have a bearing on it. It is just another angle to the India 'development' story. As for Polavaram, I have sought an answer to the question as to why Polavaram does not evince such a sustained interest in national media or politics, when compared to the issue of mining and the steel industry projects. Is there politics to it? Media seems to arrive when an iconic figure takes up the cause, or visits sites of struggles that have been ongoing even before the icon visited them. It has come to a state that even local struggles find themselves obliged to call at least 'one national figure' to address a meeting they organise with their limited resources. For they do not believe the media would cover a meeting, however important, without these 'media pullers', as I have witnessed in case of meetings organised by political activists on Polavaram. Something more 'eye-catching' than everyday resistance - especially if it is resistance by tribals, dalits and other marginal communities – is considered paramount for TV or print space. This is exactly how during our freedom struggle(s) we find mention of a few 'national' leaders and that local struggles against imperialist forces had to somewhere merge with the 'national' in order to gain respectability and a place in our history (thereafter).

In my case, I write this book as a human being, disturbed by the sequence of events I see happening all too frequently, leading us to an illusionary nowhere land. Many seem to be happy with the status quo while those who resist are branded 'insurgents' – as though no ordinary person, outside of political groups, or with no 'Party' affiliations, could ever be 'political' (outside of the now 'regularised' shades of Left – 'centrist', 'extreme', 'traditional', classical-theoretical, or even what is called the 'Common-person' politics) and protest against a world-view built on a 'development' paradigm particularly dangerous to the larger questions of life on earth. In between is a world of protest itself as a professional exercise, and just as footfalls in a shopping mall, counting footfalls on an open ground seems the only guarantee to unending coverage on national news. Is there a middle path which even the Buddha did not envisage? Just like the seasons, there is a change with a non-changing regularity, or change with ever-so-subtle nuances and conditions sometimes as dramatic as a tsunami or as gradual and impending as the climate change process. If there is one non-change, with a pathetically historical 'recall' factor to it, it is in the matter of 'exclusion'. Today's development is significant: the cement and mortar of it is replacing, displacing and dislocating the nerve, flesh and blood, and the human or animal or life-form of it. Godavari within the impending Polavaram dam is one such instance.

Notes

1. In then State of Andhra Pradesh, now in Telangana.
2. 'When Godavari comes'; 'If Godavari comes, we will go to the village'.
3. This meeting was thanks to an accident in my life, and had it not occurred, the meeting may not have happened in the way it did. A PhD that needed to be completed, search for resources for it, a chance phone call from a friend, pushing me to take up a 'job' (and a mother requesting I give it a shot) and my joining an international organisation, briefly. An accident: unexpected, uninvited. The job helped me take a trip to the field, Godavari, subduing the pain of losing my

mother within a month of joining work. I ended up meeting Muttu Rama Rao once, then twice thereafter. There are some accidents in life that impact you long after they have occurred. I left the job, and waited, till the next accident, a friend's chance mention of Prem Bhatia Memorial Scholarship, which gave wings to my innate desire to revisit Godavari, and engage with the Polavaram issue head on. Godavari stayed, and I stayed with her, through numerous other ruptures. On hindsight everything seems to have been one long-winding road to an almost providential destination, Godavari.

4. A day marked in my diary – the way Surakka quenched my thirst as well as having tasted the waters.
5. Strangely the name, Godavari, has kept recurring in my life. The first hostel I ever lived in as a student was Godavari (in JNU, Delhi). My first high-salaried job (not my calling) had the Godavari basin as its field of interest. My first journalism scholarship came for my proposal on the dam on the Godavari river. In fact, I travelled along Godavari even as I completed my PhD thesis.
6. That once belonged to my father. This camera has been equally crucial to all my Godavari journeys as it has had to compete with the most sophisticated gadgets that I did not possess but which always beat one in the media-game; it has also made one realise the power of moments that do not last forever and can never be captured all the time in all circumstances. You have to feel some of them. Nights have to be nights without the 'flash' and distances have to be distances without the 'telephoto'.
7. Though he did not 'belong' in 'submergence zone', yet may have lost his fishing zone and home to a Lift Irrigation scheme – Devadula.
8. The term, "people's history" I owe to a friend who saw in all my conversations and writings a sense of 'people's history', which I did not, at that point - being far too immersed in the experience of it all and assuming 'history' was too big a word. But at the end of that phase, I began to realise some truth in that advice. And one has to demystify History itself, for people's voices here are records of history to be retrieved at some point. Not all history is embalmed in buildings called library and archives.
9. As an ITDA project officer said to me – 'it is time to get them out of these forests and develop them.'
10. Government of AP, I & CAD (Irrigation & Command Area Development) Authority, Note on Indira Sagar (Polavaram) Project, Office of the Chief Engineer, Indira Sagar (Polavaram) Project, Dowlaiswaram. Unless otherwise stated, all the components mentioned here are taken verbatim from this report.
11. The elaborate details are appended in the book for information.
12. Shot up to Rs. 13,000 crores even within a year. Some reports say the number of villages may be more than 300 and not 276 (which was based on an old survey conducted several years ago).
13. At the time of the release of this report the proposals for dereservation of forests for the project had been sent to the Forest Department (as on February 2005) and approvals were awaited. But even before gaining approvals, forest land near the spillway site had already been 'encroached' by the companies for works for the tunnels. Especially near Mamidigondi, Devaragondi, and surrounding areas in West Godavari district.
14. Also from the DPR.

15. Report of the I & CAD, cited above.
16. I retain for the sake of continuity these District names as they existed in former Andhra Pradesh, when my journeys had begun. Needless to say, East and West Godavari districts have now become part of Andhra State, which combines the Rayalaseema region as well as some of the districts that were earlier part of the Telangana region.
17. Shared by the Andhra Pradesh Rythu Sangham – the CPI(M)-affiliated farmers' organisation.
18. Nokia 6600, one of my richest 'exchange' purchases in the year 2006 which pre-empted the circumstance of its use in my journeys.
19. Later came the voice recorder and a basic digital camera that was handed down to me by a friend who may have been irritated with my outdated technology and the money I was spending on taking photographs, washing and printing them – thousands of them; nearly a hundred-odd from each journey.
20. The 2006 and 2007 journeys covered here are excerpted from my own report for the Prem Bhatia Memorial Trust Scholarship for Journalism - "Taming Rivers, Drowning Hopes: The Polavaram Project: Irrigation or the Deluge of Contracts" - with the support of which I was able to make these journeys. But a lot here is additional to what was contained in that shorter report version, based on my field notes, reflections and research thereafter. Some developments that were not covered in that report are mentioned here, and in the 2007 Godavari journey, following this.
21. One of the phone calls I received, not very long ago, was to learn that many who received cash for their lands and bought vehicles (motorbikes, mainly) were selling these off these because they had no money for petrol. After losing their lands, cash was fast dwindling, too.
22. Podu patches of land are usually in the midst of forests which have been under cultivation by tribal communities across several generations. Prior to colonialism, the idea of 'ownership' of these patches was an oral memory, recognised and sanctioned by community traditions, but with colonialism the idea of 'records' of ownership put a physical, rather 'closed-up', fixity to these lands; even today, even with the FRA, the onus of establishing 'ownership' of these patches lies on the communities, again rewarded a piece of paper that is more a record for the government than an aspect of traditional understanding of land.
23. Hemingway, F.R., *Madras District Gazetteer: Godavari*, Reprinted by the Superintendent, Government Press, Madras, 1915, p. 74.
24. Ibid, p. 78.
25. Ibid, pp. 71-72.
26. For instance, I found reference to a case in the "High Court of Judicature at Madras" ("A. No. 26 of 1906", with judgment date of "Saturday, 17th December, 1910"). Appellant(s) was "Sri Satoda Bihara Martapatra Kondamodalu Linga Reddi alias Sullee Abhoyee versus Respondent(s) Sree Rajah Kocherlakotah Venkata Krishna Row Bahadur Zamindar Garu, Proprietor of Polavaram. The case concerned "Land Tenure, Permanent intermediate tenure, adverse possession, tenant holding ever Limitation Act, Art. 139." There is yet another case mentioned: "Case No. A. 183 of 1905" between "Appellant(s) Kondamodalu Linga Reddi minor under the protection of the Court of Wards

represented by the Collector of Madras versus Respondent(s) Alluru Sarvarayudu of Annadevarapeta". Remarks – "Limitation Act, S. 19 – Acknowledgment by Court of Wards or by Collector as its Agent – Madras Court of Wards Regulation V of 1804, Ss. 17, 32." [Source: http://mljlibrary.com/. All material is under Copyright of MLJ Library, Comprehensive Electronic Library, Reporting Judgments of Supreme Court of India and Madras High Court. But you have to register as a member and be a subscriber after you log in for further details of each case on this website, and these are shared only after your registration is approved, when you log in. It is a pity I did not follow up on this. But I mention the numbers and the website so that someone interested – with better credentials and subscription – may perhaps browse through these old land related conflicts of Kondamodalu and other villages of the Madras Presidency during the colonial period between tribal and non-tribal communities and bring these out in public domain.]

The Godavari Journeys

The 3 am fisherman-Godavari Gattu, Rajahmundry

Godavari Journey 2006[1]
The 'Site', 'All-this-will-be-lost' Feeling, and Other Accounts

With the help of the Prem Bhatia Memorial Scholarship, I made around six visits – two were extended - between June and August. This is the journey where the metaphor *lived* itself, literally. Godavari came, and I made my second visit within a few days of her coming, though I was there barely days before she arrived. I was gathering just about everything – information, sights, sounds, voices. Somewhere, I felt there were several intriguing historical aspects to the place.

June 2006

First Visit, Godavari, Up Close

East and West Godavari and Khammam Districts (Rajahmundry, Kondamodalu, Somarlapadu, Kathanapally, Polavaram, Chegondapally, Mamidigondi, Devaragondi, Kunavaram, Rekapalle...). The context this time was the Supreme Court order to stop all work at the Spillway site of the Polavaram dam, pending environmental clearances.

2nd June

I watch the beautiful Godavari at dawn from the Gautami Express minutes before arriving at Rajahmundry railway station. My journey, then, starts with Godavari, and I wonder, would she be there with me through the entire stretch? How long? I head to the Head Post Office near Kotipally Bus Stand. I had gained access to 'Suite 1' (!) thanks to a friend's husband who was then in the Postal Services. "Suite 1" was to be my safe haven (even if for half a day) often until 2010; I got to stay there for a bare minimum cost (Rs. 80 per day), or for free.[2] Mr. Radhakrishna, an employee who sometimes seemed the 'all-rounder', welcomed me at the HPO. It was a spacious old-world suite, with two rooms (and AC and a TV, if you wished). I was to start my actual travels the next day, after meeting some people whose names and contact details were given to me in Hyderabad by some activists who were then in touch with developments on Polavaram. Everything about the room suggested I could set it as the base for a day or two to contemplate and write in peace or even sleep peacefully. But I ended up sleeping at 1 am to wake up at 2 am, not enough time for a bath. Mr. Radhakrishna came by, to drop me off at the *laanchila revu* (the boat anchor-point). It was time to proceed on those paths for the first time. I was feeling fresh with excitement in spite of an hour's nap.

Diary Notes

In my diary-notebook (to accompany me on several Godavari travels)[3] I had written, with the excitement of a child – "At Rajahmundry Head Post Office! Guest house, that is. Nice, clean place. People a little too eager to take care – a problem, of course. But it is okay, may be, to let oneself be taken care of for a while…Met an activist advocate here, Nagaraju, from the AP Rythu Coolie Sangham[4] who suggested to me an altogether different route plan, the best of all. So tomorrow, for the first time, will hop on to a *laanchi* (a motorised boat) carrying provisions, at 3 am (first time on a river that early) to visit a village fair (*santa*) at Kondamodalu. He said we would reach just in time for it. Am excited, needless to say. But I see that one is not allowed to show childlike excitement in one's attitude, in these moments, to everyone. It seems to betray (to them) a kind of non-seriousness. But one wishes, at times, that these happinesses were also part of the larger world-view, where just to take in some moments of nature in silence was not considered abhorring…"

***

Mr. Nagaraju's advice and itinerary was a perfect way to commence my Godavari journey: taking me from the point of contestation, Godavari river, through Polavaram dam spillway site and the villages en-route, and all the rest of the places I ended up visiting, which gave me first-hand knowledge of situation on the ground and what people in the city believed they knew but did not. I was all the richer for taking roads different from the usual. References, suggestions, dos and don'ts (that so many people in Hyderabad, engaged with the Polavaram debate in some form or the other at that point in time, suggested to me), were all thrown aside once I plunged into the travels. Roads, routes, acquaintances and 'sources' just happened on their own, from one journey to the other. At a level of philosophical contemplation, though, every individual's journey is hers (or his) alone; and the truths you come to, are your own truths, based on your own experiences. That is the only true journey – where roads just form on their own, after you take the first plunge into the unknown, shedding yourself of all the knowledge and claims to be the knower. The river became my first 'road'; and how perceptive of that man to have understood my moderate budget, even without my saying so. The luggage launch, although an exclusive experience, actually cost me just Rs. 12 for one person; in all Rs. 24 for a luxurious cruise, with my own 'guide', Ravi. The long ride on that boat, an entire day, gave me a very important insight into life in these parts; about the centrality of the river and its meanings. Travelling on a provisions carrier, *luggagi laanchi,* meant travelling with everyday ware and with people's monthly stocks of life's 'supplies', as it were; it also meant taking an absolutely slow, non-rushed view of things, taking in the river with its own pace and rhythm, travelling with/in the flow.

***

Making Hay While the Project Lasts, With CAD Drawings: Stepping into an Internet café at Rajahmundry

Must mention, of course, that this other new angle to Polavaram that happened by sheer chance. At Rajahmundry, a day before I was to take the river boat ride, I quickly went in to check an email, at an internet café by Kotipally bust stand, close to the HPO. There I saw an auto-CAD print-out of a map highlighting the Polavaram Dam Right Bank Canal. Intriguing! So the work, which was officially supposed to have been stalled at the dam site, thanks to the Supreme Court order, was ongoing elsewhere! And I found out that the consultants for this project were a group called 'Progressive' (couldn't help smiling at the unintended pun!). I asked the woman at the internet café for a copy of the same, saying I was ready to pay for it, if needed. The woman was initially willing but somehow got a little suspicious, it seemed, and told me she would need to ask her boss. The boss came in, and said he would have to ask his 'clients' (Progressive). Fair enough, I thought. In the meanwhile, I also observed quickly some other maps, one of them Papikonda Wildlife Sanctuary, and this was drawn by the Irrigation Department (not the Forest, as one might believe). I managed a phone number of the Irrigation department Divisional Engineer, who said over phone that he would send me a copy of the Wildlife Sanctuary Plan with his staff to the Post Office. He never did. I never got my hands on that work plan ever again, but I also realised, by then, as to my naiveté in believing I would. NGOs manage to get these kinds of information far more easily than ordinary mortals do, I thought. But what was more interesting to me was the career graph of the internet café owner here. He set up appropriate infrastructure to take large printouts of maps and related information. He also entered into digitisation jobs. I was sure he was involved in the Polavaram dam project in an indirect way, at some end of the line. Ram Babu (which was his name) told me the first year he had set it up, he incurred heavy loss. But "with Polavaram project coming up, I am making good business, since I am the only one doing this kind of work, in Rajahmundry." Ram Babu asked me to pay him a visit whenever and offered to get me some tea, which I refused. I wish, instead, he might have shared the CD of those maps.

***

On Godavari's Bosom - 3rd June'06

So, at 3 am, I took my first trip on Godavari downstream, on a launch. It had poured the previous night and was still smelling of fresh earth after rain, with a mild breeze blowing. The clouds above were waiting to rain again. The launch was ready to leave, the owner and his assistant loading the last few supplies (bananas, diesel cans, kerosene cans, rice bags, etc as far as I could see) on the deck and a few people, just like me, were waiting to get in, on the banks of Godavari at what is called '*laanchila revu*' or *Godavari gattu* at other times. A lone fisherman's silhouette was becoming visible, starting his day at work. The launch was a virtual provisions store – provisions meant to be delivered along the banks at the villages with handwritten notes identifying their recipients and who might come to pick them up. I perched myself atop the deck and the launch revved up. I understood for the first time how beautiful *slow* can really be. I was 'switched off' to all other aspects - including all other sensations, except to feel the Godavari air, and listen to the sounds of water lashing against the steamer's belly, if you paid attention, ignoring the loud rumble of the motor. I was to do all that always did in my journeys – try to capture moments in film, on tape, in my head. Only, this time, when I had taken in enough of silences, I would get into the record it all mode with the single worry in my head – is this all going to disappear, when the dam comes up? The launch ride was beautiful, sleeping, briefly, under the skies almost on the lap of Godavari, an expansive lap. Lightning, slight drizzle, strong breeze, but the river seemed unmoved. Seemed to say something to me along the way; it was like getting a love that calms many angers. By the village Tallapudi, as the launch gently roared on, the Sun was just coming out lazily, and as ever, missed the perfect 'sunrise' for my camera. At Tallapudi, where the boat anchored and Ravi and I had some tea, I saw the deity carved on the bark of an old Tamarind tree (so intricately carved it could be missed for the tree itself, just as old tamarind trees have a way of expressing their age in the shapes of their base trunk, just as wrinkles on a person's face hides the stories of life). The launch made way for the steadily increasing number of women coming down the river to wash clothes. When Godavari flows, in the normal pace of her movement, she establishes relationships with the big and the small, with the everyday people, as with the once-in-a-while people. But anyone who watches the flow of her body can tell - it is in the smaller - the smaller movements, smaller transactions, smaller activities - that Godavari has her essence as a river. She bonds with the everyday people with each phase of her life, when in spate and otherwise. This is the only river I have seen that seems to believe she is the sea! Vast, voluminous, all-encompassing. Is she beginning to possess me, Godavari? I wonder. Is it possible to work in a non-attached context, as a 'professional' journalist with this kind of an idea of the river in me? Does not one need to go beyond metaphors? But perhaps it is in the metaphor that lies the depth of the issue that I am beginning to feel responsible for exploring? Different colours appear in the sky, and it looks like it might rain again. There is some movement, meanwhile, in the village by the bank that we are approaching. This boy, Ravi, a Koya, has been rather quiet so far through the journey. We both had tea at a shop in the village, where I noticed the most intriguing figurine by the tree, which seemed almost like an extension of the tree itself. But then I noticed a Siva *linga* (the phallus symbol) placed most incongruously by its side, of course, a later addition, not part of the tree. Still further, I could notice a very old stone image of the deity Ganesa and that of a Yaksa. That was all there was time to observe and return to the *laanchi* again. I am told it will be seven more hours before we reach Kondamodalu (literally, 'where the hill begins').

7.50 am – The launch just anchored at Polavaram. The last passengers (who had come from Vijayawada – a family of an elderly woman, a child and a young couple) alighted here. So it was just going to be me, Ravi, the boat driver and his assistant. At a distance one can hear voices, people talking – the Godavari *yaasa*, or dialect, almost feels like the river itself, the way the words sway up and down, musically. The tone bends a little, just a little. And then I hear that sentence, with the metaphor, my first in this journey, *"Godavari occhinappudu"* (when Godavari came), as Ravi, the guide suddenly broke his silence to tell me we had just crossed the Sivalayam (Siva temple) at Pattiseema and there was a vast stretch of sand by the banks, and he was telling me how high up Godavari can go when she usually comes, enveloping this vast stretch and almost up to the base of the hill where the temple rests. Then he said there was still time for her to come, may be in a month or two. These smaller *laanchilu* (launches) support the smaller rural livelihoods - everyday, movements, of goods, people, basic needs and basic essentials, to lead a life. Once the project is constructed, and Godavari stands still (or the state hopes it will) there will no more be the movements of these kinds. Wonder if someone ever actually calculated the volume of transactions (economic) that happen each day in these smaller movements, from village to town, village to village, village within village? Do these ever find themselves worthy of being added to the economy and GDP? The launch owner, the boat driver, and his assistant – what could they be thinking of now? I asked the 'captain' as to what will happen once Polavaram dam is built. He says "Godavari will become bound within the canals – how can she support movement of launches like mine then?" He makes around Rs. 1,500 a day transporting provisions, and other goods between Rajahmundry and these villages. He is well aware that this route may not exist once the dam is built. I had drifted off to sleep when suddenly an angry woman's voice broke the silence around. We had stopped. The launch owner had not delivered what had been promised. Small village traders depend on stuff arriving on time for their shops. This woman shop-owner and her daughter shouted at the captain using the choicest of abuse terms prevalent in the region. She held the long rope that anchors the boat and would not allow it to move until she had made her point. Meanwhile, an Andhra Pradesh Tourism launch playing a film song passes us by – a song on Godavari from the new film titled Godavari (by filmmaker Shekhar Kammula, where cinema and tourism connived). Right thereafter, one can sense the immense depth of silence despite the motor boat chugging along and the gurgling river.

About Godavari as a 'waterway' Hemingway in his *Madras District Gazetteer* (Godavari districts) makes note of different kinds of navigating mediums. He mentions the "stern-wheeler" (which) "touches all the ferry stations on both sides of the Godavari between Rajahmundry and Polavaram and has recently been run experimentally as far up as Kunavaram to provide communication with Bhadrachalam ..."[6] And he mentions Kovvur, Arikarevala, Kumaradevam, Tallapudi, Sitanagaram and Gutala as the ferry stations. He writes, "a great deal of goods and passenger traffic is also carried on the river in native sailing boats. They are generally 'dhonis', which run up to 35 tons capacity..."[7]

Many people depend on Godavari – the tribal farmers, the fishing communities – seasonal and itinerant – among others. There are fishermen's families along the banks of the Godavari. They put up their huts on the sand banks and stay here most of the three peak months of fishing. As we move in the launch I spot thatched hutments of fishing families. Smaller fishermen from Vanapally, Kapileswaram, Thotipally, Chevur and other places set up their huts on the banks of the Godavari – their peak season is between March and first half of June and they leave just in time before the Godavari's levels rise. But a few of them do stay put, and it is these fishermen's boats that come in handy most times when state machinery fails during the Godavari's coming. The fishermen are not seen as 'project-affected' incidentally though they lose out on a livelihood they have practices since centuries now.

Hemingway (1915) notes that, "Among the great rivers of India the Godavari…runs nearly across the peninsula, its course is 900 miles long, and it receives the drainage from 115,000 square miles, an area greater than that of England and Scotland combined. Its maximum discharge is calculated to be one and half million cubic feet per second, more than 200 times that of the Thames at Staines and about three times that of the Nile at Cairo…The holy waters of Godavari are said to have been brought from the head of Siva by the saint Gautama, and the seven branches by which it is traditionally supposed to have reached the sea are said to have been made by seven great rishis…It is customary for the pious (especially childless persons desirous of offspring) to make a pilgrimage in each in turn and bathe there, thus performing the sapta-sagara-yatra or pilgrimage of seven confluences. The traditional seven are the Kasyapa or Tulya (the Tulya Bhaga drain), the Atri (the Coringa river), the Gautami, the Bharadvaja, the Visvamitra or Kausika, the Jamadagni and the Vasishta. The Bharadvaja, Visvamitra, and Jamadagni no longer exist; but pilgrims bathe in the sea at the spots where they are supposed to have been."[8]

In a contemporary text you will find a technical description (of the river's expanse): "The Godavari…rises in Sahyadri mountain ranges of Western Ghats at an altitude of 3,500 ft above mean sea level near Triambakeshwar in Nasik district of Maharashtra and flows West to east for 910 miles to join the Bay of Bengal. For the first 30 miles it passes through the high rainfall zone of Western Ghats and then enters into average and below average rainfall areas of Maharashtra. After traversing 431 miles, it enters AP…The river flows in AP…for a distance of 189 miles in Nizamabad and Karimnagar districts…and Adilabad…The Godavari becomes a mighty river only after Pranahita joins at mile 620 and Indravati…at mile 650…Godavari at its mile 650 has a total flow of 2,370 TMC...The Godavari river after its confluence with the Indravati, northern side becomes the new State of Chhattisgarh for about 15 miles then it is AP on both sides. The last major tributary Sabari joins Godavari at mile 788. The river enters the narrow gorge of Eastern Ghats hill ranges and finally emerges out and flowing through the delta regions joins the Bay of Bengal at mile 910. The Polavaram project site is at mile 829 and Sir Arthur Cotton Barrage at mile 852." [9]

Reflecting, on the River…

At the larger level, it is about the river, Godavari, about water, and about conflict on a number of complex issues, though displacement will be the immediate 'after-shock' of the dam after construction. Who really 'owns' the river, or 'distributes' its largesse? Who controls the Godavari? Who expresses this right, through state power? In a country where river waters have become a site of contestation, what could be the implications of large, larger, largest dams – in which is engrained the language of state, power, prestige backed by interests of commercialisation? Polavaram dam or any talk of Godavari cannot be singularly about displacement of the tribal communities and dalits (largely) – but also about it – but at a deeper level about conflicts between agriculture and forests, forest-based agriculture, land ownership and landlessness, among others. Land alienation has been an important issue in the east Godavari region with occupation of tribal land by non-tribal settlers. There is also conflict between dalits and tribal communities in some areas. In a sense, the present government in AP would 'complete' a task begun in Arthur Cotton's times.

At Kathanapally, 4.30/5 pm – Finally, we reached Kondamodalu. Then it was a long way up the ancient steps leading the village on the hill, followed by another long walk (seemed the day wouldn't end; if only one could sit for a while) till Kathanapally. It is a small Kondareddi hamlet, in Kondamodalu panchayat, East Godavari district, beautifully nestled in a forest. It was around 5 pm. Slept like a log left in peace, absolutely dead to the world. I am put up in this house here for the night - I do not know whose. Early next morning, I heard, and went in search of, a bird that sang every half hour, a long note, exactly at half hour intervals! Finally noticed a tiny black bird, up on that tree (perhaps it was the *ippa*[10] tree), but could not make out what bird it was. Its song was the perfect note, ending the immense silence of the early morning hamlet every half hour, starting from the base and reaching a crescendo at the end of it. Perfect start to the day. I could not record the sound on my cassette player though I tried. There is something about birds and animals of the wild – they just sense it when you wish to freeze a moment with them, on a camera or a sound recorder, perhaps on account of years of being conditioned to the guns and weapons that 'froze' their bodies forever? Bathed under the open sky, in the morning, just thatched reed walls around me, with warm water that Varada provided, perfect for a tired body, though anxious if somebody might watch – of course they wouldn't, just a matter of urban conditioning. People never have supper/dinner here without a bath. I realised now why they kept asking me to have a bath late yesterday night. Most of them bathe twice here; early hours of morning can still be missed, sometimes, but the pre-supper/dinner bath cannot. It had perhaps disgusted them that I ate without the evening bath the previous day. Little later, I met Illa Rami Reddi[11], the former Sarpanch of Kondamodalu panchayat, but even now addressed as sarpanch or *president*. He was a short, stocky man, with a booming voice and spoke a lot, with great conviction.

Kondamodalu in History Records

The colonial Gazetteer mentions Kondamodalu - "Twenty seven miles west of Chodavaram (population 332). The head quarters of a mokhasa estate at the entrance of the gorge on the Godavari. The present owner is the grandson of the Linga Reddi who assisted Government in the Rampa rebellion...His services were rewarded by the grant, as a mokhasa, of the village of Ravilanka, which is held on the condition that the grantee attends the Collector with peons when required to do so, and pays a quit-rent of Rs. 300. Linga Reddi had previously, in 1858, been granted an allowance of Rs. 50 a month to compensate him for the withdrawal of his right of collecting fees on goods passing up and down the Godavari. This grant is conditional on good behaviour. Linga Reddi had just then earned the gratitude of Government by holding aloof from the fituri of his partner Subba Reddi. Kondamodalu comprises four villages and pays Rs. 110 annually to the zamindar of Polavaram. Its precise relations with the latter are at present the subject of a law suit."[12]

Kondamodalu, Post-Independence

And this is what's written about Kondamodalu in a volume edited by Haimendörf – "The state government became aware of the need for implementing protective legislations in the early 1970s when Naxalites began increasing their influence among the Reddis in this region...In 1969 they encouraged the Reddis, Koyas and Kammaras of Kondamodalu village and its 13 hamlets to harvest the paddy fields occupied by non-tribals. This they did and carried away 680 bags of paddy. The police immediately swung into action, arrested a large number of tribals, and filed cases against them. Many tribals were kept in jail pending disposal of the cases until 1975 or 1976. However, the government has since filed *suo motu* suits against the non-tribals of this village for occupying tribal lands in contravention of the Land Transfer Act."[13]

Sitting with a few members of the Agency Girijan Sangham here I learn of some of the intricacies of the tribal land issue and "RR" (Rehabilitation and Resettlement) - a term that has unfortunately become part and parcel of the language here, ever since the Polavaram project commenced. This was the first time I was to hear some of the names of the villages that were to become part of my vocabulary soon thereafter. They told me that tribal villages Chegondapally and Singanapally (West Godavari) are affected on account of the spillway of the dam, while the non-tribal villages affected by the spillway are Ramaiahpeta and Pydipaka. With the canals being dug at the spillway site, the danger of the latter two villages (being closest in line to Godavari's coming) getting submerged during the coming monsoon season is even higher. I was to find out how true this prediction would be, within two months. The tunnel of the project affects the tribal villages Devaragondi, Mamidigondi and Thotagundi, near Polavaram. And while the village Ramannapalem is also affected similarly, it is not being considered as "project-affected". Then Illa Rami Reddi and others also enlightened me on the paradox of *patta* and non-*patta* lands. Many of the tribal families here do not have *pattas* declaring them as official owners of their lands. Without *pattas* they will not be eligible for the *RR packagi* (package). Rami Reddi took me back to a past I was to refer countless times again in my research in between the journeys – the 1969 tribal peasant agitation in these parts, through which the tribal communities, under the leadership of the T. Nagi Reddy (of the group that broke away from the CPI following the end of the Telangana Armed Struggle) reclaimed hundreds of acres of tribal land from non-tribal upper caste landlords. In the course of this conversation I heard the name of one particular NGO which was also involved in negotiations with the Government over the displacement of tribal communities in the Surampalem reservoir project in Rampachodavaram and Kovvada project (both these were minor irrigation projects) in Gangavaram mandal. Rami Reddi pointed out that the contract for building one of the R&R colonies in these projects had been given to a contractor who was a blood-relative of the founder of this NGO.[14] Some NGOs have been talking of better compensation package for Polavaram while others are against the project as of now. I was also informed of the panchayats – in the submergence zone – dominated by tribal communities and non-tribal upper castes. Among the former were Manturu, Chappagonda, Tunnuru, V. Ramannapalem, Kondamodalu, and the latter (dominated by upper castes) were Chinna Ramiahpeta, Pudipally and Devipatnam. Among the problematic points in the present RR package, Rami Reddi said "rehabilitation is not in terms of one hamlet; they are talking of resettling us in different settlements. Agency Girijana Sangham has demanded Land for Land, Forest for Forest and to retain the cohesiveness of the tribal hamlets in each of the R&R colonies. Let them show us all this first, and then we will think of moving. These are not our demands, but conditions." Incidentally, the Kondamodalu Gram Sabha had passed resolutions to this effect, and these were quite contrary to the idea that the government officials spread through the media that all these panchayats had unanimously given consent to the Polavaram project. I was handed a copy of the handwritten Gram Sabha (held in accordance with the Panchayati Raj Act 1 /98 on 3rd June 2006; wish I had been here two days earlier) resolutions.

They also shared a pamphlet brought out by the Agency Girijana Sangham and AP Rythu Coolie Sangham, titled "Chalo Polavaram", calling people for a protest dharna and public meeting at Polavaram on the afternoon of '25.02.06'. The pamphlet read:

> "On 8.12.2004 the Chief Minister (Y.S. Rajasekhara Reddy) came for the bhumi puja (ritual offerings to the land) for the Polavaram project. Thousands of girijans[15] (tribal people) demanded that the Polavaram dam not be constructed without assuring land for land and forest for forest. It was also assured that the government authorities would repair the breaches that had developed in the Kothuru cheruvu (freshwater tank/lake) on 24.6.2005. But that was not done and instead they are filling it up with soil. They have been dumping soil and rubble at the spillway and canal works sites into this *cheruvu*. There is danger of many villages drowning in the coming rainy season if this continues."[16]

And I was shown a pamphlet dated '23.7.2005' condemning the arrest of their comrade Sunnam Raju (Chegondapally village). It said, "Sunnam Raju was taken away forcibly and to prevent the tribal people from East Godavari, West Godavari and Khammam from moving in to support him, the local officials stopped launch services on Godavari. On 25th July more than a hundred people gathered and the Joint Collector, Ramanjaneyulu and the RDO tried to make a bargain with us saying, if we promised not to obstruct the Venkatreddygudem canal works, they would release Sunnam Raju. This shows that the government will go to any length to crush any tribal resistance to the project…On 13.7.2005 tribal people from six villages in the submergence zone demanded stopping of the project unless the land for land issue was resolved…In June and April girijans protested at the district offices and at the office of the Joint Collector. One of our people, Made Siramayya died and others were wounded on the way[17]...We want to know as to what is the status of our cattle and livestock? Where do you promise grazing land for them during the fallow period? You mention 150 square yards of house site. Where do you expect us to keep our cattle and livestock in this kind of housing? There seems to be no other way than to kill them or sell them." There were other petitions too which I had to read (or get them read out to me) in a short time, including an Open Letter to UPA Chairperson, Mrs. Sonia Gandhi. The situation on the ground seems heady, and I hear the term *package* often. I was happy to find people sharing all this with me within a moment of introduction. And this was true of whosoever I met, not just the political groups.

On the morning of the 4th June, I also met some women of Kathanapally village sitting by the tamarind tree. Most of them were Kondareddis. When we got talking, the 'Project' was upmost on their minds, of course. Reva, a young good looking woman, said, "We shall neither leave this land nor this village…"

Joining in, Vidya said, "Since 30 years we have been requisitioning to the ITDA for a hospital in our village or the panchayat. They have not given us one till date…what is the guarantee they will give us what they now promise if we leave this place? This project is not for our benefit… The government officers came here to conduct a survey; they never spoke to us, or our elders. They did not seek our opinion." The elderly woman, Appayamma, pitched in, saying, "We will lose this kind of land and these homes. We will be economically worse off 'there'."

I asked them what would be different about the land they were being promised. Appayamma said, "Well for one, the soil will be different. Our land reaps gold. Our forests give us other things as supplement. We get pulses; we have the *konda podu.* If we leave this village, we will lose everything." Appayamma and Vidya said, one after the other, "every tree, every stone here is akin to gods…Our festivals are celebrated differently; we all gather at one place and decide a festival date. We bring pots and place them under a tree. Someone assumes the role of the *poosari* (priest). And we offer prayers. We don't eat a mango or any other fruit without offering worship to them. The first fruit of each season is worshipped..."

And then they sang together, since I asked them to (feeling later like a colonial anthropologist, asking them to sing out of context). I was anxious in that first trip that whatever I was seeing now may not all be the same, or even there, very soon. I may not hear this song again, and god knows if would go there again or where would they be. Strangely, I never heard that song again.[18] The women moved around in circles, placing their hands over each other's shoulders. That too for me seemed a metaphor, human (feminine) bonds. And of the life that the R&R colonies will force them into. These kinds of bonds will never understood in these colonies that they will be 'resettled' in.

They sang,

"Rela relayaaara mula ela.....
rabo ra muttellamma ravo muttellamma......rela rela madhavolu...
grama talli, na ochetappudu..."

The men, who were until then simply squatting, watching all this action, suddenly asked them to sing a wedding song. There was general laughter all around and the children laughed. By now I was feeling like the perfect outsider, or an insensitive filmmaker asking her subjects to act and behave a certain way to capture them. I did not intend this. But on hindsight, I am glad I captured their smiles, for never again in that same village did I manage to find another collective moment of cheer in the midst of the that impending doom. I never found Reva, Appayamma, Vidya together in a context in the same village, after that.

Meanwhile, the song, and the bonding over it, continued for a while longer:

"kodaalaiyo.....achho kodalaiyo...
kodalaala...teere....rama totha pelli....
sarkakaaya konelu katte....totha pelli...
chelamma devi.....amma kontaa...
pellikoduku..peddavada...
pelli kuthudu chinna daani...
cintha puvu tecche....tippegatte
chinta puvu te pope mida ...
niku naku....telepodam...neeku naaku ...ayte ...."[19]

While they move around in a circle, two women sing the lead lines and others join in chorus. There was perfect synchrony in movement and in what they sang. Post-singing, as we moved into our respective homes, Reva said, "Here we share rice or food. We share all that we grow. The whole village gives a share. Everyone has to give a part of their harvest, which helps the families that don't have enough."

Vidya said, "We believe we will not be so together 'there' (RR colony)… During the monsoons Godavari comes here and stays for a few days. Though you might feel Godavari is way down there (about 2 kilometers downhill), she comes up to this point" (*ikkada varaku ostundi*, she said, showing the tamarind tree). She added, "We have wells, and Godavari water and we have motor pumps. We catch fish in Godavari. Here we get food from our fields; if we move to the RR colonies, we will have to buy food."

Women, more than men, I realised in the course of all my journeys, understood the more profound truth of it all, reflected in this last statement – the distance between growing and buying food; between self-sustenance and market-dependence. All the women I spoke to here were convinced they did not want this project. Before we broke up, the women 'broke' into an impromptu song, and this time there was no prompting from any side; no convincing, and no recording in anxiety. Just a general feeling of camaraderie and peace. I had a long journey ahead of me. But I left with the promise to myself that I will come again. And shall sit down and ask them what these songs mean…Women seem to have little or no place in the discourse both for and against the project The economics as women see it, is very much part of the environment they live in. They can not dissociate the two. If they were to evaluate the land they are living in today, the value of the land would be far dearer than it is in government records.

Meanwhile, **at the village Somalapadu**, I was introduced to Kondla Rami Reddi, 90 years old, one of those who participated in the 1969 peasant agitation in Kondamodalu to reclaim tribal land. He said, "People had neither food nor land they could call their own. We did not take pattas. The landlords used to expect us to work for three years on a measly amount of Rs. 50, and would charge interest. We were all perpetually indebted to the landlords. We have lived on struggle. Comrade Simhadri Subba Reddy, other leaders (from the party of Communist Revolutionaries, later coined CPI (ML)[20], came here. We continue to struggle even today. We will struggle against this project too. We stopped police from coming here in those days (1969). Now no policeman ever comes here. We are looking after ourselves."

***

Some memorable moments in my Cell-phone Notes [21]

4th June'06: On Godavari…first time from Rajahmundry; first time downstream. 3.30 am! Rain, small drops, breeze, lightning, drizzle. Me, covered in my raincoat, lying on the deck…a little scared of the lightning though….Perhaps Godavari is saying, "It's like forcibly injecting the seed and establishing paternity, this dam!"

The old blind man and his wife, a couple I just saw. So very frail and old. How on earth will people like these, with no younger members in their family, move to an unknown, new environment? After all, aren't these sounds and smells their way of establishing a connect with the place? And what will they do? There is no mention of elder citizens, and physically or mentally challenged people in the R&R package.

***

Reflecting on Loss of History, Perceptions of Tribal Communities

The submergence zone in East and West Godavari and Khammam districts is home to Kondareddis, Koyas, Koya doras. Konda Reddis also call themselves Reddi Rajulu, kings among the Reddis, virtually 'kings' of the forest, about whom Haimendörf wrote, "The Konda (of Hill) Reddis of Andhra Pradesh are one of the tribal groups which depend to a great extent on slash-and-burn cultivation…Traditionally, the economy of the Reddis is based on the periodic felling of forest and the cultivation of various millets, maize, pulses, and vegetables in the resulting clearings. This type of tillage, in which, the axe and not the plough is the primary instrument, is in Andhra Pradesh known as podu...The Reddi of the Godavari region broadcasts all small millets without so much as scratching the surface of the ground and dibbles the great millet (*Sorghum vulgare*), maize, and pulses into holes made with his digging sick. It can be safely said that Reddi agriculture represents as crude a form of cultivation as may be found anywhere on the Asiatic mainland. It is by no means efficient, and at some times of the year when their stores of grain have run out, Reddis subsist on wild forest produce, eating the sago like pith of the caryota palm or the kernels of mango stones. They also hunt with bow and arrow, and those living on the banks of the Godavari add to their food supply by fishing often from dug-out canoes…The sense of unity based on a group's common ownership of a tract of land finds expression in joint ritual activities..."[22]

In an unrelated (to this discussion) context, speaking of 'national' pasts and what makes for 'national heritage' (the archives of the old territories and geographies constructed in a specific context) David Ludden says, "All histories of the peoples of the world currently appear in the cage of some national past or another, but some need their own space. It is a pressing challenge to imagine at least some history in non-national terms..."[23] He speaks of the "geographical grounding of social identities inside agrarian territories in India's southern peninsula" [24] I mention this to point out as to how the tribal communities and their historical geographical spaces have been perceived and in a sense frozen in time in all talk of tribal 'development'. For instance, even in Indian historiography, the tribal communities/adivasis do not have an identity as agrarian communities in any sense of the term. Agrarian necessarily refers to settled cultivation in the plains and a certain socio-economic context of agrarianism that has been the more dominant historical paradigm in India for various reasons (which one will not go into here).That the tribal communities/adivasis practise agriculture, that they cultivate and are among the earliest food producers in the history of human civilisation, is a fact that is more or less negated, ignored. Archaeologist M.L.K. Murthy (Professor Emeritus at the University of Hyderabad) who had visited this region (all the three districts) with a team in 1984 has noted, "We found rainforest adaptations. From 5000 BC onwards, the use of rice is traced in this area. The Polavaram submergence area indicates presence of a rice variety, a crossing breeding one [sic], between wild rice and domestic rice (cultivated rice) …At Koida and Isunur in Khammam district near Papikondalu, several materials of Iron Age were found. Here we recorded dolmens, three storeyed tombs belonging to 800 BC to 200 AD. Redware pottery too was found here."[25] The team found, "a variety of…material which established beyond doubt that human presence in this area is of 1.5 lakh years old...The tribals who inhabit this area too are of significant value, as they practised the age old lifestyle and farm practices...."[26] I have a problem with this last point as it again ends up treating the tribal communities equally as 'artifacts' of history. But that is a matter of political and historical perception and more important to the present context is the point of the archaeological heritage of the region. The tribal communities have been confined and frozen to a context of forests, again with forests having the sense that the colonial rulers propagated and maintained for their own benefits (many a times giving a few 'concessions' to these communities in return for labour they extracted out of them when needed); thus the identities as well as the locales of tribal communities seem frozen in time. Their limited access even within these locales has doomed them forever, it would seem. Their minimalist agriculture, shifting cultivation, seemingly small-time when compared to the high input-based voluminous agriculture, does not count for policy purposes. The perception of their agriculture being 'primitive' is unmindful of the process of change from hunting-gathering to food cultivation and innovation in agricultural practices. The perspective of the Forest department has continued since colonial times, especially the idea of forest conservation, on the face of it, but commerce in timber, and either ways, it meant seeing the forest 'apart from', or outside of, the tribal societies who had lived here for centuries. Even post-Independence there continued the propagation of the notion that destruction of the forests was on account of the practices of tribal communities and hence a need to 'educate' and 'sensitise' them of the need to conserve and preserve forests. While terminologies, referring to the tribal communities have emerged seemingly on the basis of how one perceives them, in the context of a nation's history – thus, 'adivasis', 'tribes' each supposedly defining one's understanding of this segment of population –

in a larger sense they have remained confined within narrow spaces and constructs. Thus, even in discourse on tribal communities and any debate on the nature of development occurring in areas dominated by them, one finds a certain ahistorical sense and exoticisation of their spaces. Moreover, their spaces seem to be seen as spaces that should/could not have changed, altered; that they must have remained pristine. The forests, and forest spaces and conservation of these, by extension, became the battle-cry post-Independence. At one end, since these communities had been perceived to be part and parcel of the forests they lived in, or around, the 70s model of environment conservation and wildlife protection perceived them to be a threat to the propagation of forests. Their access to the forests, and their being confined (perpetually) to that identity both need to be seen as two ends of a spectrum, each important in understanding their levels of socio-economic progress. Podu was allowed in some regions, but even for that they needed permission from the Forest department. What is far more important is the fact that here are the few communities who practise agriculture in synchrony with the changing rhythms in the forests; and these may well be the last few pockets left of this kind of forest-agrarian economy that will soon be wiped out. They were also making innovations in agrarian practices throughout history.

I just found one instance of one such innovation in technology recorded as early as 1932. "The locality where this instrument is used is in the interior of the Polavaram Taluk, Godavari District. The people who use it are Kois, a Dravidian hill-tribe. Oil-seeds are placed in a shallow flexible box made of palm leaf, and are then crushed between two heavy logs. The nuts crushed were of the *Bassia latifolia* (Mowha). A shallow dish on the ground under the logs receives the oil. The logs act like a nut-cracker. A fulcrum is obtained by roping the logs together at one end…A stake passes through the two logs just inside (to left of) the fulcrum. The object of the stake is partly to keep the two logs in position, and partly to support them at a convenient height above ground. In order to increase the pressure on the nuts a second lever is brought to bear transversely on the distal end of the two logs (to left in the photograph). The fulcrum for the second lever is a horizontal piece of wood secured through two uprights; against this the end of the lever pressed upwards. In the illustration two men are shown pressing down the second lever. The chief point of ethnological interest in connection with this oil-extractor is the use of the double lever. The Kois use the double lever again and again in their traps. They seem to have thoroughly understood its advantages, and have varied its application in a number of ingenious ways…."[27]

Their identity as cultivators, farmers, was never understood, nor acknowledged and they have remained in records as those dependent on forest produce largely, and needing careful handling etc. In the name of development whatever has happened with the tribal communities has never come from them but from the top, or others who claimed to represent them or their interests. They would not like to remain in a state where they have no access to things that mainstream society has. But they also need the true freedom to make the choice that they wish to. An angry man in Kondamodalu had once questioned the officials visiting them in the wake of a calamity (that I will shortly come to), "when we wanted roads, you gave us boats."

***

World Environment Day, But Does It Matter?

5th June 2006 - At *The* Site, Polavaram: Recording Both Sides of Arguments on the Way

"Polavaram: Head-quarters of the Agency Deputy Collector (who, however, is temporarily located at Rajahmundry) and the deputy tahsildar. Population, 4,455. It also contains the office of a sub-registrar, a local fund dispensary (established by Government in 1880), a police station, a traveller's bungalow, a Government Girls' School and an English lower Secondary School for boys. It was formerly the chief place in the important zamindari of the same name, which formerly embraced the whole of this division and much of Yernagudem and Rajahmundry taluks, but now comprises only twelve villages paying a peskhash of Rs. 6,713"[28] records the colonial Gazetteer.

My journey from Kondamodalu towards Polavaram dam site started with a ferry ride from Kondamodalu to Sivagiri. It is the World Environment Day, I realised, when I reached the Polavaram dam spillway site. The Supreme Court has ordered a temporary stay and stoppage of all works on the dam site. But I can see that the work continues, if not so much during the day, then twice as much at night. Here, I am led to the site by a few people from Polavaram and Chegondapally. Meanwhile, I meet (alone, without the others around me) some of the workers settled at a makeshift camp at the site. It so turned out they were speaking in my mother tongue, Tamil! Mobility of workforce, or height of distress, I thought to myself. They were working as welders, simple mechanics, security guards, etc on daily wages for the BGR Company, a Nellore-based Company, which was sub-contracted by Madhucon Company, for the spillway work. There were ten of them who had come all the way from Tiruchengode in Tamilnadu, as contractual labourers. Four machines are seen busy in the lining works. One of them, Vijay Kumar, around twenty years old, a Helper, tells me, 'We are working every day. Contractor Ravi from Amman Dealers brought us here. ...We will be here till the work is completed.'

Even as I was speaking to them, the Project-in-charge (from BGR

Company), someone who called himself Verma, walked by and told me I was not to talk to the people working here without 'permission'. But he offered to guide me around the place if I wished. He told me his side of the story, the Company's point of view, obviously. He said, 'Many (from the TV channels) come here. Our work has been stopped because of all this. We stopped work thirteen days ago. But we have to dump this loose soil from blasting (dynamites, to break through the rocks). Fifty per cent work must be completed within the deadline given by the government. Who can be sure about this government? They might ask us for proof of work later though they are asking us to stop work now. As of now, there are around 350 labourers in the camp. We are paying our labour force Rs. 150 per day even when they aren't working. We already paid them an advance. We have been dumping soil on the fields we have taken on lease, given to us by the Revenue department. We are losing lakhs of rupees per day. It will cost us at least Rs. 50 lakhs to shift our machinery if we are asked to leave here. We have already sent back 70 per cent of the vehicles we had. What will happen to our investment if we go now?'

I found out some more details of the work being done at the Spillway site by BGR Company.The Engineer-in-charge of the project, Mr. Venkayya (a retired engineer of the AP government) was not too happy about the 'stop-all-works' order recently issued and he had his reasons for it. He informed me that BGR Company pays Rs. 20,000 per month for electricity (three-phase) connection. They shifted out six excavators so far and hope to be paid by the AP Government some 'idle charges' as compensation; the Company was planning to move the court of law to claim it. They had incurred an expenditure of Rs. 40 lakhs for shifting heavy machines so far and they had to pay salaries to the skilled workers, though they were not working. They had acquired land for dumping soil and rubble with the permission of the Revenue Department, he said. Madhucon Company had deposited Rs. 1 crore to the Revenue Department nine months ago for 70 acres of land. The department, he said, has not settled rates of fields till date. The Revenue Divisional Officer (henceforth RDO), Jangareddygudem, the Joint Collector, West Godavari district (at Eluru), are in the loop. Apparently, the Company also deposited crop compensation of Rs. 3,500 for the Pydipaka farmer whose field is being used for dumping. Around 600 acres will come under the spillway, of which 400 acres are revenue villages. He then said, '960 MW electricity (may be less) will be generated from this project. It will be used for industry and navigation. This project is essentially one of less expenditure and more benefit.'

A security guard working here said he was from a village nearby, perhaps one of the only locals employed here (I was to see mention of this man in another context, of conflict, later). He earns Rs. 2,000 per month. I also saw a welder at work. Everyday - as I witnessed - trucks and heavy vehicles keep moving up and down through the day and more so at nights, when on paper the works have been stopped. I also spotted a few children employed. I managed to take some pictures.

An Argument on the Road

Walking back from the Camp office, I was witness to an interesting argument between the Project-in-charge (Verma) and Posi Babu and others from Chegondapally and Polavaram. The point of argument was two small pipes laid to ostensibly direct excess Godavari water (during rains) away from the fields of local people. I keenly recorded the argument:

Posi Babu – "You have dug up all this soil and dumped it on our fields already. One of these days when you blast those dynamites, somebody will die. The other day our women went to the fields and a huge boulder fell close to their feet. Who will take responsibility if it falls on human beings? People use that path everyday to fetch forest produce."

A farmer who walked past just then joined in, saying, "We are unable to bear the smoke. The walls of our homes are all cracking up."

The Company Engineer retorted, "We keep issuing warnings to people using loudspeakers in auto rickshaws to move away before we begin the blasting."

Posi Babu (tells me) - "We are unable to bear the stench the smoke (*tattukoleka potunnaamu*); the ground literally trembles....Everybody is being told the work has stopped, but it hasn't! Blasting goes on night and day."

I asked Posi and others if they would meet the Company officials again the next day on this issue but Posi says, "We have already been threatened and false cases have been booked against us when we tried to stop work the other day (in February 2006 they held a dharna at the site). What is the guarantee they will not do so again? It has been this way ever since we started objecting to the project. If we get agitated because our lands are going, the police mark us out. Some have been booked under cases of '*rajadroham*' (literally, anti-state activity/basically Sedition)!"

Varma butts in, to tell me, in Tamil (intriguingly), "They will not come."

A woman meanwhile (whose name I could not find out) said, "Now they sweet talk; the moment we ask them questions they (the police) will put us in a vehicle and take us away!" This was the first time that I heard of the Sedition cases having been filed against some of the people who protested against the Polavaram dam. It was time to move along, as it was getting darker and we needed to reach Chegondapally.

Chegondapally village (West Godavari district, Polavaram Mandal): An Argument at the *Ramalayam* (a Rama temple)

It was twilight when I entered the village, and a local photo studio owner kindly lent me a flashbulb as my camera didn't have that accessory. Later that night, I was to go pillion riding on someone's motorbike and take pictures of the work carried out at the spillway site at night. And we had to be careful that none from the Company spotted us. Varma from the Company continued to follow me even into the village and caught himself in the midst of an argument with the people. The Koya peasants showed me a written undertaking the Company gave to them promising to remove the debris from their fields by April (2006). Till date (June) loose soil and rubble remained on those fields, as I could see.

The Company men, I was told, also dumped soil on the *cheruvu* (small lake, natural water body); if it rains there will be no way for anyone to utilise this pond. The only way for that water to be drained out was a channel that the Company workmen had blocked. They also blocked all the channels to the *cheruvu* putting in danger this whole village. It will be under water if heavy rains come in the next two months. (Sure enough, during the 2006 Godavari, that is what happened, and some villages were under water longer than was normal in the heaviest of rains). Sakkubai said, "This Project is a loss for us. They have not been compensating anything so far - they are dumping rubble on our fields. We have asked the MRO, RDO so many times. But we have no answers. Suramma was forcibly held. How can they (policemen) hold a woman's hands that way? We are criminals in their eyes. But there is no compensation given to the farmer who has lost the land due to the dumping of loose soil...."

***

Cell-phone Notes

5th June'06: World Environment Day.[29] 3 km *kutcha* road; 1 PHC (primary health centre) for 13 villages, at Kondamodalu. A boat needed to take a sick person to a hospital.

On an auto - bumpy ride. Sivagiri, towards Polavaram, crossing many of the to-be-submerged villages. Next time, will stop by. Stay in many of these villages. Punnam Mutyam, the ferryman, ferries people from Sivagiri to Kondamodalu. Just as our boat rocked, when a huge, garish tourist launch passed us by, a big project is about to kill several small people in its path. Punnam Mutyam has a beautiful broad smile. Where will he be, when it will all be water here and his boat will not longer ply? About the Polavaram project, he said, "the world may want the project, I don't." So many metaphors in a single crossing-over, half-hour boat ride across Godavari from Kondamodalu to Sivagiri.

Debate at the bus-stand now (Polavaram). Tribal agriculturist from Ramiahpeta, "if the project does not kill us, the dynamites will! We will leave our homes if this continues. May be that's what the government wants?"

6th June'06: This bird, sounds like a barking deer almost! Calls out as night approaches.

Punnam Mutyam…![30]

6th June, in my diary - at 9 pm, I am sitting at home in one the most beautiful villages in this part of East Godavari district, Manturu, under a moonlit sky, quiet (not *so* quiet – wish I was left alone for a while – even if I go for a walk, someone accompanies me!).The house I am put up at tonight overlooks the Godavari and just next to it, is a beautiful, intriguingly placed (naturally, of course), pond with a green-blue hue. (I try to recreate it in my diary with my hand - not so good a sketch). I heard this bird call. They said it was called Kabbilam. I could not really record the sounds of the night and the bird, what with all the men around having their own serious discussions.

7th June'06

3.30 am: Roosters crow together, all at once, after my cell-phone alarm goes off (a rooster's call!). Have upset the quiet? By Godavari, at dawn, 4 am; quiet, under the moonlit sky. The river, a lifeline, also a handicap; timings of the smaller launches, bad road connectivity and for some a matter of life and death, to reach the hospital.

The fishermen angle: the Yanadis, the 'bestas', as they are sometimes referred to, are not recognised as 'project-affected'. The term, 'project-affected' even before the project (has been) completed! Some NGO in future might perhaps have a 'project' called 'participatory rehabilitation of project-affected communities'?

Sitting in a packed boat. Seems like Godavari might just overwhelm! But no, she is the benevolent one now. The boat is overfilled. From Kondamodalu to Pocharam (Pochavaram).

Tourism launch - to see these *kondalu* (hills) devoid of people. With music unconnected…. The 'Sujala Guest House' at Polavaram mentions "State of the Art" housing, and they have a model of the same. Incidentally, there is no hospital in Polavaram, which is the mandal headquarters?

7th June, in my Diary

If you were to touch my heart at this moment you would feel the immense silence within. There are the words, of course, under construction, but the depth has silence written all over it. At present, am back in Kondamodalu. This part of the journey - the first phase - began and ends here. Like coming full circle. Sitting in this very nicely done home with its own garden and Gladioli and a flower creeper called Radha-Krishna (beautiful!) in the front yard. Sparrows abound. Quiet abounds. It (the house) belongs to Kumari (a non-tribal – seems to be among the better off from the rest here; of course there is this history of the non-tribals once having dominated the 'girijans' in these parts and the tribal revolt against them). She has been introducing me to others as her kin (*chuttam*)! She has put her sons through the best of schools and colleges in Vijayawada and Kakinada, and they are employed now, one in Chennai and the other in Hyderabad. My 'guide' Ravi, has left, it seems. A teenager with his set of problems… Ravi told me later he cannot afford EAMCET entrance fees even if he wishes to join the engineering stream, in spite of the Reservations (for ST candidates).

Godaari, or Godavari, and terms of endearment! *"Pedda Godaari occhinappudu Kondekkutaamu, adi vellina tarvata kindikostaamu"; "ika Godaari raad-andi"; "Godaari Bhadrachalam varake ostundi, daani tarvaata raadandi; Bhadrachalam ninchi, routu"; I Godavarini nammukoni unnavalandi; em ayi povaali"?*

The statements I had heard over the last few days, translated as (respectively) – "When the big Godavari comes, we climb the hill tops; when she leaves, we come down"; "Now Godavari will not come"; "Godavari comes only until Bhadrachalam, not after that; after Bhadrachalam, it is the 'route' (you have to take the road)". They use the term route for road sometimes). "What is to happen to those of us who believe in (depend on) Godavari?"

***

Kumari and her Home

Kumari's home, where I stay, for a while, at Kondamodalu, before I move on further, is something to see. There are Gladioli and red Hibiscus spread across her garden. Her kitchen, a separate one – within the compound (thatched, bamboo reed) of her home. That is how people make kitchens in these villages. Everything is neat and in its place. There is way of keeping the soap boxes on a bamboo pole. The garden has flowers of different kinds. There is one called Radha Krishna (with a red flower and a white one on the same stalk, hence the name, she says). It is a beautiful house. And I am wondering, as I sit here sipping tea (she drinks a lot of decoction, black tea, mostly, but since I am there she is giving in to the luxury of milk tea – thankfully for me, thirsty for tea anytime) I wonder if all this will go. Or, what will be. But who cares about how Kumari keeps her home, or where she keeps her utensils, her soap box, her how her kitchen looks. Or for the fact that she, like many other women here, tend to keep their belongings limited. For they know Godavari will come, and they will then have to pick up their basics and go uphill, wait until she retreats, to come back and resume lives as they did before. Who will care as to how these people have made their lives in tune with the coming and going of Godavari? When the water submerges all these villages from the dam, women's way of making their homes, each differently, each a unique home in itself, will be gone too. Or, who will care about what a few women of these villages say about what they think of the project? For the project has been designed, under implementation, mainly, by men. Men can think of the numbers, the expanse of the area, money given to the displaced, the compensation, the design, and the technicalities. But women's ways with their land, forests, fields, and their homes is least of the worries. At least, till date.

***

A Fishing Boat – in Continuation of the 'Smaller' Things in Life

At 4 am today I hopped on to the fishing boat of Varaprasad and Gangaraju. Varaprasad belongs to the Yanadi community and resides at Ganesh Nagar in Polavaram. Incidentally, the Koyas and Kondareddis, I learnt, consider themselves of a higher social group than the Yanadis. This was an important dimension. Varaprasad starts his day at 2 am at Polavaram to dock at 3 am at Singanapally. He buys 10 kilograms of *royyalu* (prawns), at Rs. 100 per kg and makes a profit of Rs. 10 to 15 (per 100 kg). There is a partnership (profits are divided) between the boat owner and the trader. Rs. 100 per day is paid to the boatman. Gangaraju and Varaprasad collect fish from the fisher communities in East Godavari and West Godavari villages - Kondamodalu, Sivagiri, Chevur, Nedapuru, Talluru, Kachluru and, Tuthigunta – take them to Rajahmundry to be sold. He does this work from the months of March to May/June (up to 15th June) every year until "Godavari comes". He says, "Bigger merchants go to Khammam (town). They can invest more in this trade. Once the dam is built, this route and these small traders will suffer."

They have been in this trade for generations he says. Small fishermen from Vanapally, Kapileswaram, Thotipally, set up huts here in the banks, and fish in these three four months. They will lose their livelihoods. Varaprasad happens to be a graduate student at the Arts College in Rajahmundry. His summer vacation coincides with this work. He says "the Yanadis (*Yanadulu*) on the Godavari banks come from Kapileswarapuram and other places in these months, set up their huts and hunt fish. Godavari's coming and going is linked to their coming to and leaving from these banks. They fish alternatively upstream and downstream, timing it with the movement of the river. Once the project (Polavaram) comes up, well, Godavari will never come nor go and there will be no banks on her lap! For they anchor their boats or pitch their thatched huts on these banks. Where they will vanish, god knows!" Varaprasad will drop out of college this year and is sad about it, but still asks me, *"akka, i Godavarini nammukunna chaapala pattukunnavaallu emavutaaru"* ('what will happen to these fishermen who bank on Godavari' once the dam comes up?) He was thinking of others, not himself, when he said that.

What would happen to boatman Mutyam, who transports people from one bank to the other for Rs. 5 a person? Mutyam, with a heart-warming smile? When I asked him his name a second time, failing to register it the first time he mentioned it, he pointed to my nose-ring and said, "Mutyam, Punnam Mutyam"! *Mutyam*, meaning a pearl or a gem. We took Mutyam's boat from Kondamodalu to Tekuru via Sivagiri on the 5th morning. As we were in the middle of Godavari, suddenly the boat rocked a little. Two AP Tourism launches – garish looking, with blaring music and people playing *tambola* (a kind of number game) within (as we could hear from the loudspeakers) passed us by. As Mutyam's boat rocked, I wondered, wouldn't this be the way a 'big' project would rock all the 'smaller', of everyday people? Living by the day on the munificence or otherwise of Godavari? And then again when I got off the boat at Sivagiri and gave Rs. 20 to Mutyam, he returned Rs. 10. Unable to figure out (since I was paying for Ravi as well) I asked him, "*enduku*?" (Why). He said, "*inta chaalamma, devudunnaaru naaku*" ('this much is enough; I have god with me'). And again I thought, how apt was this metaphor for whatever was happening? Here are the smaller things that make life. There is contentment and honesty in people like Mutyam, happy with their state which is also a life of struggle, which they also acknowledge. And there are, on the other side, governments and commerce that could always do 'with yet another', 'some more', 'something bigger', 'something better'. Small, there, is a crime. And as I reflect on things 'smaller' (though really mighty in the sense of the philosophy of human life and living), I see this sparrow here and this small black singing bird and the tiny million life-forms that these forests in the foreground and background nurse and I wonder, wouldn't they all go too? Disappear? Die? Where will their alternate rehabilitation packages be set up? Is it a crime that they exist at all? And in these dwellings, homes, that I have been fortunate to stay in, there is a place for all of them (including the lizards who I am yet to make peace with); most importantly, without much ado, without token sloganeering. Everyone, simply, just lives along with every one else. And of course there are problems. But those are problems exactly of the state negating, ignorant of, the smaller. These villages have no roads, no electricity (or have apology for electricity in one pole that may fall every time there is a storm or a shower), no hospitals, PHCs without doctors, schools where one teacher manages four sections of students. These are villages without drinking water taps; though few have dug wells. These are villages where everyday living is indeed a struggle but people have learnt to survive with two, three or four options. Thus, when crops fail, "*kondakelli testaam*" ("we go to the hills and bring – forest produce, etc); or fish in Godavari. No, they are not sad about basic things – food, clothes, nor do they complain so much. But in acknowledging their rights as citizens, the state has failed and in fact the state complains that 'nothing we give is enough for you'!

7th June – a Long Walk through a Different Forest

At 12.30 afternoon, we (a young couple from Kathanapally and I) take a boat at Kondamodalu, to Pochavaram. It is packed to capacity, though. It will reach us in time for me to get the bus at 3 pm to Bhadrachalam. A slight variation in the boat timings and you miss the last bus to Bhadrachalam. Similarly, a bus starts at 5 am from Bhadrachalam to reach Pochavaram at 7 or 7.30 am; time enough to get a boat to Kondamodalu by 12 noon. The boat passes through Tummileru, Kondapudi, Kolluru, Tekubaka, Kakisanuru, Perantapally, Telladibbala, Sirivaka and Koraturu. So when the public hearing was arranged in Bhadrachalam as a 'central place' being very 'near' to the villages in the submergence zone, as the government pointed out in its defence, it says something. And this is only about Kondamodalu. There are other Panchayats even farther off.

In Search of Elusive Sleep, an Adventure Mid-way

In my diary - "Five kilometers (?) trek, from 4.30 pm to 5.30 (or was it 6?) pm through a beautiful forest patch between Pochavaram and Ippur, on my way to Kunavaram. Something totally unexpected, but something part of a deep desire. Wanted to spend 7th all by myself in parts of Papikondalu Wildlife Sanctuary. It happened, but 'almost'. Was not alone. Couldn't be alone; was not left alone, but thankful this trek even happened, so unexpectedly. The three hours boat ride was something - luggage, people, jam-packed, capsized almost twice with the weight. Seemed like Godavari might just envelope me!"

I learnt later, this forest was part of the Bhadrachalam South Division forest range. I was to find out later from people at V.R. Puram that it was most certainly not less than ten kilometers, though they would not believe that I had walked all that way. It was a fine trek through a very beautiful forest patch. My 'guides' this time, a couple from Kondamodalu (one of them being Varada), came along too. We had taken a passenger boat from Kondamodalu that day around 1pm, to Pochavaram. It had stopped midway (was packed beyond capacity and would have all but capsized twice) to ask people to make some adjustments for some balance on the deck. So, by the time we reached Pochavaram, it was way too late - 4.15 pm. We missed the last bus of the day to Bhadrachalam which would pass via Kunavaram, my next halt. I was to start my Khammam district tour from Kunavaram either the same evening or the next morning. It wasn't to happen. No problem. We had some *soda* at a tiny shop at Pochavaram. The couple had to continue on their journey to another village (Chintur, perhaps?) in V.R. Puram mandal, so they had missed an important link as well.

At Pochavaram, we met the elderly Jogi Reddi (a Kondareddi, must have been in his eighties), once a leader, and participant in the peasant struggle of 1969, I was told. He was the first and only Kondareddi man I had seen, through this entire journey, who wore the traditional loin cloth, and had a tattoo on his arms. Though he permitted me to take his picture, his son insisted he wear a towel for it! I was to meet him just once more, on my next trip (in 2007 and gave him a copy of his picture), but regretted later not having taken a full interview with him, for he passed away soon thereafter. I lost out on his personal story of resistance.

Meanwhile the couple and I could get no means of transport, not even a bicycle to carry on. They seemed to know no one in this village for us to spend a night at. I wasn't carrying very light luggage with me either - books and notebooks I 'might' use. It would soon get dark in this forest. So Varada and her partner walked with me all the way up to Ippur from where we managed to find a rare shared (by how many!) auto rickshaw to Kunavaram via V.R. Puram. In the meanwhile, at the only spot where I found 'network', I managed to call a friend in Hyderabad, who said I could stay at Kunavaram (though the original idea was to go to Rampachodavaram) Forest Rest House – thanks to her contacts there. Later I was to know how easy it would have been for me to get off at V.R. Puram instead of taking another long ride in another auto to Kunavaram. There was a place to stay in V.R. Puram as well! But then, this was my first journey, after all. The walk was beautiful, through pristine forests, amazing quiet and the sounds of various birds. The kind of green just readying for the monsoons, but glistening in the light showers of the past two days, all the way. Fortunately, we found a little boy on a bicycle mid-way. Must have wondered later why he did! His name was Posi Reddi and we coaxed him into carrying at least my rucksack on his cycle to Ippur. It had been a long day - waking up at 3 am and walking through the villages, journey on a launch that missed an accident and finally the walk through the forest. All one needed was a bed to crash on. So, I got off at Kunavaram, right at the bridge where the Forest Rest House stands (down below you could hear the sound of the river gently flowing past – this is where the Sabari meets Godavari), and in that dark, nothing could be seen. It must have been around 8 or 9 pm. I was already dreaming of the bed. But the Sub-Divisional Forest Officer had kindly arranged for a meal I didn't have the energy for. I had to oblige. They showed me the room for the night. I saw nothing but the massive bed. The officer and his attendants took leave of me. Shutting the doors after them, I dumped myself on the bed. There was utter darkness of the night, sounds of night crickets and a gentle river. But things didn't remain so dreamy. What happened thereafter is a memory for a lifetime, to be recounted later.[31]

Kunavaram, in Colonial Records - Incidentally, Kunavaram is mentioned in the colonial records thus: "Stands at the junction of the Saveri and Godavari rivers...Formerly the station of the Special Assistant Agent and now the head quarters of the District Forest Officer, Upper Godavari. It is an important point for the river borne trade, as it is beyond the Ghats and the unabridged Saveri, and carts can travel from it to Bhadrachalam."[32] It was the Sub-Divisional Forest Officer at Kunavaram who, in a reply to my question said, "Animals will just flee when the dam waters come down, and the birds will fly away" But will all the animals be able to flee when the waters come gushing? He said, "of course not, but can we help it? We cannot transport all of them." What about the forest as a whole? "We are being compensated for the trees, and compensatory afforestation area has already been given to the Forest Department." And where might that be? "Ananthapur district." The Irrigation Department has been drawing forest boundaries now, and I can see the reserve forest boundaries shrinking when it comes to the question of compensation! New alignments are being shown to prove that not that much of forest land is going to be submerged after all. SDO says, "only 13 hectares" (in his division) will be submerged. And showed me the alignment "newly made, by the Irrigation Department."

Ananthapur district? I thought - we are losing deciduous and semi-evergreen forest of more than 3,000 ha which will be compensated with an artificial plantation in a dry-land scrub jungle habitat? Are foresters trained so as not to feel pain for these losses, but all the pain for loss of small measures of fuel wood that small communities need for truly smaller uses? Incidentally, the SDO apologised for the condition of the room which was quickly readied on account of my sudden arrival. But more importantly, he said it was not being repaired since it was 'anyway going to be submerged under the Polavaram dam'. You read - why waste money on it? Even if it is a historical monument (to colonial exploitation, nonetheless!)?

***

Khammam District

8th June – Valairpadu Mandal – the Jinxed Scheduled Castes

Babji, an active relentless worker representing the AP Vyavasaya Vruthidaarula Union (APVVU[33]), and the dalit cause in the Polavaram displacement issue, says, "Since many dalits do not have the BPL ration cards they will not be taken into consideration for the Polavaram dam R&R compensation. The dalits have a peculiar disadvantage. They have been working under the tribal farmers (on their lands) or the non-tribal (upper caste, few BC) landlords in the Agency Areas. They are denied the constitutional rights guaranteed to them elsewhere in the Agency Areas. They cannot seek special rights or concessions in Agency/Scheduled Areas…out badly....APVVU has tried to organise dalits in this context but some groups/organisations alleged we were trying to divide the communities threatened by displacement. But we are merely all displaced marginal people should come together on a single platform, including dalits. Most meetings on Polavaram dam until now have been held unfortunately in groups, so Agency Girijana Sangham talks only to tribal communities; Adivasi Aikya Vedika does the same. They refuse to include others. Each group is speaking up for one community and its interests. Till date the question of dalits has not been taken up by anyone. The movement is fragmented..."

Karakagudem (Kunavaram Mandal) Dalits (Madigas)

Gampala Ramulu – Agricultural wage worker. He says men are paid Rs. 40 to 50 and women get Rs. 35 for work on the fields of non-tribal landlords. "We do not belong to this place. Our village is Potlavai, on the banks of Godavari. We faced around three Godavaris some years ago; and some officials, including the MRO there told us we should move and brought us here. Now even that village, Potlavai does not exist! We have settled down here. We had rights as SC in Potlavai but ever since we moved here, we have lost those too as we are in Scheduled Area. Every government scheme is aimed at the *girjans*. Nothing comes to this hamlet. We are all landless agricultural labourers... Our names are definitely there in the voters lists! But many of the young adults here do not have a ration card. We work for six months and are jobless for the remaining six… I built my house with my hard earned money here. It is incomplete because we don't know when we will be asked to leave….We were not even chosen for the Food For Work schemes. The only (primary) school here is 3 kms away at Marrigudem. At least some *girijans* collect beedi leaves and are better off than us. We are dependent on non-tribal landowners….We will lose out even on this work…."

Sujata (Mala/SC) works as an agricultural wage-labourer. She says, "We have around 40 families of Malas and Madigas here…Women earn around Rs. 900 per month and men about Rs. 1500. For six months a year there is no work and we take loans to make ends meet. Some men go to Bhadrachalam and other places and work in tea shops and so on. Few educated ones work in STD booths. Unless we get assured work we will not leave this place."

A non-tribal (from the Raju community) landlord from Marrigudem says, "We are demanding at least Rs. 1,10,000 per acre (not the present rate of Rs. 50,000 per acre)." He happens to be a BJP activist (a local organiser). If we get a good package we are ready to leave. But we will not get this quality of land anywhere else. At present we get an income of about Rs. 1 lakh a year, even without adequate water (for crops)…Godavari ('floods') is not one of our fears..."

At which point, an agitated and angry 70 year old Sarojinamma says,

"You landlords will get your land. What about us? Where will we go? Drown in Godavari? What do we eat there? These packages will kill people like us, who have no land anyway. Will your government be happy then? May be Polavaram project will be happy if small workers like us die. We just have to cry holding on to these trees may be."

The strange case of Pydigudem in Suchirevula Panchayat, Kottur: it will become an island - This village is not in the list of submerged villages under the Polavaram project. Yet it will be affected since the surrounding villages will be submerged rendering this an island. A few dalit families here spoke of their problems. Kurnelli Venkateswarlu (Mala agricultural labourer) said, "We came here from Garibpeta around 35 years ago...We used to work in Bhimavaram on farms of upper caste landlords (*pedda kolastu vaaalla bhumi*). Here we do not have pattas since this is ST land...This village is not on the *barrage list* (Polavaram submergence list) we are told. But we will be stranded nonetheless - this village will become an island! For all the villages around this village are in the submergence area. Where will we go then when the land around our village will be under water? From Garibpeta we had to move because of the Kothagudem canal project. We did not get compensation even then. Then we went to a village near Palavancha. We were moved from there too, because another canal under Kinnerasani project would submerge that place...."

The project affects some of these people indirectly. But since they do not have rights in the Agency Areas on tribal land, they are more vulnerable than the upper caste non-tribal landlords on whose fields the dalits work.

At V.R. Puram Mandal, I met people belonging to the Koya (and a few Kondareddi) community. Korram Lakshman Rao, Ward Member of the Gurrampeta panchayat, said to me, "Our lives will be destroyed and forests will be destroyed if the project is built. We survive on these forests. We fetch small forest produce (NTFP) and have small lands which we cultivate. Girijans here own between five to twelve acres of land. There are also rare cases of those owning 20 acres. We grow *jonna, pesaru, minumulu, mirchi, pogaku* (respectively, sorghum, pulses, types of grams - usually black gram, horse gram - chillies and tobacco)...Instead of building a barrage somewhere for welfare of a few people, they can improve the lives of lakhs of people in these areas...We all went to the public hearing and protested. We did not allow the meeting to happen that day. We are not being compensated for toddy trees. Each toddy tree gives us at least Rs. 1000 for a year...They never held any panchayat meetings in our panchayat. They just conducted a public hearing at Madapally where ten people from our village had gone to say that we do not want the project..."

Korram Ramanaiah, a Koya, owns 10 acres of land in Kotaragummu village, in Jeedigupa panchayat, and has Toddy and '*ippa*' trees. They use *ippa* oil for its multiple properties, and the flowers, fermented, make for a fine drink as good as wine, slightly bitter and more potent. He said, "We will not get this cool and pure air (*challati gaali*) anywhere else. In fact when we go to Palavancha (in the plains) as labourers for the Forest Department, in the months of May-June to uproot stumps, for which they pay us Rs. 500 per tonne, we yearn to get back...We fetch *kalapa*, broomsticks, *jiguru* from Papikondalu forests to which we will loose access and we shall loose our cultivated fields…We shall stay right here (with) our *grama devatas* (village deities), such as Gangalamma, Mutyalamma, Lanka Sitamma. We celebrate *kolupu* once in three years for them, with sacrificial offerings. We celebrate *bhumi panduga* in June; we go hunting for five days in these forests. Once this project comes, most of these things will disappear…Since the project has commenced, people from West Godavari have stopped entertaining marriage proposals with our children. No marriages have happened for months now…"

Vallam Krishna Reddy from Kotaragummu in Jeediguppa Panchayat (V. R. Puram Mandal) is a farmer and agricultural labourer. He said, "The Collector and PO (ITDA) came here; we *gheraoed*[34] them. Even the Velugu (a Government NGO) staff came here that day. We didn't let them leave our village that entire night. The MLA (Sunnam Rajaiah) and MP (Midiyam Babu Rao, both from the CPM Party) also came here. We threatened the officials that they shouldn't come here in the name of 'Polavaram survey'…People from my father's generation were brought down here from the Marempudi forests, on the pretext that they were destroying forests by doing *podu* cultivation. Now they want us to move again…They were going to build houses for us under the Indira Awas which was stopped mid-way because of the dam. Our people will die and fall to death like birds if they go elsewhere leaving these forests (*'akkada pittalu sachchinattlu sacchi potaru"*)…."

Katakala Venkateswara Reddi remarked, "Even if they give us one crore rupees we shall not leave. We do not want the project."

M. Satyanarayana (Ward member of the Panchayat) remarked, "The CM's *Jalayagam* will only benefit the crorepatis who anyways do not need this water. This water will be used for electricity generation, drowning all of us. We told them we will give up our lives but not allow Polavaram dam to be built. But do you think they are listening to us (the government)? *"Em ayina sare, dunnapotu mida varsham kurisinattuga prabhutvam nadustundi"* ('whatever happens, for the Government it's like water off a duck's back'). In the name of Jalayagnam the government is wiping out the 1/70 Act."

Rekapally, in Colonial Records - As we conversed, I realise Rekapally is close by (and you have the office of a local NGO, ASDS there). Rekapally and surrounding villages (with Godavari) had once participated actively in the Rampa revolt. To this day, the tribal communities in this part seem strongly opposed to exploitation at one level, yet vulnerable at the same time. "Twenty eight miles east south east of Bhadrachalam, and below the junction of the Godavari and Saveri rivers. Population 617. The name means 'wing village' and is explained as referring to the abduction of Sita which tradition locates in this taluk. It is supposed that the wings of the bird Jatayu who tried to oppose Ravana's flight but was killed by him, fell here. Rekapalle is still important as the chief village of the most considerable of the inferior proprietors of this part of the country... This country joined in the Rampa rebellion of 1879, and at one time gave a great deal of trouble to the authorities....Under the Central Provinces administration, podu cultivation had been almost unrestricted, and the assessment on it had been only four annas an axe. The Madras Government almost trebled the assessment, excluded the cultivators from certain tracts, and levied a tax on the felling of certain species of reserved tree... (It) was this reason that the Rampa leaders found adherents in the Rekapalle country. On the tenth of July some Rampa insurgents under Ambul Reddi, aided by a number of Rekapalle people, attacked the Vaddigudem police station. They were driven back, and a party of armed police was directed to proceed up to the river from Rajahmundry in a steamer and launch. The steamer...was attacked a little above the gorge, and was taken by the insurgents. A force of 125 sepoys was then sent up the river, the Godavari and Saveri were patrolled by steamers, and posts were established along their banks...The Rekapalle country was again disturbed by an incursion of Tamman Dora in October 1880..."[35]

Meeting the Forest Officer at Bhadrachalam

I met the Division Forest Officer, Mr. Chalapati Rao, at Bhadrachalam on 8th June. He informed that 13 hectares in Bhadrachalam South division and 80 ha. in Palavancha are to be submerged in the Polavaram dam. Timber will be sold to compensate timber extraction. According to the Supreme Court guidelines, net present value of the forest is determined according to the density of the forest – in this case it comes to Rs. 5 to 9 lakhs. [I wondered, 'only'?] Physical loss (of trees) will be compensated by sale of timber by the forest department, in the market. They 'prioritise' species and areas; valuable tress, prone to smuggling, will be chopped off. The rest of the trees will be left behind. Afforestation (compensatory) will be done in Vizag and Kakinada forest divisions (not Ananthapur, as I was earlier told). In all, 2200 hectares of forest area comes under submerged area (according to his estimate) which includes 13 ha of Wildlife area of Papikonda Hills. As I came out of that office, a strange feeling, mix of anger and pathos. How do they 'value' trees and 'prioritise' species or trees? He never mentioned birds. Nor animals in this entire talk of compensation and 'net present value' of the forests. Good somethings are not valued in terms of money, but bad, they are not valued at all! So the Forest department has accepted its role as being merely custodians of marketable timber for sale to private contractors; they have no role in ecology or conservation (even in its narrowest definition). Incidentally, as I was leaving his office, he informed me, the V.R. Puram Forest Rest House is "better"! Embarrassing – the news has spread already! I hope I do not have to take shelter in another Forest Rest House given up for good since the Polavaram dam is coming. But, sure enough, situation led me to try out that Rest house as well, much later, in 2010. Speak of 'never again'.

9th June – Meeting the local MLA at the CPM Party Office

I met the Bhadrachalam MLA of the Communist Party of India (Marxist) Party, Sunnam Rajaiah, who is from the Koya community and a farmer too. Incidentally, the Khammam CPI (M) cadres had been (and continue to be, as of today) actively opposed to the dam, relatively more so than their counterparts in other districts coming under submergence. This is because Khammam district has been one of their strongholds with a firm grassroots cadre base. Organisations (NGOs, activists) working in this district however quickly point out that the CPM (which, indeed, in the initial stages demanded reduction of the height of the dam, and study by an 'expert committee') changed its stance mainly on account of the growing public displeasure in the submergence villages of Khammam district (one of the greatest impacts of the dam would be felt here, with 7 out of 9 mandals to be submerged), which it could ill-afford. Moreover, the party has its MLA and MP (Midiyam Babu Rao), both Koyas, from this constituency. Both personally opposed to the project too. "We want the project stopped", said Rajaiah, "since more than 300 villages are to be submerged. An expert committee should be set up to investigate the issue. We stopped the Revenue department from surveying the villages and forests for the Polavaram project. Just yesterday in a few villages, people told them they do not want survey teams in their villages to come in the name of the Polavaram project. The Centrally Empowered Committee (CEC) asked the Government to stall the works. But the Government is adamant. We intend to intensify the opposition the project area…We are not against other irrigation projects, but this one displaces so many people at one go…More than 300 schools and tribal welfare Ashram schools will also drown…We had opposed the project at the Tribal Advisory Council meeting in February (27th) 2006. We asked for voting, but the Government did not relent. There is no consent from the President, Governor, for this project. It is in violation of all rules/ Acts 1/70, Panchayati Raj (Extension to Scheduled Areas) Act and so on. Why drown two lakh girijans for generating mere 5 lakh MW power? In the Panchayat elections our campaign is to defeat pro-Polavaram lobby parties, and to ask people to vote for anti-Polavaram parties, candidates."

10th June - At the Gurrampeta dalit basti in Khammam district - the same old story. Landless agricultural workers with no hope.

10th June - in My Diary, Reflecting on This Journey

In terms of a journey embarked upon with neither too much nor too little 'preparation' (though with no expectations of any kind, and also with a new path) this I would place as a complete one. The sheer length and breadth of it all - in terms of places seen, roads traversed, the overall route, the cross-section of people one met, the multiplicity in modes of transport, food eaten (or *not* eaten), lack of sleep, or the nature of sleep...adventures...the list could go on. Most important of all, this was a true journey along, across...Godavari – and Godavari just hung on to me. In some way or the other there was always a 'Godavari' somewhere close by – in name, metaphor, words people spoke, or physical form.

***

July'06

12th July – At Delhi

I attend a CEC hearing on Polavaram in Delhi. I am taken to task for trying to record the proceedings on my cassette player, for it is not 'legal'. I stopped recording. Yet I share the general tenor of the opinion-crossfire that happened there (since many of the points raised there have now officially come into the public domain through the CEC order). Refrain from quoting verbatim. The first thing that I noted, of course, that I was the only woman present. I learnt of the meeting by the sheer chance through a journalist friend (male) who happened to be one of the petitioners in the Polavaram case (against the project). It was also then that I began to wonder about the concept of 'representing'; none of the people who would be affected by the project were present. Some of the angles of the discourse were redistribution of resources, the Orissa[36] and Chhattisgarh objection to the project, etc. Andhra Pradesh Government was represented by the Advocate General, C.V. Mohan Reddy, Satish Chandra (Secretary Irrigation), Venkateshwara Rao, Chief Engineer, Polavaram Project, and some representatives from the Forest and Irrigation department. The petitioners against the project included Sivaramakrishna, from the NGO Sakti, activist, J.P. Rao, advocate J. Ramachandra Rao and others. The loudest voices were of those in favour of industry. There was some discussion on what is a 'forest'. I wondered, who decides on what is the meaning of forests? Where are the people who actually live there? What is 'public property'? Who delineates on interests that so many people's basic lives depend upon? I heard, in the course of the hearing what I thought were funny words, such as "Government River". The CEC noted that they had "serious reservations" on implementation of

R&R, Scheduled Area and monetary implications. They also had "reservations" on the Wildlife aspects. On behalf of the AP Government, Satish Chandra said the AP Government was ready to conduct a joint visit on the issue of submergence in Orissa and Chhattisgarh and they had meetings with the Secretary of both the states, but Orissa had shown no interest. One of the questions was about the tiger habitat, which lies in the submergence zone and this is one of the few habitats left for tigers and how would the AP government compensate. Satellite imagery of AP shows dense forest but many aspects of flora and fauna were not covered, it was noted. The wildlife study should have been done for the entire catchment area covering the submergence area. The CEC also requested Orissa to give "complete and exhaustive" reply as per the AP affidavit on Bachawat award and whether Orissa had taken steps to challenge the provisions of the award and if the Orissa agreed to the AP Government's offer to build "protective embankments" to address flooding of their villages. Orissa had submitted that no land in Orissa should go under submergence. Satish Chandra submitted that action had begun on the R&R and Gram Sabhas had been conducted in all the panchayats (which was of course, not true, as was also represented in some of the petitions). He said the government had changed 3 cents (for housing sites) to 5 cents (242 sq. yards) and they estimated an expenditure of Rs. 3600 crores for R&R.[37]

13th July, YSR Addresses a Press Conference at Andhra Pradesh Bhavan in Delhi

Dr. Y. S. Rajasekhara Reddy (Chief Minister of Andhra Pradesh) made a point at the conference, which to me was important, that 5 per cent of water from all irrigation projects in AP will be earmarked for industries. Am wondering, how much of Polavaram dam is for industry and how much for agriculture. How much will Kakinada and Vizag take of the dam in terms of percentage? He also spoke of subsidies in power and water for industry being a priority for the government.

***

The CEC (Centrally Empowered Committee) team visits the Polavaram Dam Submergence Zone in Andhra Pradesh 29th and 30th July

People in the Polavaram-affected area were truly looking forward to the visit by the Supreme Court Empowered Committee. For them, it was definitely the crucial moment when a mere word of the Committee would put a full stop to the project and their fears. There were several petitions readied for this visit, by people from different panchayats, NGOs, other activist groups, political parties that were opposed to the project (and otherwise). Even I had hopes about the visit, considering all that I had heard so far from everyone I had met in the villages. But I had fears, and questions, as well.

What Will they See? CEC and the Visit to the Project 'affected'[38]

In the recent past (say, in the last two months) many things have happened it seems, favouring the government of Andhra Pradesh. For one, the Chief Minster Y. S. Rajasekhara Reddy had a fairly successful (as per reports) trip to Delhi. A successful meeting wooing industry big-wigs to invest in AP. AP was showcased (in bright looking brochures) as the "Best Business State". Responding to a query at a press conference in Delhi, the Chief Minister reiterated the government's will to offer the best incentives to the industries willing to set up shop here – water from irrigation projects, subsidies in power and so forth. And he especially stressed on development of the coastal industrial corridor. A video CD handed over to media persons at the Press Conference shows the CM, Y.S.R inviting investors to Andhra Pradesh, saying, "in Andhra Pradesh, we have a friendly government, a proactive government and this government wants industry to come in a big way. That is why we have created APInvest, a corporate-like single-window agency which hand-holds the investors and sees to it that their goals are achieved; their meticulous planning goes without any hindrance." The CD itself is symbolic of how the state looks at itself and what is the politics of the state; in the entire run-time of about 30 minutes, Andhra Pradesh, seems to mean only Hyderabad and its people seem to mean only the engineers and IT professionals and few industrialists from the IT sector. And there are the white faces of a multinational banker and of course former U.S. President, George Bush who, in one of his usually staccato media interview sans homework, says he is going to meet the "the future CEOs who will drive the engine of economy" in Hyderabad (when he visited the city). The CD also gives some startling facts, on the nature of investments already pouring in and the AP government signing MOUs on the same. The key investment areas are shown as (in this order) – "Hi-tech Manufacturing, IT/ITES, Food Processing/Horticulture, Tourism, Textiles/Leather, Education/Health Services, Pharmaceuticals/Biotechnology, and Retail/Real Estate." "Major projects under implementation" are shown (with the CM signing MOUs) as – "HPCL, Vizag (Rs. 10,000 crores), Sem India Fab City (Rs. 13,500 crores), VSP (Vishakhapatnam Steel Plant, Rs. 8,200 crores), ONGC (Rs. 8,200 crores), ITC Bhadrachalam (Rs. 2,000 crores), Nano Tech Silicon (Rs. 2,600 crores),." And the CM goes on to say, "We are willing to take care of the needs of all the prospective investors in the state…We are future-ready…"

The Press Note circulated at the press meet in Delhi on 13th July 2006 mentioned some important points in the context of YSR's meeting with the Prime Minister, Finance Minister and Agriculture Minister, especially with regard to the Accelerated Irrigation Benefit Programme (AIBP). Since farmers' suicides had been a context in the previous regime (under Mr. Chandrababu Naidu), the YSR government was playing up the card of addressing the issue through several programmes, including loan waivers, etc. But at the same time, the real intent of the government was on securing foreign capital investment and towards this end, work had begun full-swing on massive irrigation projects in Andhra Pradesh within the next few years. Polavaram dam is just one end of a long linkage in this powerful game of capital investment from the manner in which these reforms are being pushed for. At the press conference YSR also mentioned the nature of subsidies being given to industry – "we are reducing power tariffs (for industry) every year, we are going to make it 55 paise as electricity rates for new industry, 25 paise for food processing units" and so on. The Press Note mentioned "the Government of Andhra Pradesh offers power at Rs. 2.25 per unit, the lowest in the country." Yet another development relaxed the State Government a little more, namely, environmental clearance for the Indira Sagar dam. With these two seemingly different bits of 'good' news, AP seems poised for its next leap into a future which looks at industry as its first best bet. What would the visit by the CEC in this context entail for a project that has already attained environmental clearance from the centre? It may also be worth noting that perhaps this is one state where the Irrigation department and the forest department are working in solidarity to show alignments in high-end maps of forest areas that will come under submergence.

***

Submergence Zone, or its Periphery?

The CEC team had a short stay at Bhadrachalam, the temple town and was promptly whisked away to the temple (should religious visits be part of their itinerary in the first place?)

A whole crowd is gathered at the Punnami Guest House owned by the AP Tourism department, at Bhadrachalam, to hand over requisitions and memorandums. Police is out in full strength through the length and breadth of Bhadrachalam, it seems. At least the way towards the Guest House. CPM cadre, activists of the Adivasi Samkshema Parishad, members of the Mandal Praja Parishad, a few Sarpanches from the panchayats affected, and a few 'pro-Polavaram' (read, Youth Congress and Congress workers) are all gathered here. The CEC members are surrounded by the people when they make their appearance before leaving for the temple and the field. The policemen try to stop people from having a word with the CEC members. But a few people manage to hand over their memorandums. The CEC takes the road route from Bhadrachalam town, visits the Rama temple, moves through Burgampadu and Kukunur to Koyda; leaves out the more 'appropriate' routes, namely, the one with more number of villages to be submerged and a vast, thick forest patch, namely, VR Puram, Ippur, Pochavaram (which are also motorable). I reach VR Puram, in order to take a boat to Koyda. But all local boats, ferries, launches, one is told, have been put in 'service' for the visit, and local people, and I, are stranded, unable to cross over. At Kunavaram, waiting for any local ferry to take us across to Koyda, on the other side of Godavari, I can spot them, the cavalcade of cars, and hectic activity of some kind. Within fifteen minutes the CEC team is on a special launch and off across the Godavari, through the Papikondalu sanctuary area (that is, only from the launch). They go via the same route that most launches take, that is, via Perantapally (perhaps they will get off there again to see that famous ashram?). The police hire seven cars and 'Tata Sumo' vehicles. Waiting with me at Kunavaram to cross over to Koyda, were two villagers who had come here all the way from Motu (Orissa border) to meet their relatives here. Motu Mallayya (from a village on the Orissa border, near Chhattisgarh) said, 'We will lose everything. I have 3 acres of land. I grow *jonna* (sorghum) and paddy. We do not want this project...In the year that Godavari does not come I get about 20 bags of paddy and 2 bags of jonna (sorghum). We supplement that with income from farm labour. We men earn Rs. 60 per day and women get Rs. 40. We work on the lands of the '*lankollu*[39]' (the non-tribal landlords who own the most fertile fields on the alluviums of Godavari river, or the delta farmers). We do not know where this Promised Land is which they say they will give us. We haven't yet seen it. We are not in too good a shape in our own village so what do we say about a land we know nothing about?"

Katani Venkateswarlu, a Koya farmer, who has ten acres of land here, said, "We wonder where they will throw us. Our fields and livelihoods belong here...What about our children? Since generations we have lived on our fields and lands. Will we get all that we have earned in so many years? Polavaram *barrage* shouldn't be built."

***

And What They Saw: The CEC Visit to Polavaram, its Mandate and Many Questions[40]

The Centrally Empowered Committee had visited the Polavaram dam site, and affected areas (besides Kolleru) between 29th and 31st of July 2006. But what was exactly their mandate? Or what did they come to 'corroborate'/ substantiate in terms of the affects of the dam on the area? These are questions that will remain until we have a report from the team at some point soon. In Hyderabad there was no access to the CEC team for some people (especially media and activist groups) but according to sources, some agencies did gain access to the CEC team when they arrived here late night of 28th July. In the field, should the CEC have actually 'not' met local people, affected by the project, merely because their mandate is the Wildlife Act, and the Forest Conservation Act? In fact, do the Forest and Wildlife Acts 'spiritually' include people (communities living with forest areas, since centuries)? And does the CEC team have the mandate of meeting communities intrinsic to these forests and biodiversity? The CEC in its short stay at Bhadrachalam town visited the temple of Rama – now, is a committee empowered by the judiciary mandated to pay visits to religious places of worship? Surely, if the Rama temple, well outside the purview of the submergence zone (official records do not say it will likely submerge), and the Forest Conservation Act, can be accommodated in the CEC team's itinerary, the affected community could, too. The team took the road from Bhadrachalam town, visiting the temple via Burgampadu, Kukunoor, to Koyda, leaving out the more/most appropriate route – in terms of the forested area, affected villages, and most essentially, the villages that have been consistently protesting and vociferously so, against the project. Karaka Lakshmi, a recently appointed MPP in the Panchayat elections from Karakagudem in Kunavaram mandal said, "We came here to tell the CEC that we do not want Polavaram. We are just about getting some facilities; our children are wearing pants and shirts, and going to school. It has taken a long time for us. Now we will lose all this if we move from here..."

Their memorandum to the CEC had this to say – "our village is drowning under Polavaram project. Indira Sagar – forests, our lands that give us two crops...which will be drowned under Polavaram."

Large numbers of Mandal Parishad and Zilla Parishad members (mostly affiliated to the CPI (M) had gathered at the Punnami Guest house in Bhadrachalam on 30th July to show their opposition to the project and give their representations. Sonde Veeraiah, Convenor, Adivasi Samkshema Parishad believed that "the administration and government have designed this tour of the CEC in such away that they do not touch any of the affected areas. They are hiding the real facts this way...." Strangely, a motley crowd put together by the local All India Youth Congress was raising slogans - "we want Polavaram." On closer scrutiny one realised quite a few of these were from villages that do not come under submergence. Obviously they had gathered to present the government's point of view. One of these, a trust board member in the local Bhadrachalam Rama temple, said, "a little injustice (*anyaayam*) for a few is valid for development of the state as a whole..."

People of Burgampadu had their own 'representation' to give. It was fairly clear they were from Youth Congress – they waved the Youth Congress flags. The local administration had on a previous day warned local groups and activists against raising slogans, or bringing flags or any kind of banners at the Punnami Guest House. Local groups who had come to give representations to the CEC team against Polavaram dam were incensed about this favouritism shown to the Youth Congress in this regard. Payam Venkaiah of Kunavaram mandal, member of the Polavaram Project Nirvasutula Sangham (displaced people's association) said, "the culture, livelihoods, of people, mostly tribals is getting lost....86 panchayats have passed resolutions and filed petitions...We will give our memorandum to the CEC."

Lakshman Rao of the AP Girijana Sangham point out, "people talk of wildlife loss, environment, and not about tribal people who are losing their culture and livelihoods which are based on the forests..."

Between 30th and 1st, it was a clear signal that people were not welcome to meet the CEC team. Does a team empowered by the judiciary to look into the conflicts in the project not empowered to demand from the local administration the time to meet the affected forests and people? That they had the entire district administration geared to serve them was clear from the number of officials appointed for their visit. Security bandobast included. The CEC team visit happened like a visit sponsored by the government in power. From the choice of routes, time the CEC spent on the field, in the forests, to time they had to receive representations and memorandum from people, it was the administration's show. In Polavaram on 29th July, nearly 1500 tribals had gathered (under the local activists of the Agency Girijana Sangham, Rythu Coolie Sangham, and CPI M) at the spill-way work site from early hours and forcibly got the CEC to listen to their opinions. But at Koyda, police force was deployed to prevent villagers of Kasavaram, Tekupalli and Kosuluru from trying to meet the CEC team. The CPI (M) MP (Bhadrachalam constituency, Midiyam Babu Rao) and MLA (Sunnam Rajaiah) were also present there with their support groups. Where would people like those in Kondamodalu gram panchayat and their representation figure in this?

For these are people who proclaim, in their petition, "This land of ours has glorious history of struggles. Our forefathers waged many struggles against British colonial rulers in defence of our light to land, forest and self rule. Korukonda Subba Reddi of Tuthigunta village near Polavaram and seven of his followers were hanged to death by British Government on October 7th 1858. It is our long experience and hence our belief that it is through struggle only we are able to survive and live. Sacrificing lives in defence of our common interests is not new to us. But we hope wise council shall prevail and so we fondly wish that we will be sparred or another arduous struggle in the context of Polavaram Project... We fervently hope the Empowerment Committee understands our agony..."

***

And Godavari Comes! August'06

9th August

I am in Bhadrachalam. Godavari has come, after all. And I am here unfortunately a day or two after her coming. Look how nature does its thing! Thankfully, found myself a very cheap and clean room in a *chatram* (one of the pilgrim dormitories built several years ago by a philanthropist in these parts, meant for those who cannot afford big hotels), thanks to a kind manager (in an otherwise troubled time, as they fear Godavari will rise further and the Bhadrachalam town will be in danger). Do not know what this journey is about to show me. Right now, am happy to find a clean bathroom and bed and a clean toilet, and happy to get tea for a tired soul. Within an hour or so, must take to the road. Meanwhile, Bhadrachalam readies itself for the visit of Congress Party President Mrs. Sonia Gandhi – a 'one in several years' kind of visit (at least to Bhadrachalam). There is tight security all over the place. Before she lands, several helicopters can be seen doing sorties in the sky. This is her visit to the 'flood-affected' area. After a short visit to the relief shelter in Bhadrachalam she would be taken to areas in and around Amalapuram. Most of the Polavaram-dam affected sites are also, incidentally, Godavari–affected sites, but they do not form part of her itinerary (prepared by the State Government). Local MLAs and Congress supporters have placed their banners (their photograph next to Sonia Gandhi's picture) to show their contribution towards flood relief. All along the way, to the relief shelter where she is supposed to participate in a relief/charity programme (feed flood victims – a huge cauldron of food is being prepared here), there are numerous advertisements of the measures taken for relief.

But...Nothing New![41]

The CPM MP from this constituency, Midiyam Babu Rao, who I have been in touch with, offers very generously, to take me along with him for a tour of the flood–affected area around Bhadrachalam. Following comes from my visit through these localities. Later that evening, I am accompanied by a local CPM organiser, Madhu, to a relief shelter near Bhadrachalam. Every year, almost, the Godavari affects the SC colony in Burgampadu (mandal headquarters and not in the list of sites Sonia Gandhi visited this time round) near Bhadrachalam. Every year water level at 43 feet at Bhadrachalam is enough to inundate this colony. Every year, the SCs and Muslims (few) here, mostly daily wage earners – some of them working in the factory of the Bhadrachalam Paper Mills Pvt Ltd, locally referred to as BPL. Few work as agricultural labourers. They have been requesting the authorities to shift them elsewhere. But the local administration never heeded their request. Today, five days after Godavari, these people are without drinking water and few have been given only 7 kgs of rice so far. Surveys have been faulty. They do not count nuclear families (that is, adult married son not counted as a separate unit); they do not count migrant labourers, who do not 'own a house'. Nothing is new, in any of this, of course. At Reddypalem, Muslims and SCs are living in a makeshift shelter in an empty market yard. They managed to fight with the administration to get an electric bulb just yesterday. There are no toilets and cause a problem for women. Infants and children have no milk. Bleaching powder has as yet not been spread by the authorities in this area. When Godavari comes these people fend for themselves when the system fails them. Worse still, hard working, dignified people are rendered 'beggars' with 'charity' doled out. International NGOs with expertise in disaster management are still busy 'assessing' the situation and still 'monitoring'. Some have "money to release" but would rather wait for the 'right' time. They are still 'coordinating' with local NGOs. Nothing new ever happens here. The MP's visit gives the people the opportunity to just pour out all their woes, not all connected to the floods, or the present state they are in on account of it. These are issues that will come up no matter when a local political representative is accessible: old age-pension doesn't arrive on time, says one old woman; no ration card, says another; blind people have their own severe handicap...

***

Pay up or Perish! The MFI Loan Recovery Sharks[42]

At the Zilla Parishad school, Sarapaka, one system reached the affected families in the most unfavourable circumstances: Microfinance Loan Recovery. At Sarapaka near Bhadrachalam, in Burgampadu mandal, one of the worst hit, local staff of a few MFIs operating here braved the slush in the wake of incessant downpour, seeking out their borrowers from amidst a large number of women housed in crowded relief camps. Rice, milk, drinking water, food, even bleaching powder to clean up, almost everything the people needed most urgently was in short supply. As were government officials. But the MFI staff came to recover their loan installments. Never mind that their women borrowers had lost homes and livelihoods. The MFIs such as Swayamakrushi, SHARE, Spandana (with head offices in Hyderabad for most of them) - with a large customer base in Khammam district, reached the day they promised (or had warned) they would. Even local moneylenders (said the women I met) were better since their homes were flooded as well (and they were not linked to global finance capital, anyway). At the Zilla Parishad High School building relief shelter (for people from hamlets of Sarapaka, Talagummuru, Masid Road, Gandhinagar) a group of very worried, anxious women started narrating their plight almost as if they had been waiting to do so all these days. A local CITU organiser, Mr. Madhu, informed me that while these women have been sharing with him the regular hassles faced by loan recovery agents, the MFIs this time have degraded to the most inhuman state. Rahimunissa of Talagummuru said "Our village is inundated...They will come tomorrow. What will we do? We took loans from Swayamkrushi sangham. They came to us even on the day the village had flooded. We have even sold our household things to repay. When Godavari came, we requested them to wait just a little; we said we would pay later. They told us repayment cannot wait, even if our village is flooded..."

Dhanalakshmi, another borrower, said, "On Monday Godavari came into our homes and yet they asked us to pay on Tuesday. There was no respite...They have promised to return tomorrow to this camp to recover the money…" Many women had taken loans to start their own little enterprises – an Ice cream and candy shop, a provisions store and so forth. They say, almost in unison. 'Even if we die in the Godavari waters they will seek us out... Is there no humanity in these circumstances?"

Most of these women belong to SC, BC, Muslim communities. They have to pay a fine of Rs. 15 per Rupee if they don't pay the installment in time. What they did not expect of course, was a situation where even a disaster of this nature would keep the loan recovery guy on his feet. Shakuntala said, "I am a labourer. My husband had a kidney operation. I had to pawn my *mangalsutra*[43] and pay them before we moved here, unable to bear the abusive words that man (staff of the MFI) said to me in front of so many women…I wanted to drown in Godavari…"

I checked with one of the staffers of SKS in Hyderabad, Kiran Kumar (AP Operations), who said, "only the first week was a problem in recovery, but we had no problem recovering the amounts later in some areas. Our Bhadrachalam portfolio [!] was affected badly. But even that was in places where we could not reach the members because of bad roads and where there was no way to reach them. Or we would have been able to collect the money. But we are also distributing relief to our members – in places like Sarapaka, Burgampadu – we have given 7 kg rice to each of our members as relief. We are also giving relaxation time for those who could not pay...."

But I saw no such distribution at least in Sarapaka. But the MFI guys did go the next day, as warned, to collect loan installments, as I got to know.

***

10th August '06[44] - Polavaram

I reached Polavaram in not such a good way this time round; standing in a bus from Bhadrachalam, most of the way. Had time enough to dump my bags at the Head Post Office, Rajahmundry and head to Polavaram in an autorickshaw. How different it all looked! I reached Saikrishna's home whose backyard has also been visited by Godavari. We both then left for a survey of the area around.

Human-made 'floods' or a Natural Coming? Discussions on a Fisherman's Boat

We spotted few volunteers (one Manikyala Rao and his friends) from the nearby towns even as we waited, on a hillock, for a fisherman's boat to cross over to Ramaiahpeta from Polavaram since there are no government boats in sight; people say the government relief boats did not ply all day. The volunteers just dumped the food packets (*pulihora* - Tamarind rice) there and left (which was to be carried by others in the boat). The boat arrives. It is mid-afternoon. The fisherman belongs to one of these villages. People pay him either in cash or kind (grain, etc). I am with Banerji, Saikrishna (Prajashakti newspaper agent), and others in and around Polavaram. People express opinions on government's relief work and Polavaram dam. The conversation (many joining in) in the boat we are in, wading through what was a road the last time went this way (as I recorded) –'Godavari is receding (*teesindi)* but backwater levels are rising because of upstream river waters and breaches in canals badly maintained. There was no timely warning this time. Overnight, we were caught unawares.' One of them said, "It is the Polavaram project. People here are strong enough to face the sea. Why talk of Godavari? Godavari came earlier (previous years) too. The government has hatched this conspiracy to displace all of us for Polavaram project. When Maharashtra, Chhattisgarh and these states released water, Dowleswaram barrage water was not released on time. This is their only way to get us all out and blame it on Godavari." Someone pointed out, "even smaller lakes are filled up, or the water would have receded in three four days. They have built projects upstream..."

Whether or not the reasoning was 'scientific', it was clear that people here thought each and every moment about Polavaram and about the inherent contradictions of irrigation projects. People in this boat had come to Polavaram to fetch some essentials few days after Godavari's coming. Conditions hadn't improved. They were saying, "We had no boats; no government boats were visible all these days here, except one boat in the morning. Mamidigondi and Devaragondi girijans have climbed up the hills (*kondekkaaru*)." The bus stand, the hospital and fields and even the Polavaram dam spillway site were all under water. Nothing had been left. A local boat carrying people with provisions had capsized just the previous day, killing two. The atmosphere was tense in this boat as well. The only disturbance was a snake that tried to get on to the boat and caused some flutter. Reaching Ramaiahpet one realised this was the first time any representative of the media was stepping foot there, what does one say about government officials, people tell us. Boatmen had their own grievances. Local boats have been put in place by an office peon at the MRO office. Boatmen have been promised Rs. 500 per day for their services. But have only given a bare minimum as advance. Most important, though they belong to the same villages and have lost their homes too, they do not get any relief supply.

2006 Godavari: Many Images[45]

5 or 7 kilograms of rice instead of 20; 3 litres of Kerosene instead of the promised 5, was a common sight almost everywhere. Some of the tribal hamlets in Chegondapally, Mamidigondi, Devaragondi (some of whom where in relief camps for the first three days, and a majority on hills in their self-devised shelters) were not even accessing relief packs. When I asked the MRO at Polavaram, Sai Kumar, as to how they plan to reach them, he said helplessly, "*alaage undaali mari*" ('well, that is how it will be, can't be helped!)" The Mandal Parishad President[46] K. Lakshmi (Polavaram mandal) who one caught up with at relief shelter (through her husband) said, "The MRO refused to give us a *laanchi* (motorised steamer) though I have been begging I want to see my panchayat. Even former CM Chandrababu Naidu (who was there that day) was refused a launch. We managed a private launch on hire and went up to Singanapally.... As a mandal president I do not get a launch even to assess the damage!"

At Ramaiahpeta Barla Venkateswara Rao, member of the village welfare committee said, "For three days we could not go out with the rice bags the government gave us. We made our own distribution method. We got rice for 350 cards, of those in the "barrage list". But we shared rice even with those who did not have cards..." It was clear somewhere even the tribal families did not figure in this list. The 'barrage list' - what exactly did he mean by it? A local CPM activist (and Venkateswara Rao, too) confirmed it was the list of Polavaram dam oustees. No wonder then, all over, in villages in Polavaram, Devipatnam, Burgampadu mandal the constant refrain was lack of ration cards and fewer ration cards than people. Perhaps only those in the "barrage list" were supposed to get rice and kerosene. Obviously this list would show lesser numbers! At Burgampadu Zeelanbi's house is in ruins. She works as a daily wage labourer in BPL paper factory. Since she does not have a card, she will not get rice. She lives with her grand-daughter. Though Burgampadu is the Assembly and mandal headquarters, yet, the SC Colony there is in a sad state. No helicopters surveyed this part, despite the fact that Sonia Gandhi was barely few kilometers away that day. There were false rumours of another Godavari coming. But people had to get to their homes to clean up, to see what they would make of their lives now. This is when they need the most crucial help of all, conveyances to reach them to their homes. Well after the waters have flown, it will be time to assess whether Godavari has helped the government lay its plans for Polavaram project – despite the questions raised by the CEC.

***

'2006 Godavari' and the Polavaram Dam Submergence Areas: Some Issues[47]

Almost overnight, Godavari rose risen in nearly 30 villages of V.R. Puram mandal, 20 (approximately) in Kunavaram mandal, 12 in Chinturu, 12 in

Velairpadu mandals of Khammam districts; around 14 villages of the Kondamodalu gram panchayat in East Godavari district; including Devipatnam (mandal headquarters), nearly 13 panchayats in Polavaram mandal in West Godavari district, as sources reveal based on rough, instant estimates. If the river recedes people can still sow some seeds provided the government intervenes and distributes seeds free of cost to farmers in these villages. According to local sources, this extent of destruction could have been avoided had there been timely information. No timely communications were issued by the local Revenue officials – as used to be done in earlier floods. For instance, in Bhadrachalam, the temple microphones used to warn local people to move to safer places and shelters but this time round this was not done. According to Ravi, member, CPI (M) Bhadrachalam Division, "Bhadrachalam revenue officials failed us completely this time. Overnight Godavari entered people's homes and fields. No warning was given...More rescue teams are needed and more local boats and launches and medical teams."

Mutyala Rama Rao's home (V.R. Puram) is one of the many homes washed out. Currently taking shelter in the MRO office in VR Puram, he said "they told us on the evening of 4th August (at around 4 pm) that Godavari is coming. By then, half our homes were already under water! We had power cut from 3rd onwards. Government gave us 5 kg rice packets just today."

But the sub collector, Khammam district, Yogita Rana said, "We have been issuing warnings since long. The last one was the third warning given by us. We issued the first warning in the 1st week of July. By the month of June we already had 12 launches especially for this (possibility/eventuality) in place. Administration is well-equipped to tackle this situation...."

Most people wondered what they should expect when the dam comes up, if this was the situation in a 'normal' flooding. The Polavaram spillway site is now under water too. When the CEC team had visited, almost all local boats were hired for them and the police officials. Today, when local conveyances are required, the administration is not harnessing them. M. Venkatesh, of VR Puram said, "Government never told us anything. Only on 4th August they asked people to move. We do not even know what happened to the tribal families... There are no boats to rescue them. Jeediguppa, Kotaragummu, Chintaregapally are cut off... There are no boats in sight..."

Even in Polavaram, with the entire nation's attention focussing on this once small town, a local stringer of a Telugu Daily, Sitaramachandramurthy says, "No one knows where the girijans are. We are assuming they have climbed atop hills as they always did in the past. But others here at the MRO office in Polavaram are living under fear that any more breaches in local canals and water bodies, and Polavaram town will be consumed...Autos and other means of transport have been used by the government officials to shift their own families, even before they shifted local people..."

The local telephone exchange at V.R. Puram closed down owing to non-availability of diesel. There was no communication possible for at least two days. Even the BSNL cell phone networks have just about started functioning. Illa Rami Reddy called from Rampachodavaram to tell me, "Food did not reach us. People of Kondamodalu, Manturu, Tunnur panchayats have climbed hills and been staying there. Some are surrounded by water all around and stuck in small islands. There is no drinking water. Around 20 villages here are under water. We have come here – walking through hills and forests – to the Project Officer, ITDA at Rampachodavaram. We cannot find a single official. They say they have gone surveying. But the response of the officials has been very slow. In Devipatnam some people got stale '*pulihora*' thrown down from helicopters but many packets drowned in water. Official jeeps are out of access to local people. Only one launch was made available to us after we raised a hue and cry. We are all truly stranded...without conveyance, without water or food and without medicines and doctors..."

10th August - At Rajahmundry HPO, Reflecting in Solitude

While some things were the same as I last left them, many things weren't. For instance, look at Polavaram. What a change! To go on a boat where there was a road the last time! Roads have merged with Godavari. In fact, one couldn't make out where the 'true' Godavari was in all that backwater-inundated areas. Visited Ramaiahpeta on a boat that I got luckily. Though I was told nobody from the media had visited them so far yet local stringers from Telugu dailies have been duly reporting on the situation here. At the same time, one also heard of tales of a few stringers who wrote a story and took 'commissions' from people – from their relief money which they received later on! Local Telugu print and electronic media covered the situation relatively extensively compared to national TV media. Unfortunately, much as I wanted I could not make it to Chegondapally this time round, nor to Mamidigondi, or Devaragondi. Tomorrow I hope against hopes that I will be able to go to Kondamodalu, by road may be[48]? There were some familiar faces on the boat trip. Including one of those farmers of Ramaiahpeta who had a heated argument at the bus stand in June! Yes, this time, no BGR Company staff followed me! They left in time before the floods arrived, taking with them their contractual workers from Nellore and Tirunelveli, but left behind the local workers here. Apparently, people of Ramaiahpeta helped the local security guards and others by sharing their relief supply with them. And yes, the artificial tunnels that the Company officials placed, ostensibly to avert flooding of fields, did exactly what the people had warned they would: blocked the natural flow of the water into the local tank (*cheruvu*) inundating even those fields which none of the previous Godavaris ever left inundated! People kept repeating, even *enabhai-aaru* Godavari (1986 Godavari) did not 'stay' this long, all of five days... Incidentally, the R&R colony built near Polavaram for the Devaragondi tribal people was also flooded! There is no saying if it will not be submerged post-dam (if the dam is built). Wonder sometimes if this entire episode was a government conspiracy?

***

11th August - Godavari Forces another Route on Me

I note in my diary – "Road to Kondamodalu – one of the most adventurous so far in the three journeys down Godavari." I hired a jeep, with the help of Mr. Radhakrishna at the Post office, to take me to Kondamodalu from Rajahmundry. The road leads us through Devipatnam, Rampachodavaram and Karakada. The driver is an old man and the jeep looks really run-down, but for my kind of money, even this is a luxury I can ill afford. Yet, I can't *not* visit the people of Kondamodalu and the river route is next to impossible. I shall also get to see this historic road. It would take us four to five hours, I am told, mainly because the old jeep can't be driven too fast. On the way, I am thinking about my visit to villages so far. After five days people go back to their homes to clean up; to retrieve things, etc. In this time, it does not help to focus only on the relief shelters; in stead it would be important to look at means of communication. More boats are needed, but the government is not providing them. The focus is on the immediacy of the floods and immediate relief activities. Thereafter, people fend for themselves. People also need utensils, fresh vegetables, doctors, ANMs and delivery dimensions of rural health systems are most crucial in this time rather than packets of *pulihora* sent by good samaritans from the city. And they do not like being in relief shelters beyond two days. At Rampachodavaram I find a relief shelter set up at the Velugu (Government NGO) office. There seem to be no problems here on the face of it, as people tell me. The Kondareddis from nearby villages have been eating here for the last five days.

All about 'GAFs' and 'PAFs' – Interview of the ITDA P.O, Rampachodavaram (East Godavari District)

I stop by to meet the Project Officer, ITDA, Mr. J. Murali, at his office in Rampachodavaram. And this meeting ends up being quite significant. I make sure he knows he is on record. He says, "P.O, ITDA happens to be the administrator of the R&R component of Polavaram project. The Polavaram project occupies 70 per cent of our time."

I think to myself: So, 70 per cent on 'integrated tribal displacement'!

Meanwhile, Murali goes on, "The Polavaram project will be like rehabilitating flood affected families. Rehabilitating of GAFs (Godavari Affected Families). Anyhow they will get affected. Survey of the PAF (Project Affected Families) is happening."

For him it was a matter of an altered acronym.

"A list of objections has been published. It is being enquired into. By the way, the list (of households) prepared for the Polavaram project came in handy for relief. It was easy for the teams to distribute rice and kerosene."

Does this mean, until then, there was no proper list of households, or status of the area with the ITDA? And this list was made only for the Polavaram project?

I asked him, "What is ITDA's mandate in this (disaster such as this)?"

Murali: "Advanced Action plan – asking departments to be alert. We have four months worth of stock of essential commodities for people of Kondamodalu, K Gonduru, Manturu (inaccessible areas). Girijan Cooperative Corporation depots have four months of ration – rice and Kerosene. We asked the Kondamodalu people to open up the depots and collect their supply. They will get five kilograms per family. Since the last two days we have been trying, but are unable to get to the people. ITDA does not have boats, since maintenance is a problem. During floods we keep five to six boats on the stand by. We had an unofficial warning at 4 pm. By 5 pm we got four launches readied. Next morning they were pressed into service, at 9 am. Besides, thirty to forty boats were pressed into service. We sent a medical team that treated 130 patients. Floods are a recurrent feature here. On 5th (August) we dropped supply from helicopters. Since 170 tribal villages in Khammam were affected, all the helicopters were in service there. We could get the helicopters only by the 5th. We sent a few things by road but these did not reach them. We had prepared 850 food packets (prepared between 1 and 4 pm that day). We dropped them around 4.30 pm. We sent seven to eight biscuit packets, too. We couldn't supply water. Drinking water sachets had come from Rajahmundry but we didn't have enough time to send these. But floods are not uncommon for tribal people here. They have survival strategies. They have methods of cleaning up water and drinking – they drop *ingupu* seeds in water to cleanse water. The girijan depots have a good supply of these seeds."

I ask him, what about the road?

Murali: "On 6th we engaged labourers to clean the forest patch (roughly 40 km stretch) between Rampachodavaram and Kondamodalu…In the evening we sent another team. They carried the packets as head loads and reached there in the morning. I went on there on the 8th. It took me five and a half hours to reach (by walk from a point where the jeep could go no further). Nobody came to us to take these provisions. We sent five jeeps with labourers on 9th (August) morning. There was a sanction for the 30 km road between Lankapaka to Kondamodalu. Since these villages (in Kondamodalu) will get submerged under Polavaram, they diverted the money. But I revived the proposal. In a year we will lay this road. I have given them my own reasoning for it. I said if these villages get submerged, anyways we will have to transport people, shift animals, etc., from there. The project will anyways take three to five years to complete, At least people can use that road and it will be comfortable for the administration as well. We are laying the road under Food for Work programme."

On why is the road as bad as people mention it is, Murali says, "There are some very precious trees in these forests. Species like *egisa*, and *irugulu*[49]. These are smuggled by the tribals. When they drag these trees on these roads they get damaged."

(Incidentally, the Range officer had told me that this forest has many Rosewood trees. People had told me that private paper mills used to bring truckloads of labourers to chop these trees for paper)

I asked him what was ITDA's role in Polavaram since the ITDA is meant to protect tribal people's interests.

Murali: "Tribal people are mostly encroaching on government land. We are giving them pattas. Even for encroached lands they are supposed to get compensation. In order to include them in revenue records we carried out an "enjoyment survey" of patta and government lands. If it is not objectionable to us we will give them pattas. And then compensate land for land. In a way, the Polavaram project is a great opportunity for tribal families affected by floods on a permanent basis. It provides compensation for land, structures, housing and wages. In normal times, we cannot build pucca houses in flood-prone areas. Kuccha house compensation, as of now, is Rs. 1,500 per house."

I ask him, how many conveyances are in place right now?

Murali replies, "ITDA has 30 boats. Four launches and ten boats. In Devipatnam there are seven launches and fifteen boats. Kondamodalu has thirteen boats…We have 46 lakhs hectares of forest and two lakh tribal population in Agency Areas. Four per cent of tribal population is enjoying 2.18 lakh ha. forest area. Laying infrastructure here is difficult. At Kondamodalu, for instance, we had a doctor placed at the PHC there. He got *Malaria falsiparum* each time and after the third attack he refused to come back. Every Sunday we have a doctor who attends to that PHC (!) And we have a launch at 7 am everyday from Kondamodalu to Devipatnam or they have to come by road to Rampachodavaram."

***

At Kondamodalu Panchayat Office

I reach Kondamodalu by late afternoon.

A crowd gathers around the government officials who have come here with half the relief material, after two aborted attempts to reach them. People are agitated, angry, as they question the officials. It is still pouring and some people have umbrellas while others do not mind standing in the rain. Kondla Gangaraju, former sarpanch speaks agitatedly, "People sent numerous requisitions for a road from Kondamodalu to safer places. Each panchayat must have a (motor) launch. We have been asking ITDA since many years now for launches for our panchayat...Since they are building Polavaram barrage they feel these developmental programmes are unnecessary."

Revenue Deputy Tahsildar Radhakrishna tells me, "39 villages in Devipatnam are affected. Right now we are supplying food and provisions; later we will assess house damage, crop losses. We started a week ago, but had to go back; we started again two days ago, and again returned because it was unmanageable. Finally we reached today. But the provisions are arriving in another launch tomorrow. We brought with us whatever was left in our stocks. We couldn't come to Kondamodalu because there was no way. We have to deliver the remaining stocks of rice and kerosene – some were delivered earlier. We brought twenty litres of kerosene (for 750 ration card holders) and thirty bags of rice. There are two relief centers in Devipatnam and one at Rampachodavaram at the P.O.'s Office."

Gangaraju continued, angrily, "Fishermen come here every year. Had there been no fishermen, this place would have been a graveyard today. They helped save our lives."

The Tahsildar says they are giving compensation to fishermen too (in Polavaram, fishermen were not being given any relief); he said, "This flood came without warning. So we were not prepared."

But Tatikula Abbayi from Kukkarigudem screamed, "It was not sudden; it was already raining heavily in Orissa and Chhattisgarh; our government failed to take steps to prevent this much damage. They should have been prepared for it...This Government has no conscience. Didn't we say earlier we wanted more launches and more roads? Nobody listened to us. Now they are building a road. What is the point? They said they will give us Rs. 50 as daily wages. We are giving our labour since we need the road. Will the road be of any use after we all die?"

***

The Road to Kondamodalu[50]

Few roads, like the road to Kondamodalu, cruise so perfectly through life's close hits and misses. If you fall ill in Kondamodalu, the minimum price to pay for life is a three-hour boat ride on the Godavari – survival not guaranteed during the monsoons with the Godavari in turbulence. Sometimes, children have been delivered mid-way, unable to wait till they arrived at the hospital hours away; as this child did, during the 1986 (*enabhai-aaru)* Godavari; and he was promptly named Jalasandra Reddi, to remind him forever of the circumstance of his birth, with water all around![51] For, nothing but a myth of a road exists between Kondamodalu and the nearest government hospital. It may take lesser time to reach a doctor far away, or bring the doctors closer, were this road not one folklores were made of. The distance between life and death on the road to Kondamodalu, is the distance between the myth and reality of 'beneficiaries' of democracy. The road to Kondamodalu is both a metaphor and a reality; or, one where metaphor and reality of a state and its systems converge. It marks an almost imperceptible distance between relief and rehabilitation and displacement. The pathos unleashed by the '2006 Godavari' lay in the fact that, at least for the 12 out of 14 villages of Kondamodalu panchayat, the road to Kondamodalu was the only alternative source of surviving it. Some point that calls for a re-look at the way modern technology perceives floods and reacts, in response to it, was one of the points that the road to Kondamodalu and the 'relief' mechanism, raised. 2006 Godavari will be an important turning point in the official response to floods. And a sheer paradox. For now, these "Godavari affected families" have turned into daily wage earners building their own road towards their own displacement. Making it convenient for the government to smoothen out at least one rough feather in the long battle to build the Polavaram dam.[52] Kondamodalu has a population of 1,960 ST families (approximately), with one Primary Health Care centre for 14 villages and an apology for a road, among other handicaps. The way to reach people in normal times is through the far simpler and more beautiful Godavari on everyday launches. But during monsoons most of this part is out of access. Since years, this has been the case. This time, of course, with 'flood level' at nearly 70 feet at Bhadrachalam, there was no way but to hit the road to meet them. The road to Kondamodalu starts from a point near Rampachodavaram, and is near about 35 kilometers long stretch of a 'kuccha' road, inlaid with boulders and stones and passing through streams of water at intervals. And a pristine forest with (older) Rosewood and (younger) Teak trees, among other local species. Raju of Somarlapadu, near Kondamodalu said I was the first journalist to set foot on Kondamodalu village (or Manturu, or Kathanapally, for that matter) and the first to take that road. And he thought it was good I experienced the road and their plight during monsoons 'first-hand'. Perhaps the last point where media stopped until then was Devipatnam mandal headquarters. For almost the last two decades (or more) people of Kondamodalu panchayat have been requesting the ITDA to repair the road. Godavari's coming and going is nothing new to these people. But each time that happens they are stuck with their fates to climb atop hills to live in thatched huts. People here are forced to a tragedy of existence that can hardly be exoticised as the 'tribal survival instincts'. On 11th August, passing through this road, one realised the indifference to these people in matters of flood relief. It was literally the case of 'build, or perish'. For accessing the relief rightfully their due, the men and women of flooded Kondamodalu village panchayat were employed by the ITDA. People from Somarlapadu, Kokarigudem, Kathanapally are working on it. Contractor Kattam Rajasekhar of Kothagudem said, "This is to reach relief commodities in jeeps. The President of the Panchayat and Madi Muttamma and others wanted the road...Because of this calamity it has to be repaired..."

Accompanying me that evening at 7 pm, on the way back from Kondamodalu, were five young men from Kondamodalu, Somarlapadu, Kokkarigudem and other villages. They had come to the forest clearing to collect food which was air-dropped hours ago. The jeep, they surmised, would take them closest to the point from where they could walk back to their respective villages. Woefully for them, they ended up walking me down to Kakavada since the jeep refused to budge, not necessarily unconnected to the fact that the driver got drunk on *ippa* toddy. After the near-fatal miss (that I recount later in this book)[53], almost pushing us deep down into the valley, we all walked from 7 pm that evening to 4 am the next morning through intermittent downpour. For those men, it meant paying a price – all for relief. Having to walk back a day later to their homes, all for a loaf of bread, literally and figuratively. Needless to say, the loaves were consumed all the way down that road; besides the bananas and water and dry dates I had carried along. The walk from Kondamodalu that night made me realise the value of a tribal person's life – to the Polavaram project. If there was no road until then, it was because these were tribes with 'survival instincts'. And if there was going to be a road now, after all, it was to push them into fringes of existence through their own labour! Images of the 2006 Godavari seemed like a sneak preview of what these villages would look like when the Polavaram dam is built. Local people are seeing this as a conspiracy by the state that renders only these villagers victims to the floods, so that it is easier for the government to intensify efforts to shift them prior to the dam construction. And it is easier to move the 'flood affected' rather than the 'dam affected'. 9th August happened to be the International Day of the World's Indigenous Peoples. The day the indigenous people of Andhra Pradesh – Koyas, Kondareddis, Koyadoras – were struck off the mainstream. Government, NGOs and media either did not reach them or were unable to. They remained virtually cut off for more than five days from relief. Food, water, basic needs had failed to reach them. As did communications and communication channels. The road to Kondamodalu may have an answer. Or a question forever. The road, at the moment though, does not seem promising; though the woods that it courses through are lovely, dark and deep. Especially when seen by night (as one did). Little do they know this might well be the pathway to a nowhere place, without the ado about 'reaching the unreached'.

***

18th November - Changing Landscapes in Hyderabad (Reflections in my diary)

At home. I wonder, even as irrigation projects by the dozen are in progress, the nature and form of Hyderabad, the city, change, by the day. Is the money from these projects building this cityscape? Sprawling bungalows, Spanish villas, 'gated community' projects, (how I hate that term), malls, shopping complexes – are all these inversely proportional to the foundation stones laid for each of the irrigation projects? And deals thereof? As Hyderabad expands, in this manner, agricultural rhythms will collapse, by and by, just like the trees in my neighbourhood visited by the axe of the electricity department (leaving, thankfully, the tree that rises by my window), to give way to more and more electrical connections and rockfaces giving way to deeper borewells and higher skyrises? Then, there is no place for the simple things of life – plants thriving on mere sprinkling of drops of water, the *koels* waking you up at dawn, sometimes 2 am, the call of the street vendors (which my neighbourhood thankfully still has, each with a specific intonation, tone, *sruti*, timing of the day), each in a different *raga*, a different voice modulation, each distinct, each perfectly timed at different times of the day. So there is the spinach man shouting *'akulamma!'* at 11 am, *'samose'* man at 3pm, the *'moggu'* (used to be made of rice flour earlier, but it is chalk powder that is sold for the patterns that decorate the entrance of homes) woman at 7 or 7.30 am, the *'mouz'* (plantains) man at 12 noon… If I do not record these, would these disappear in a few years from now? Aren't so many things in life interconnected? Though they do not seem so? Just how many complexes and complications will devour how many simple things? When life consists of distances between two Godavari journeys, these things too fill up the space.

***

As This Journey Ends, Is this a Work in Progress?

In a budget announcement in February 2006 the AP Government allocated Rs. 10,000 (crores for irrigation (major irrigation, essentially) projects. The Chief Minister Dr. Y.S. Rajasekhara Reddy also publicly claimed that to be the "highest allocation" ever by any other state (in post-Independence India) towards the irrigation sector. The Annual Plan outlay for the year 2006-07 for irrigation sector in AP was fixed at Rs. 19,551.91 crores (near-about 200

Billions). The issues of access to river waters, for irrigation and in recent times, industry and power, unequal development, the Telangana angle (as to why the TRS as a political party has been involved in the anti-Polavaram discourse) are all important issues of long-standing. There is the question of coordination between states about opening and closing of flood gates – in the wake of 2006 Godavari.

***

Meanwhile, the Land Acquisition Progress as per Official Records[54]

Note: At the end of each year's journey, I present the hard-hearted reality of land acquisition that had been and continues to be progressing in spite of all. It is a 'meanwhile' activity or a harsh backdrop that each year people realise the ruthless power of state when it is hell-bent on proving its point against logic itself.

Land acquisition 2005-06 (When the project had not yet got basic clearances for projects of this nature)

July 2005 – Land surveyed = 88.92 acres
Land Acquired/"Handing over Land to Department" (as it says in the chart, to show consent) = 41.38 acres
Award value –Nil
(Dept here, means Revenue Department)

December 2005 – Land Surveyed = 3715.98 acres
Handing over Land to Dept = 2965.69 acres
Award value = Rs. 11.45 lakhs

2006 March – Land Surveyed = 1231.13 acres
Handing over Land to Dept = 1348.10 acres

[Which means here land is acquired in excess of land surveyed, or even compensated for; award was passed only for 8904.68 acres in this year.]

Cumulative Progress for 2005-06
Land Surveyed = 15,355.28 acres
Handing over Land to Dept = 8147.60 acres
Award Value = Rs. 29,186.75 lakhs

Land Acquisition 2006-07

June 2006 – Land surveyed = 286 acres
Handing Over land to Dept = 152 acres

August 2006 (This is important; the year in which Godavari came in spate or 'flooded')

Land Surveyed = 203 acres
Handing over Land to Dept = 2066.63 acres (again in excess of land surveyed, in effect without true consent of people)
Award value = Nil

[Note: September 2006 the Government was using Godavari in spate as a bait to convince or coerce people to give their lands for the project while resistance was relatively strong in that period. They specially targetted the tribal villages where dissent was the strongest, to delay flood relief and quality and quantity of relief as I was personally witness to it in that moment. Look at the Figures below.]

Land Surveyed = 1082.67 acres (This was the survey that was conducted ostensibly as Flood Level –FRL – survey/assessment)
Handing Over land to dept = 39.78 acres

December 2006 – Land surveyed = 88.63 acres
Handing over Land to Dept = 93.59 acres (in excess of land surveyed)

[My note: This is the period of relative resentment, resistance and dissent among people and the Godavari in spate limited the land acquisition work.]

***

Notes

1. The 2006 and 2007 journeys covered here are largely excerpted from my own report for the Prem Bhatia Memorial Trust Scholarship for Journalism - "Taming Rivers, Drowning Hopes: The Polavaram Project: Irrigation or the Deluge of Contracts" - with the support of which I was able to undertake most of the journeys narrated here, barring few, which happened after the scholarship tenure. But there is a lot more here which is additional to the (shorter) report, based on my diaries, some reflections and research thereafter, etc. Some developments that were not covered in that report are mentioned here, and in the 2007 Godavari journey, following this. I am grateful to the Prem Bhatia Memorial Trust for the support, as well as to Ms. Asha Rani, whose initial comments on the draft PBMT report, helped immensely.
2. And there were other similar safe havens in each of the three districts. After a point, the only thing to do was pick up the bags and go. Someone would always be there. A family would always open its doors to this woman.
3. I shall be chronicling some of my entries in my diary (a notebook for the personal) for a good part in the book, as a witness to the developments happening each moment, and a personal reflection.

4. Which owes its allegiance to the small splinter group formed along the Charu Mazumdar line, with Tarimella Nagi Reddy as one of its founder members, politically perhaps quite inconsequential in Parliamentary democracy terms, but important in issues it has been taking up, with a strong presence in East Godavari V Schedule (or Agency) villages.
5. Photo-credits in the entire book – mine, unless otherwise mentioned – taken from a Konica Autoreflex T4 (SLR, 28mm wide angle), Nokia 6600, Sony Cybershot 5 mega pixel.
6. F.R. Hemingway, 1915, pp. 127-8.
7. Ibid, p. 128.
8. Ibid, p. 6.
9. S. Prabhakar, 2005, p. 1.
10. Referred to as *mahua* in other places (*Madhuca indica*). .
11. I would meet him quite a few times during all my subsequent journeys.
12. Hemingway, op.cit, p. 269.
13. Jayprakash Rao, P, 1982, p. 284. Though popular history in Kondamodalu, with supplementary history from the Party records do not owe ideological allegiance to the "Naxalites" mentioned here. They were a group that broke from both the Srikakulam movement as well as the CPI post-Telangana Armed Struggle.
14. I miss till date not having followed up on this story even if they mentioned to me the name of this NGO, as I was this in the rush to meet as many people as was possible in each subsequent journey, with each leading me on to several other pertinent issues. Hence, without having substantiated this information though my own 'leg-work', I refrain from mentioning the name of the organisation. I also wish to add here, that it wasn't just the Kondareddi communities in Kondamodalu who mentioned this NGO, but also the Konda Kammaras in Gangavaram mandal (East Godavari district, not in the Polavaram submergence zone), and a few others I had met in Polavaram in the course of my journeys. But more than the name, it is the nature of development discourse that brings to fore these paradoxes.
15. Incidentally, I have preferred to retain the term girijan (literally, forest dwellers) when used by the tribal communities referring to themselves and tribal groups/ communities (tribal as a sociological, historical construct with distinct kinship patterns and other culturally distinct aspects that differentiate them from the caste-based social systems), communities elsewhere, and not adivasis (in these contexts). There has been a debate on the term adivasi not taking into consideration the marginalised groups within the caste construct (the officially termed Scheduled Castes and other Backward Castes).
16. The pamphlet mentions names of, among others, Illa Rami Reddi (Girijana, Girijanetra Pedala Hakkula Parirakshana Samiti - Committee for Protection of the Rights of Tribal and Non-tribal Poor), B.D. Sharma (Retired IAS officer, Chief Guest), Jeevan Kumar (Human Rights Forum), J.P. Rao. The 'invitees' included Manish Kunjam, Kunta (Odisha, former MLA), Sunnam Rajaiah (CPM, MLA Bhadrachalam), Midiyam Babu Rao (CPM, MP), Lakka Raja Rao (MLA, Paderu).
17. When the tractor they had hired had skid due to the slush on the way to the office.
18. I was able to record bits and pieces of that song on my mobile phone and

freeze some moments in my slr camera. And of course, record some of their voices, and the song, in my old cassette player.

19. I did not get all the words of the song right; and just about managed to gather that the song is about the young woman and a man getting married in a grove (*thota*), the bridegroom is a big man (*peddavada*) and the bride is a young girl (*chinna dana*), and about bringing the sweet tamarind fruit (*chinta puvu*).
20. The organisation that worked in these parts owed allegiance to the Kanu Sanyal group, and it may be a question for historical engagement as to why in these parts – barring the initial phase of organising, during the 1969 peasant agitation (to reclaim tribal land, when the idea of forming tribal guerilla squads (armed with bows and arrows, essentially) – the notion of armed struggle gradually faded out. Speaking to some of the older people associated with the movement here, I felt it important to look at the element of the local contexts within Communist movements (of any leaning); the elders told me they felt it was important to take with them all the people in their villages and which was not possible if they had continued with armed struggle. They would, of course, be prepared for it (in the 1969 phase and few years later) but that was not the main focal point, on which they differed from the Naxals in the Srikakulam tribal movement. They believed in Communist politics and managed to bring a lot of young people (which continues to be the case) because, they say, they did not follow the Naxal line of continually arming themselves against the state. Politically, of course, the CPI (ML) and Agency Girijana Sangham (which is affiliated to this Party), do not have a significant presence though they have been on the forefront of several issues of workers rights, tribal communities, displacement, among others.
21. On many occasions, there wasn't an occasion to write, or to hold a pen and a notebook. Cannot say in how many ways my mobile phone journeyed with my emotions of the moment in these situations. Some of those points I feel were worth recording as moments, here. Many a times one typed in these notes on my mobile sitting on a boat precariously, or pillion on a bike, or in auto rickshaws amid bumpy rides. The anxiety to just record the thought of the moment!
22. Christophe von Fürer-Haimendorf, 1982, pp. 7-10.
23. David Ludden, 2004, p. 216.
24. Ibid, p. 217.
25. M.L.K. Murthy, 2006, p. 135.
26. Ibid, pp. 136-7.
27. L. A. Cammiade, 1932, p. 84.
28. Hemingway, op.cit, p. 280.
29. Found just the spot to call my nephew on his birthday. Perfect timing, he said; perfect spot, I thought, for my phone to find its network! Not coincidentally, at the office of the Madhucon Company which is overseeing the spillway construction!
30. The old boatman from Sivagiri, my regular 'ford-maker' on most journeys, crossing me over, across Godavari, had in a dramatic manner, told me his name when I asked, pointing to my nose ring, saying 'Mutyam' ! Punnam Mutyam, always blessing me, the boatman who saw Mahatma Gandhi when he was a child!

31. See *Personal Memorabilia...* later in the book.
32. Hemingway, op.cit, p. 263.
33. One of the very few activist-mode organisations still working on the premise that the dam is to be opposed, while some others have either become quieter or changed their 'vantage point' in the discourse. I met him first in June last year, when we visited the dalit bastis/hamlets in the submergence zone. I thank him for highlighting the dalit angle to me in the course of my visits and making it possible to tour the villages and meet the people (he says) the media did not meet.
34. Surrounded the officials in protest.
35. Hemingway, op. cit, pp. 263-65.
36. It was still Orissa then. Today it is officially Odisha. I retain Orissa to retain the historical truth of that period.
37. If Telangana State ever becomes a reality, there will be a new inter-state issue in this.
38. A Telugu version of the above was published in the Telugu daily, Andhra Jyothy, on 30th July 2006. The English version was published in www.indiatogether.org
39. *Lankollu* originally referred to the landlords who owned or rented out rich alluvial fields which were used for growing tobacco in the colonial times.
40. A version of this was published as an op-ed piece in Andhra Jyothy Telugu daily.
41. I credit this phrase to a response I received from a features syndicate if I can send them about Polavaram, the 2006 Godavari 'floods' and displacement. They said, it was not "new" and rejected the story.
42. Thankfully, I found space in the Times of India, Hyderabad edition for this story; however, the names of the MFIs were promptly removed in the final printed version that appeared, incidentally, many days later. The above is my original version, unedited. I heard of this case from a CITU (CPIM- affiliated trade union) organiser, Mr. Madhu. Our idea was to 'catch them red-handed' as they had threatened to come back that evening to the relief shelter. We waited for a long time, but perhaps word got around and they did not show up.
43. A chain worn around the neck as a signifier of marriage.
44. Some of my experiences here were published in The Times of India, Hyderabad edition as "2006 Godavari: Many Images".
45. A Version of this was published in the Times of India, Hyderabad.
46. Representative of the local self government, the Gram Panchayat.
47. A version of this was published in the Times of India, Hyderabad. Times of India, Hyderabad published four of my articles – during the visit of the CEC and 2006 Godavari.
48. Sometimes hopes do realise!
49. He could not tell me the scientific names of these.
50. A version of this published in the Frontline – September 8, '06 under the title, "Road to Nowhere". I have not reproduced the article in its entirety here. I wrote this piece after coming home and suffering a blackout, besides fever. It felt like I was reliving, as closely as possible, the pain of all those from whom many of these things have been, unfortunately, a matter of routine. On

hindsight, there was no other way to have written that piece, but in pain.

51. He is a Kondareddi man from Kokkarigudem, and Saroja had told me of this story. Apparently some volunteers who had helped these people in the 1986 Godavari suggested that name. Like the many things that end up incomplete in life, I could not meet him or his mother, now no more. I had told myself, I shall meet him the 'next time' which did not happen in this case.
52. The journey that took well over 12 hours – from Kondamodalu back to Kakavada and another hour's journey to Rampachodavaram over ups and downs and rubble and slush – smacked of the sheer apathy towards the Kondareddis by a system (ITDA) meant to integrate them with the 'mainstream'.
53. See the section *Personal Memorabilia*...
54. http://ppms.cgg.gov.in/

3rd June'06

On Godavari's Bosom

Inside the Luggage Carrier

Brief Halt at Tallapudi

A Tree but a Goddess -Tamarind tree at Tallapudi

Fishermen's boats on one of the banks

People's Ferry

Evening - Relaxing on the calm waters!

Evening - Mother and Daughter catching seer-meen at Kondamodalu

Kondamodalu village from afar

4th June'06

Tamarind by the house at Kathanapally

Vidya and Others (an image of cheer)

The Dance of Camaraderie

Children at Kathanapally

Kondla Rami Reddi of the Peasant Agitation of '69

Photograph from L. A. Cammiade, 1932, p. 84.

At The Site - Polavaram, 5th June'06

Towards the Spillway Works

Welder at work *A Child at work*

Argument at the Bus Stop (Posi, others)

Chegondapally - 5th June'06

Ramalayam at Chegondapally

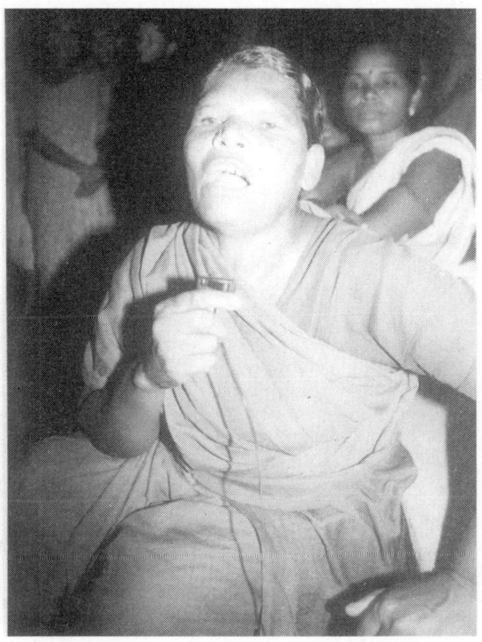

Mutchika Suramma, the fiesty Koya

People watch as rubble from dynamite blasts falls on fields

6th June-In my Diary-Manturu by the blue-green pond and river

7th June'06 - Kumari at her home, Kondamodalu

7th June - 4 am. A Fishing boat-'smaller things in life'

Boatman Mutyam- always blessing

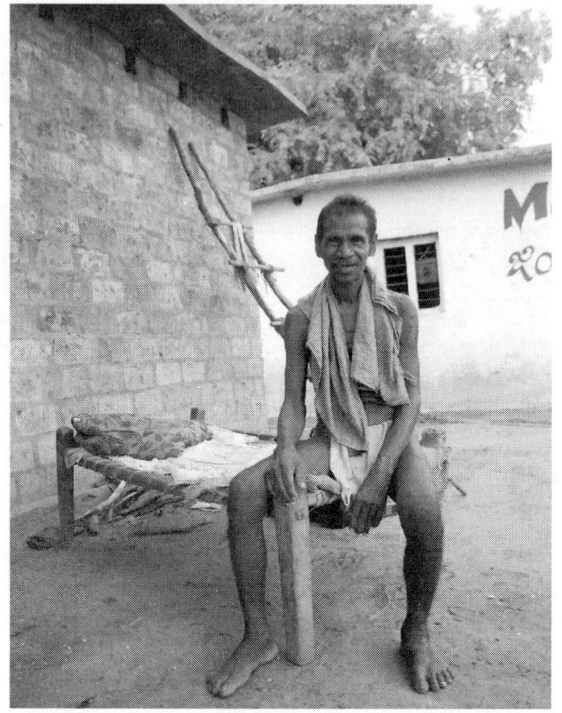

Jogi Reddi at Pochavaram:a Resistance Story I missed

Dalits of Khammam District, 8th June

Gampala Ramulu's family, Karakagudem

Karakagudem Sujata and others

Angry Sarojinamma

Strange Case of Pydigudem-Suchirevula

At Jeediguppa Panchayat

Ancient Tamarind Tree at Kotaragummu

Korram Ramanaiah (Koya), Kotaragummu

M. Satyanarayana, Ward Member, Jeediguppa Panchayat

9th June, Meeting Bhadrachalam MLA Sunnam Rajaiah

29th-30th July, CEC Visit

Posters Welcoming the CEC team at Bhadrachalam Punnami Guest House & Some of the Political Activists Waiting with Petitions

And Godavari Comes (August)

Polavaram Inundated

Houses under Water

Backwaters at Kondamodalu Panchayat

And nothing new - Old man lined up for relief at Burgampadu

Zilla Parishad School, Sarapaka - Relief Shelter

Women Harassed by MFI Loan Recovery Sharks (at Zilla Parishad School, Sarapaka)

A Young Child at makeshift shelter at Reddypalem (Bhadrachalam)

Waiting atop a hill for any boat in sight (Polavaram)

Discussion on a fisherman's boat towards Ramaiahpeta

On the Road to Kondamodalu

A Board with a picture of the Wild Gaur welcomes one to Papikonda Wildlife Sanctuary

Gherao at the Panchayat Office

An old man, with his belongings

On a fisherman's boat on the backwaters

A woman feeding her baby in a paaka (thatched shelters Kondareddis construct)

Some of the paakas uphill

The road

Women labouring on the road

People from nearby villages waiting for the relief packs to be air-dropped

Trying to move the Jeep

Godavari Journey 2007
Of Marginal Lives, Unique Landscapes, Prison Terms and Tourist Gaze

Sometimes a postscript can be providential. That was a brief "P.S." to my scholarship report to the Prem Bhatia Trust, which ended with, "Does this journey end?" For I did not know then, as I wrote, that if I were to continue on these paths (a desire I could not deny), where would the moneys come from, after this resource ends? But they did, after all, for a short while and the journey continued through the entire stretch of 2007.[1]

In my first visit within this journey, I see new crops, new life in the fields, a new season. I visit villages in all the three districts in this time, East and West Godavari and Khammam. Some places for the first time, of course. Something else is 'new' - I left my 'dairy' (notebook) behind at home. So this time round, my work and personal moments will happen together, in moments of relative solitude. There may not be many this time, considering the number of places I plan to pass through.

New Crops, New Life in the Fields but Simmering Discontent

29th January - 5th February'07

Bhadrachalam, Khammam District, 29th January

I am at a public protest gathering organised by the CPI (M) in Bhadrachalam, by the Gandhi statue, held in protest against the police firing on the CPI (M) activists who staged protest marches against the dam. Several were injured in this firing, including the MLA, Sunnam Rajaiah. One of the speakers (I think the District Committee Member of the Party) is saying, "Over two lakhs will be displaced (due to this project). We are asking the government to think of smaller projects and stop Polavaram. They arrested our MP (Midiyam Babu Rao); lathi-charged activists and slapped cases against us. The previous government (under Chandrababu Naidu) behaved this way and lost power. This government too is going down the same road. One of the party activists had to get his leg amputated..." A bullet pierced through his leg on the day of the police firing. He left for his village by the time I reached the hospital to speak to him.

3rd February'07 - The Dalits (Scheduled Caste) of Polipaka SC Colony, Gunduvarigudem, Kunavaram Mandal[2] (Khammam distirct) and the Onus to 'Prove'

Errapotu Chinipiri, small farmer, tells me, "Surveys are being carried out only for STs[3]. The Government will give pattas only to girijans. We have one to three acres of land. Some of us have even less than one acre and there are many landless. In 1978 we were given Settlement pattas (for assigned lands) and later pass books in 1986. We have been cultivating these lands since then. Now they are saying these lands will go under Polavaram barrage. What about us? But the MRO and Collector only meet the girijans. They do not consult us on matters of R&R package. Three MROs visited us so far. About twenty-six years ago we got pass books, which we used all these years, thinking these were proof enough that these lands were ours. Now suddenly they are saying these pass books are not valid and only those with pattas will get the compensation. The Government is causing conflict between us dalits and girijans. We cultivated these lands. The Government must give us answers. The girijans want pattas for *podu* lands. Why take away our lands? They are taking our land and giving it to STs. We are poor people. We have been living on daily wages after the 2006 Godavari, since we lost our crops. Since three months now we have had no regular work. We will not be given land to land compensation... We have not seen any R&R colony yet. Three Godavaris came in recent years but when Sabari (being closer to these villages, which also rose up during the 2006 Godavari) and Pedda Godavari came, we lost our crops. We had to go to far-off places to work.They gave away some of our lands which we were cultivating for last two generations to the STs."

People say, almost in chorus, that at Gunduvarigudem the government officials are asking people to move to save themselves from the 'Godavari floods'. "They are not mentioning Polavaram. They are saying we will move you to other land as flood-affected people. That is what they told us." Pada Mutyam (a Mala) said, "I made a requisition (in Telugu) to the Sub-Collector (Kunavaram, then one Ms. Yogita Rana[4])." He showed me the same; someone read it out to me – "For generations we worked and toiled on these lands. We were not given pattas. We could not go when the Settlement officers called us since we had no money for transport and someone was ill in my family. We want a patta for our land. Otherwise, we will have no other go than to commit suicide in front of your office. Signed: Pada Mutyam Survey No. 3/1 (2 acres land)."

Mutyam has not received any response so far, in spite of taking the bus journey to personally hand over the letter but there is a stamp of receipt on the same. The problem here regards pattas and dalits having to prove to the government as to how these former Estate 'Banjar' ('waste land') came under the Assigned Wasteland category handed over to the dalits by the government. This should actually be a predicament not of the dalits but the Revenue officials who have not moved on several petitions from the dalits who wish for regularisation of their ownership. For me, this was a new angle to the R&R that I had not known of earlier. I was struck by the paradox of the upper caste non-tribals actually benefitting from the 'legalising' in the form of 'declaring' their lands and giving consent to the project, and in fact earning crores of rupees in the bargain, but this 'advantage' was not available to the dalit small peasants and farmers. There was a lot of history that actually came out from conversations with people regarding these complexities.

I then moved on to some interior villages in V.R. Puram Mandal (Khammam District).

"Will our mango trees be 'there'?" Questions the Kondareddi Sarpanch of Sriramagiri[5]

They said her name was Seetamahalakshmi. It immediately conjured up in my mind an image - that of a middle-aged woman, probably one of the many 'proxy' women Sarpanch in a constituency reserved for a tribal woman candidate. Yet, for me, the fact that she was a Kondareddi (still defined as a 'Primitive Tribal' group in government records, now called Particularly Vulnerable Tribal Group) and a woman, were reasons enough to seek her out. The image was struck off with a pleasant blow, as out walked a young, wiry woman, on that warm late-afternoon, with a sportsperson's gait, dressed in pants and shirt, her wavy hair plaited behind. I was introduced to Kotla Seetamahalakshmi, the 23-year-old Panchayat President of Sriramagiri panchayat. This was an image I least expected in a village predominantly settled by the Kondareddi tribal community (having visited many other villages in these parts) in V.R. Puram mandal. Seetamahalakshmi's home (where I met her) is in the Godavari gumpu (by the river bank) in the hamlet called Reddigumpu (hamlet of the Kondareddis) of Chukkanapally village in this panchayat. It is also quite near the village named Sriramagiri.[6] She had doe eyes and skin like porcelain, I couldn't help noticing. Seetamahalakshmi's beauty is not a stereotype to bind her character in, yet, something that added to her persona, of one leader, among many, who refuses to move an inch from her stated position against the Polavaram dam and the "packagi" (the R&R package as it is referred to here). And she is single. Definitely one heartening instance in this particular trip of mine, in which I witnessed people of many of the submerged area villages gradually losing the hope to resist the power of the state. Seetamahalakshmi studied up to the tenth standard at different government schools at Chukkanapally, Chintur and V.R. Puram. She was also a Vidya Volunteer (teacher) in the MPP (Mandala Praja Parishad) school for four years. Her parents had wanted her to study further but she says, "I did not want to put further burden on my family. I wanted to help at home. So I quit. Moreover, I did not have good teachers. So I lost half the interest." She is one of two siblings; her older sister is married and lives in another village. And she lives with her parents. Her father, Kotla Kamireddi was the sarpanch of this panchayat, for ten years, following which she got elected, last year. There are three hamlets here with Kondareddi settlements (there are a few Koyas as well here) and the community owns, in all, approximately 500 acres of cultivable land (according to her). Women are more in number than men among the Kondareddis here.

The panchayats of Jeediguppa, Tummuleru (besides Sriramagiri) boast of women sarpanch. Incidentally Sriramagiri is the site of an ancient temple (popularly believed by some to have been built by the Kakatiyas), which has suddenly become a thriving tourist attraction. The colonial records note: "Sriramagiri ('holy Rama's hill') lies forty-four miles south by east of Bhadrachalam. It is supposed to have been here that the bird Jatayu, who had tried to hinder Ravana's abduction of Sita but been mortally wounded in the attempt, told the news of the abduction with his dying breath to Rama as he passed that way. The grateful Rama performed the funeral rites of the faithful bird at Sriramagiri. The god is known as Kulasa (the joyful) Rama, because he here had news of his lost wife...The temple is supported by the zamindar of Rekapalle, who devotes to its maintenance the net income derived form the village of Kunnavaram, which ordinarily amounts to about Rs. 800 a year. The neighbouring hill called Vali Sugriva is so named from the legend that it was there that Rama obtained further news of Sita from Sugriva, the brother of Vali and a king of the monkeys."[7]

From one to two buses in a day in the past, today there are at least ten buses bringing visitors from outside to this temple. This has happened especially in the last year, following the government's ongoing Polavaram works. Of course, I was told by people here that the money from this new tourist attraction is not reaching the panchayat funds, as it rightly should. The tourists (as always) come in hoards, and leave, with little connection to the surrounding reality: that this temple will also be submerged under the Polavaram dam; or that there are tribal hamlets around this place. Conversation with her reveals what her community feels about the Polavaram project and its 'packagi' and one realises these are not people to given in to any pressure tactics though all around them village after village looks helpless faced with the Government's land acquisition process. She says, "It is sad. The MRO visited all villages in my panchayat and seems to believe that the President (Sarpanch) is not necessary for any of these. One of our community men in Kalthanuru village (Sriramagiri panchayat) in fact told me, the village elders there said they want to consult the panchayat president (sarpanch) before deciding if they will give their lands or not. But the MRO told them, we do not need her. "The MRO told them, as he told us the other day, *'mundu poina vaallake tikkatlu, veneka oste tikkatlu dorukavu'* (if you come first you will get the 'ticket' – compensatory package - not later on). We don't want to go. There are poor landless people here whose livelihoods are linked to the forest, collecting non-timber forest produce. What will happen to them?"

Seeta adds, "What is point of my being a Sarpanch if I am not consulted by the officials. The MROs are big people...In our gramasabha last month (January 2007) we had resolved we do not want Polavaram. The MPP members (Mandala Praja Parishad) too passed a resolution in five mandals (Khammam district) that people here do not want Polavaram..."

Seeta's anger at not being consulted is valid as much as it points out the violations by the Government. For, the Provisions of the Panchayats (Extension to the Scheduled Areas) Act, 1996, No. 40 of 1996 clearly says, "The Gram Sabha or the Panchayats at the appropriate level shall be consulted before making the acquisition of land in the Scheduled Areas for development projects and before re-settling or rehabilitation persons affected by such projects in the Scheduled Areas the actual planning and implementation of the projects in the Scheduled Areas shall be coordinated at the State level...While endowing Panchayats in the Scheduled Areas with such powers and authority as may be necessary to enable them to function as institutions of self government a State Legislature shall ensure that the Panchayats at the appropriate level and the Gram Sabha are endowed specifically with ... (iii) the power to prevent alienation of land in the Scheduled Areas and to take appropriate action to restore any unlawfully alienated land of a Scheduled Tribe..."

It is clear in many of the panchayats in the submergence zone, this wasn't done. There are other ironies. For instance, she also informs me that "the Rozgar yojana (work) funds which earlier used to be routed through the panchayats have now been diverted into NREG (the Employment Guarantee programme of the Union Government) and Sarpanch is not getting any remuneration. Even the Sarpanch remuneration is not reaching us; we attend meetings at our own costs..."

Seeta is saddened by the fact that in the month of January (2007) the sub-Collector (the Sub Deputy Collector of the Polavaram project) and MRO (V.R. Puram) "went to Mulakalapally (a village nearby) and Kalthanuru and claimed they had convinced people to give up their lands. They did not take my permission...People who have land do not want to go; only those without land are agreeing to the *packagi*. Landless people want to go because they will get money. But how many years can they live on that money? We need fields, and food. We get our food from our lands and the forests. We cannot survive anywhere else. We will never get this environment."

And she points out, quite rightly, "For the increasing population there is no place, how will they give us land to cultivate and to live? Women are not very happy about leaving." While many men in these villages are today discussing the R&R package, Seetamahalakshmi says, firmly, "We do not want to know how much compensation is coming from all of this. They (officials) did not come to tell us. But then, we are sure we will not leave this village, which is why they do not tell us about it. Anyway, when we will not leave, why even bother about the package!"

There is yet another facet about the 'surveys' conducted in these villages post floods of 2006. Seeta says, "They used to talk about survey. They came recently. They are showing forms which read, "*varadala baadithulu ki*" ('For Flood-Victims'), and not Polavaram. I questioned the MRO in this regard. He said the CM (Y.S. Rajasekhara Reddy) ordered them to prepare these kinds of forms. She points out that tribal people in Mulakalapally and Jeediguppa were also confused about the Godavari varadalu (floods) mentioned. "At a meeting in the mandal office recently, I learnt that only the Mulakalapally non-tribals consented to leave. The Jeediguppa sarpanch told me she thought it was rehabilitation for the Godavari floods. She did not think they were giving consent to the Polavaram dam! She is educated, yet she did not think anything could be wrong. I had to tell her the forms were meant for Polavaram and not Godavari floods. The officials are telling lies, conducting surveys and convincing people. Godavari comes once a year and goes. We have learnt to live with it. It is normal for us to climb on to higher reaches of our hills when that happens. But our homes and lives were never destroyed totally. We use the Godavari water through motors to irrigate our lands."

The father, Kamireddi, pitches in – "We do not know since when, but surely since civilisation began our ancestors cultivated lands here."

Their lives are intricately linked to the forests, the land and more importantly the seasonal 'flooding'/coming of the river. Post-coming (as one could clearly see) the silt gives the most abundant crop. This is the season when the landscape is filled with some of the most diverse crops – chillies, jonna (maize), some pulses, lentils, and so forth. In very small pockets paddy is cultivated, by some tribal farmers. The non-tribals (among these the later settlers, mostly) brought in tobacco cultivation (used in making the local 'chutta' and the more posh cigarettes). In tribal hamlets, of course, tobacco is not extensive. They grow more of millets and rain-fed paddy, among other crops. Each of their gods/goddesses (Gangamma, Rajulu, Polam rajulu, Potirajulu, Chikati rajulu, Gamutyalamma, Sayilamma, Malachemma, Chalalamma) has an associated tree. The term *polam rajulu, chikati rajulu* convey the sense of association of the gods with fields, darkness, or night (*chikati)*, etc. The trees associated with these deities are never cut down. The gods are not always anthropomorphic, human-made icons or forms, but just a 'concept', placed under a tree and propitiated.

For that festival, that occasion, *in that moment*. Their festivals are related to the harvest or the onset of a particular season, a seasonal fruit also is a harbinger of the good. Seeta told me, "Before Godavari comes, we celebrate *samakotta*, offering *samalu, chintakaya, gongura, pottalu, totạkura* (Millets, Tamarind fruit, Rumux *acetosa* (?), Pointed Gourd, Amaranthus). Then there is *kondajonnakottalu* when we offer *jonna, chikkudukayalu* (Sorghum, Cluster beans) as offering to Chikati rajulu. On *ugadi* we have the *mavikidkaya panduga* (variously referred to as *mamidi panduga* or *mamidikotta,* first Mango fruits of the season) and only eat after the prayers."

Less than a decade ago, some developmental programmes have given them access to education – with residential schools and colleges being part of the welfare agenda. The Integrated Tribal Development Agency was also put in place years after Independence to work closely on employment, education and health. But displacement will only negate the essence of these programmes. If land, forests and compensation, are 'complex' subjects, there are 'simpler' questions that people like Seetamahalakshmi (and I) think about. So, while Seetamahalakshmi wonders (even as I do) about the loss of these facets of their lifestyle and lives, surely, the government's package may give them house sites and land (if at all) but what about these culturally tied expressions of community living? Can these be recreated? As she asks, simply, "*maa maavidi chettlu dorukutaayandi akkada*?" (Will we get our mango trees 'there'?) And the question is not about the 'species' *mavidikaya chettu* (the mango tree) but that of the symbolic meaning hidden within – for history, culture and society. A meaning women understand well.

***

Reflections, in My Notebook

The tribal communities are being delinked from lived histories. Mainstream discourse, and the government's attitude that codifies it, perceives them as having nothing to do with agriculture, and even if they did, their cultivation is not considered significant. Their identities have become totally subsumed under forest-dwellers and hence their sense of their agency in the agrarian history of the nation does not look legitimate enough for any worthwhile discussion. Why are lived histories and spaces of histories of communities to-be-displaced not even considered by those – on either side of the spectrum – who have caught on to the "adequate compensation"(in terms of money) bandwagon? And if we were to look at *podu* (shifting cultivation), it has surely not remained stuck in a bygone era, as human societies have almost always introduced newer methods of farming appropriate to their contemporary situation (given the freedom to do so; else in a forced context, have had to negotiate change).

***

4th February'07 - I take a boat from Kunavaram to Rudramakota (Rudramakota Panchayat, Valairpadu mandal, Khammam District).

The boat is a Rudramakota Panchayat boat but privately owned. They pay (amazing) Rs. 4 lakhs a year to the government in taxes. (And these people are never credited with being 'tax-payers' which is a middle class 'privilege', heard in vigilante sloganeering, usually). They make up for it with at least nine trips a day in the normal season between Kunavaram and Rudramakota at Rs. 10 to cross over, per person. The boat plies between 7 am and 6 pm almost through the year. The owner tells me they more than make up. At least up to Rs. 2 lakhs a year (half the amount they pay in taxes).

This Premchand loves English Literature!

At Rudramakota, we had to take a 'service auto' (shared autorickshaw) to the village we were to visit, a dalit hamlet here. As the auto driver waited for the vehicle to be filled up (to capacity), I looked at the young boy (12 years old) seated next to me. He asked me what I was doing here. Premchand was his name. He said he was the seventh grade at the government school here. I asked him what he wished to be when he grew up and was pleasantly surprised when he said (and could not help being amused by a kind of 'link'), "English teacher", then added, "I love English literature" (*naaku English rachanalu chaala ishtam*). He had read Tagore's stories; that is why! And, after a brief silence between us (having asked me as to why I had come here), he said, "*maa school kooda munugutundi ee barragi valla*" (our school will also drown due to this barrage - Polavaram). I could not meet Premchand again, thereafter, though he told me where he lived. But some moments are fleeting, and must be accepted as so. Thankfully, I captured a part of that moment in my camera.

At Rudramakota[8]

Mendam Narayana (SC) said, "On 3rd January (*modati nela*, as they refer to the months here sometimes, either as numerics, 1st to 12th month, or in Telugu equivalents) the MRO and RDO came here for survey. Our sarpanch, Uike Mahalakshmi (Koya) did not go with their survey team." Penta Rao added, "They conducted the village survey saying it was a Godavari ('floods') survey, not for the Polavaram package." Kesapaka Venkati, the former Village Officer here said, "I do not know why we elected this government. Last year Godavari affected the model colony they built in Polavaram for RR (R&R colony). They said they will build a school, hospital, etc. Now, since Polavaram (works) stopped, they are repeating the same survey for 'Godavari-affected'! Some SCs here have half an acre to five acres of land. The officials told us we could come and take cheques, but they are asking us to form a Society first. Only some families agreed to do that...."

The situation here is peculiar. This is Scheduled Area and these are tribal lands. The dalits are stuck in between – there is the non-tribal upper caste landlord who has managed to get pattas in his name for lands originally owned by the tribal communities; and there are tribal communities who have usufruct rights on some of the lands here. The third category is that of the tribal peasants, some of whom were assigned lands by the government more than three decades ago. So historically, it is important that perhaps some of the villages in the submergence zone have dalit farmers owning land and growing food crops. There are also others who have been brought in to work by the upper caste landlords, and remain landless farm labourers till date. Both the landed and landless dalits now face a highly uncertain future with the unequal package assigned to them under the R&R. Venkati adds, that "there is no preference for SCs in Polavaram package. The Deputy Collector only talks to the OCs (non-tribal forward castes) and STs. Our SCs are mostly debt-ridden people. 25 per cent non–tribals (mainly the caste group called by the suffix Chowdhary) own most lands here. They take loans from co-operative societies. Today even if they are given the package, half that amount will be spent repaying those loans! Officials came here and spoke in the presence of the karnam (village book-keeper who has all the land record details)..."

Then there are other kinds of transactions that each one, by turn, spoke to me about. For instance, there was a peculiar case of a non-tribal farmer (one Palivela Raja Rao, a brahmin) who had 'sold' his land to a missionary organisation (they said the name was GELC) in spite of the 1/70 (Act which nullifies such transactions and can even be punishable) which no longer works there. Though he sold the land to them, he retained the patta in his name through some mischief. Neither Raja Rao nor the missionary organisation in whose name the sale was finalised has been in Rudramakota for years now. Both are absentee landlords. Instead, tribal peasants have been cultivating that land (rightfully, as 'enjoyment rights' are available to them in Scheduled Area). As soon as the Polavaram project compensation package was declared, Raja Rao came right back; had a tacit understanding with the missionary organisation saying he would pay them some part of the compensation. He got the compensation for 'giving' his land to the government for the Polavaram project, so did the other illegal absentee landlord - the organisation's founder. Both made their money. Or they were about to, as they had nearly done the 'rounds' (that is, paying bribes, etc) to the MRO offices, etc. The person who has lost out is the tribal peasant cultivator. And this story is being narrated to me by the SCs.

Honestly, I did not understand this situation at that moment.[9] It continues to be mind-boggling. And there may be several such illegalities within illegalities in the Polavaram compensation scam – this just a sneak preview. The dalit peasant cultivators and others told me that 70 acres of land here was being used as enjoyment rights by tribal cultivators. But the ownership on record rests with the non-tribal upper castes, willing to sell away the land as it comes under the 1/70 Act and they can never get a better chance later. In spite of the Sarpanch and Ward Members in the Panchayat being tribal people and this being a tribal panchayat. But there was a perplexing piece of information that also came forth in the midst of our conversations, which I could not verify: before 1970, Konda Malas and Konda Madigas (SCs in these areas) were *not* considered non-tribal communites. If this is true, it raises an important question, as to the distinct identity of a different set of SCs who are not part of the mainstream SC population. And one is reminded of the presence of the Konda Valmikis, who are a tribal community, with their prefix of Konda (hill dwelling), while Valmikis within the mainstream caste structure are SCs (who were earlier denigrated as 'untouchables').

Tunga Babu (SC), said, "There are 90 families of SCs here, and about 300 voters. The MRO told us, we could identify land and they will purchase it for us but it must be within Rs. 5 lakh per acre. Where will we find land at that rate now? Later they said they will give us housing site of 5 cents and later even that was changed to 2 cents. The extent of land in Rudramakota is around 450 acres. If we had to leave, why were we staying here so long? This land feeds so many; will money feed so many mouths? There are so many livelihoods here, attached to land, crafts, and other work. We do not see that (reflected) in the compensation. For tribal people there is another problem – they (revenue officials) are telling them, if you have more than seven acres you will not get land; you will get only three acres land and the rest in money form. It is the land that ties us to this place."

The revenue officials have obviously been lying; for nowhere in the R&R note of the government does it say that the government will purchase land for the dalits if it is within Rs. 5 lakh per acre. Incidentally, this village had at least 15 cases of land disputes under the Land Transfer Regulation Act and 1/70. There was yet another case that defied logic. Mendam Muttaiah (SC) was pattedar of 1 acre and 9 kuntas land. Devella Bhaskara Rao (upper caste, Chowdhary non-tribal) bought this land from Muttaiah in 1980s. But Muttaiah and his descendants continued to enjoy the lands, as cultivators (that was the assuarance given at the time of buying which was itself a case of distress sale). They also remained pattedars (land owners with record) until 2003. The moment the Polavaram project package was announced, the government officials (revenue), instigated by the Chowdharies, called Chinni, son of Muttaiah who had since passed away, to the office and asked him to take Rs. 30,000 in compensation and sell his land to them for the project, or else they would book him under the LTR case. Apparently the non-tribal absentee (though illegal) landlords have virtually descended on the village, have formed their negotiation committees; people have been told this is regarding compensation for the 'floods'. They are taken in 'Sumos'[10], and other vehicles. Fifteen days ago, dalits like Chinni were taken to the Collector's office in Khammam. At the Collector's chambers, apparently nobody heard out the plight of the dalits. Everywhere the surveys are being done for land acquisition. But the forms say, "*Godavari Varadalu Mumpu Baadhitula Package*" (compensation for Godavari Flood Victims), where people's consent signatures or thumb impressions are taken. Most of the dalits have just about enough to eat, which they cultivate on their small pieces of land. I met Uike Mahalakshmi, the Koya sarpanch of Rudramakota, too, later. But she was a stark contrast to Seetamahalakshmi in every sense. And having me wait for an hour and a half, she refused to talk about Polavaram or the issues that the SCs of the village wanted addressed. In fact, just as I was waiting that she may speak, an upper caste landlord passed by on his motorbike and sought to know my credentials. The response was cliché and as expected – "it is good for us, this package. With 1/70 it has become a pain for us, so we have decided to sell.Anyway, ask her (pointing to the sarpanch) these girijans are getting a much better package than others." Mahalakshmi simply watched, and it was time for me to move ahead.

Enroute, on a Board Giving Details of the R&R Colony in Reddygudem

This R&R colony was under construction at that point. I saw a board saying "AP Government Indira Sagar Polavaram project R&R package, Khammam district, Velairpadu mandal, for Puchirala colony Project displaced; Ralapudi village, Valairpadu." "Houses – 122; Girijan – 104; non-tribals 18; facilities – primary school, health centre; gram panchayat office; community hall, anganwadi, water , cement road, electrification, temple, sewerage canals." Incidentally, I was to hear later in a few months following our visit, that the said colony construction had been stopped mid-way. The naxals were blamed for it, but reports of their involvement (as district editions of Telugu papers had noted) were unconfirmed. I did not manage to touch this village panchayat later in the course of my journeys.

***

A Surreal Sight, Frozen in My Lens, and in My Notebook

At 3 pm, Cheravalli village, a lovely *cheruvu* filled with Lotuses. It was a surreal sight. It was around 3 pm on a rather warm afternoon when I was passing by that beautiful lake, partly covered with lotuses, on my way to Cheravalli from Kukunur, pillion riding[11] on a motorbike owned by one of the local activists in this area. The lake (a natural lake), one of the many beautiful ones in these parts, post-monsoons, called Perantalammacheruvu, is part of the Cheravalli village in Madavaram panchayat of Kukunur mandal (Khammam district). I had wanted to take both a closer look (at the lake), and a photograph, and requested the rider to stop by, a while. That was when I spotted Chitturi Subba Rao, dark-complexioned, standing knee-deep in water, amidst a few lotus blossoms and many lotus leaves. His skin almost blended with the late afternoon sun and the greenness spread over the lake. The beauty of the image – this man amidst lotus leaves, tanned skin, the green of the leaves blending with his skin painted a striking picture but a picture of pathos.

He sells Lotus leaves for a living. He is a Madiga (the lower sub-sect of SCs) from Aswaraopeta. He has no one around him, no family. No relatives. "Had a wife once, but she went away to another village, and probably married another man", he joked. He has no children either. He must have been in his late 60s. He is illiterate and knows no other job but this. But then, here he is his own master, almost. The leaves are available aplenty in this particular season, at least (post-monsoon, early onset of summer months). He is *"Chitturi Subba Rao, oka naathi leni anaathi"* – 'an orphan without means', as he said, in the same breath as he shouted out his name from the middle of the lake.This was one of the most beautiful *cheruvu*s I had seen on this side. Incidentally, in the months immediately preceding, and following the Godavari's coming, there are several such smaller lakes and ponds with water lilies and lotuses. Subba Rao makes around Rs. 15 to 20 a day selling lotus leaves (which are used as plates to serve food) at the local mother goddess, Perantalamma's temple. He has been doing this for the last ten years, he says. He can be seen, knee-deep in water everyday, in this season, for almost six to seven hours a day. In fact the numerous wetlands in the submergence zone, relatively more in the Khammam district (so far as I have seen) are as much in need of conservation as the forests are. World over, when there is debate over the need to protect the last remaining wetlands, we have in our own midst these niches, most of which will also be submerged under the Polavaram dam. Loss of wetlands, incidentally, does not figure too prominently in most of the discussions around Polavaram dam and its implications. Sometimes in this talk of the 'big' and countering it with something equally 'big', we lose sight of the small and deeply crucial aspects. Spotting Chitturi Subba Rao one felt the need to record this very surprising element in this part of my journey. That some people earn from livelihoods which may not perhaps count anywhere in the larger scheme of things. But their survival is linked to the ecology and environment they live in; and they have learnt to make most use of it, in the ways they know best. And livelihoods such as these have a direct connection with the natural world. But at the same time, only a few of the marginalised are pushed to the extreme, perhaps to make use of this 'natural' world?

***

Cheravally (Madavaram Panchayat) – On the way to Vinjaram, we pass through Cheravally. Konda Nageswaramma, secretary of a government women's thrift saving group (Velugu) said, "Money is impermanent; if they (government) gives us land it is better. They have not shown us the R&R colony. Their package will not give us much. But we still do not know, is this project on or off?" She gave me an amazing bit of statistics: there were around 80 physically challenged people in this tiny village and 150 elderly people, far more than I had seen so far. Gondra Venkataramanamma, a young woman, said, "Do you think they will let us decide? They will bring police and beat us up (if we question). They are giving advertisements in newspapers. It seems they have given cheques (compensation) in Vinjaram. What about the landless, like us?

***

Vinjaram – Mutyalammapadu

Maganti Krishnarjuna Rao, a Kamma land-owning farmer had become quite famous in these parts for being one of the first to take the compensation cheque for the Polavaram project, even as his village became one of the first ones to give 'consent'. He was also the former Sarpanch of the Mutyalammapadu panchayat and the President of the Co-operative here. So I was meeting, in the course of my journey, the first such family who had 'taken the cheque' (*vaalu teeskunnaaru*, they took it, became sort of a refrain heard once in a while, initially, and once too frequently, later). I was meeting a rich man, and an important person in these parts. His house was the typical upper caste, higher status tiled house of the region. Sitting on the pyol behind his huge house, Rao said, "They (revenue officials) said they would give Rs. 1,30,000 per acre; but some agreed even for Rs. 1, 15, 000. 480 acres of land have already been sold (to the government). A hundred and thirty families agreed to do this last December (2006). Some land will of course remain with us, or so they assured us, that they will give back these lands if the project is not completed." (!)

I asked him why did sold his lands. He replied, "There is the 1/70 (Act) applicable here; we can't sell them anyway. I have three daughters to be educated and married off. At least the government is giving us compensation; otherwise the non-tribal pattedars can't sell land to anyone outside. We can sell only to the government." About the demography, he said, "There are 20 per cent STs here and 80 per cent non-tribals." [That explains.]

He said, "We have not been shown any R&R colony so far. People are taking money because the government is giving the cash and assuring us that we can enjoy the land till submergence time. At least we are getting some cash (*dabbu*) in our hands. I got Rs. 10 lakhs; I bought some land elsewhere. Yet, this kind of money is not sufficient to buy new land and rebuild our homes. I am just hoping some of our land will remain with us and not go under submergence."

On hope rests life, I thought, as I left. And learnt of a paradox: the SCs in this village did not get the compensation money nor did they get pattas despite so many requisitions. Around 20 households here belong to the SCs.

***

5th February'07 - I take leave of the activist (who accompanied me so far) and move ahead, alone, towards the **the MRO Office, at Kukunuru Mandal, Palavancha, Khammam district.** I was mute witness to agriculturists of the state let go of the hold over their only wealth, their fields – a feeling hard to describe. What perturbed me more than the fact of their selling away their lands was the government machinery's attitude towards these farmers; it was as if the Government was doing them a big favour. Villagers from Rudramakota, for instance sold their (rights on their) lands to the government for the Polavaram project not too long ago.

A lot of them, including SCs and other caste non-tribal landlords were at the Palavancha office of the Sub-deputy Collector Polavaram on 5th of February. They waited the whole day – many of them had come collectively hiring autorickshaws from their villages. Ideally, cheques are to be given in a gram sabha, only if it is tribal land, at the village panchayat by the SDC (Sub-Divisional Collectorate), but according to the SDC Palavancha division, Dr. Dharma Rao, people 'requested' that they be given cheques at the SDC office. According to the farmers present, this was also to avoid unpleasant situations in the village where some people are still opposed to Polavaram and there are a number of disputed land cases in this village.

As on 5th February 2007, this is the state of 'progress' on Polavaram land acquisitions: a total of 3,415.81 acres are in the hands of the government and compensation (for land only) has been paid so far. Farmers say the government officials have assured them (there is no written assurance of any kind though) they can enjoy these lands until the project is completed. The government has targetted the non-tribal concentrated areas villages first, and the 1/70 Act has come in handy for them to do this. Most of the non-tribals (barring a few dalits and OC landlords) have fallen easily to the promise of money to the extent of Rs. 1,30,000 (nearly) per acre for land that they could otherwise never sell to outsiders (in Scheduled Area) and for which they would not get more than, what they say, Rs. 25,000 (at the most) per acre. Just yesterday I was at Rudramakota where the dalit farmers opposed to the project, and were firm that they would not leave. And today I meet these very farmers here. The abject dejection in their eyes cannot be missed. It shows the cruel ground reality. One transaction that I witnessed here at Palvancha was peculiar. There was an upper caste 'negotiator' absentee landlord who had come with a few people, bringing them in vehicles he had hired. He had taken their signatures on some form and presented his case to the officials. Apparently, his case was disposed off (with a hefty 'pay'-cheque) without any problem, nor any enquiry.

Meeting with the SDC Palavancha division, Dr. Dharma Rao

He read out the Government notice, vide G.O.MS. 111, 27th Jan 2007 on land acquisition, some of the important details being as follows: "205 habitations identified under submergence out of which, in Palavancha division are Burgampadu – 9 villages, Kukunuru – 34 villages, Velairpadu – 39 villages (total of 82 villages)."

He shared further information – "So far Irrigation Department filed (notices) for six villages only – Vinjaram, Gummugudem of Kukunuru mandal, Rudramakota, Koyda, Tekupally and Kakisanur of Velairpadu mandal. All of the six villages surveyed and submitted Draft notification and Draft Declaration to Collector (Khammam) approved publication in District Gazette and local newspapers and villages as per Land Acquisition Act 1894. At Vinjaram (they) completed the process and land compensation (for 208 pattadars) of Rs. 6.11 crores (was given) towards extent of 488.18 acres. At Rudramakota the process has been completed (for 194 pattadars) and started payment to pattadars to the tune of Rs. 5.7 crores from today (5th Jan'07) for 434.05 acres. At Gummugudem the survey is completed…; District level Negotiation Committee deliberations are also over. We will start payment in a week to the tune of Rs. 3.86 crores for 331.18 acres. For Koyda, Tekupally and Kakisanuru only Draft and Declaration have been published. Award enquiry is going on. We give A/C payee (State Bank of Hyderabad) cheques in favour of pattadars. Until we had completed the process in Vinjaram we made payments directly to the pattadars. But then the High Court issued interim orders not to disburse payments without a 'no objection' certificate from Banks (in case of indebted farmers). Bankers are here today. Farmers will get NOC (No Objection Certificate) on production of which we will give them cheques. Actually, cheques should be given in Gram Sabhas but farmers requested us so we have asked them to come to our office and take the cheques… FRL survey is still ongoing.[12] We are trying to ascertain the point at which water will come after Polavaram dam is built. People are ready to give their lands. We conduct gram sabhas after the Irrigation Department sends requisitions to us. We get opinion by resolutions passed by a majority.[13]

***

When Godavari Paves Way for Polavaram Land Acquisitions[14]

On 5th February I meet the Deputy Tahsildar at the SDC (Special Deputy Collector) Polavaram project office (a new establishment) at Bhadrachalam.

Nageswara Rao, Deputy Tahsildar, says "the Government is supportive of the officials working on the project. They have promised Police protection for the government employees. Farmers are willing to give land to the government, since they have the constant fear of floods in the Godavari. They are ready to leave. We went to Kunavaram since the people supporting the project called us to hand over their lands. I was among the officials who went there. 10 people were willing to hand over land. Political parties (some of them) are opposing us. They made us stay there and questioned us. But 50-60 supporters later came and gave willingness letters to give their land."

I am shown the typed letter which is to be signed by the people showing consent to the project and giving their lands for the same. The letter states (in Telugu) clearly, "Godavari flood-affected families (*Godavari mumpu baaditulu*) seeking to shift to safer places giving agreement in writing to the Government" and the heads, viz. "name of the farmer, extent of land and survey number". Nageswara Rao says, "Polavaram (that is, as a term) is immaterial. They are anyways flood victims. We are not using 'Polavaram'. We have to acquire the land by any means, in any case. Irrigation department has been surveying land and the land acquisition process is on going. This is easier. Entire village lands are being acquired (in the name of Godavari flood affected families)."

He then informs me that in Kunavaram, they had received 'willingness' letters from people for at a total of 1800 acres of land. They were in the process of buying them at that point. He said each day at least 20 to 25 farmers were coming to the office to sell their lands. Total acquisition as on January'07 was 123 villages in Bhadrachalam division (Bhadrachalam, Kunavaram, VR Puram and Chintur). He said "the non tribal landlords would never get the kind of rate for their land (under the 1/70 Act), that is, Rs. 1,30,000 for irrigated dryland, and Rs. 1,40,000 for irrigated wetland, In normal times they would get only Rs. 25,000 per acre."

They had "opened the case" that day (when we met them) for the village Tekubaka (Kunavaram mandal). They were confirming the status of pattadars (legal owners) at that point. "Where there were disputed cases they need to be settled in court first (there are several land dispute cases in these mandals). For land acquisition process we deposit the money in court; the High Court must give a no-objection certificate. Cooperative Society loans must be cleared (we have to get a no dues certificate from the Society) before we had over the cheques."

The following details were shared –

Tekubaka – where land acquisition process had begun – 49 non-tribal and tribal landlords were being given compensation for 196.38 acres.

Suchirevula – 84 landholders (251.38 acres)

Pedapolipaka – 106 landholders (384.60 acres)

Kunavaram – 1,800 acres identified; 360 acres got consent to acquire, compensation being given block-wise.

Sitampeta (V.R. Puram mandal) – 170.48 acres.

In processes of land acquisition, normally the authorities give intimation to the Sarpanch. The Irrigation department conducts the surveys and gives intimation to the Revenue department and office of the SDC. Since March 2006 they have been busy conducting the FRL (Full Reservoir Level) surveys, but the 2006 Godavari gave them a better picture of the FRL at 45.72 level above mean sea level. The forms above mentioned, incidentally, were also being printed since the last six months, that is, post-2006 Godavari (which explains the logic of using the term 'Godavari floods affected' as a convenient means of convincing people). We were informed that the Kunavaram first Block land acquisition would be completed by 20th February 2007. In case of non-compliance to sell their lands, the Government also has a system whereby compulsory acquisition can be resorted to, under the General award. In this case (without consent, if acquisition is made) the District Collector will release only Rs. 80,000 and the rest will have to be claimed from the Court. For matters of R&R the Government has formed a Negotiation Committee comprising the Collector, the Joint Collector, the SDC, and landholders. In cases of dispute, or any other problems there would be discussions held by the Committee and on mutually agreeable terms in the above case, the Committee has powers to enhance the compensation by around 50 per cent. Interestingly, the Sarpanch is not invited, but can participate in the Committee proceedings only if the Sarpanch is also a landowner. If the Collector decides, she/he can call the Horticulture Officer, Excise Superintendent, Forest Department officials and Roads and Buildings personnel to deliberate in the Committee sessions/ negotiations, if necessary.

I met the Revenue Divisional Officer (RDO) Palavancha Division thereafter, one Mr. Vinay Mohan. And he gave the following statistics of land acquisition until date.

Burgampadu mandal – 9 villages, including habitations

Valairpadu mandal – 38 villages including habitations

Kukunuru mandal - 34 villages including habitations

27,449 acres from three mandals had been acquired so far.

The government acquired a non-tribal held land of 53 acres at New Reddygudem for constructing R&R colony for people of Valairpadu. If a tribal farmer owned 10 acres, 3.75 acres would be compensated in cash and the rest in the form of alternative land. In case of non-tribals, the entire cash amount is given. If the land acquired was a dryland, up to half acre wet land would be given in compensation; for wetland, entire wetland would be given. Mr. Vinay Mohan adds, "Actually, this is rehabilitation of flood victims (not for Polavaram project alone). There are heavy floods every year. For six to eight months people face unnecessary problems."

He also says there is actually no Stay on Polavaram project. CEC is only concerned about wildlife and environment. There were no other objections raised. FRL surveys were ongoing taking into account the floods from 1953 onwards. The benchmark (on an average) was taken as 45 feet. In the three mandals (Kukunuru, Burgampadu and Valairpadu) the total landholding of STs (patta holders) was 9,646 acres and there are 3,561 ST families (households). SCs have a total landholding of 2,915 acres, with 32 families holding more than 5 acres. Notification was sent for 500 acres land; rest was under way. Approximately 500 acres were being acquired for land to land compensation (as on 5th January '07). And until that day, only the village Vinjaram had got the total compensation money (which was given to the Kamma landlords). 15 R&R colonies were constructed. Land acquisition was being conducted as per the 1894 Land Acquisition Act with subsequent amendments.

***

Another Kind of 'Assigning' Lands Here![15]

Work on the dam site may have stopped, but work on land sites continues, day and night. There is a planned sequence to the whole process. There are also deadlines, some specified, and some not yet shared with the outside world. By 5th February this year the total acquisition of lands in this division was approximately 800 acres. The level of opposition to the dam and work of political and activist groups here seems restricted, or not to have been so influential. FRL surveys are being done keeping the Godavari 2006 floods as benchmark. There is no counter data available to find out what is the correct FRL. But one is told the FRL is based also on the floods of the subsequent years in the Godavari.

Godavari Flood Levels (as per government records) –

1986 (16th August 2006) – 74.6 feet
1990 (24th August 1990) – 70.8 ft
2005 (6th August 2005) – 36.7 feet
2006 (6th August 2006) – 66.9 feet

One of the old farmers tells me, "They arrested our MP, they aimed guns at our MLA; who are we? Can we fight this Government?"

There are several complicated issues with regard to R&R settlements, as well. For instance, a non-tribal landlord accosted me at the Palavancha RDO office that day, with the story of his land and compensation. Kovvur Dilip Kumar (his name), S/O Sitarama Sastry had land in Repakgummu in Valairpadu mandal in Rudramakota panchayat. The pattadar for the land that he is cultivating (3 acres, 36 kuntas) is his uncle by name Kovvur

Sundararamayya. He said though they had not given the land to him, and were still cultivating the same, the uncle usurped the land. In 1975, the Kothagudem magistrate's court gave a judgment in favour of Dilip Kumar, and while they were cultivating, a "rival" Kakarla Suryaprakash Rao got a temporary injunction on the judgment. Dilip Kumar did not vacate the land (till date) but wants to know as to who will get the Polavaram compensation for this land (already illegally occupied in violation of 1/70 provisions)? He was amazed to find, at the Palavancha RDO office, Chitturu Gopalakrishna, grandson of Kakarla Suryaprakash, claiming and getting his compensation! The SDC gave him only two days to sort out the issue, but Dilip Kumar looked worried because he had lost his evidence in the 2006 Godavari! He wanted to know as to how Chitturu Gopalakrishna got the pattedar passbook, if not by bribing the officials at the RDO office.

***

V.R. Puram (Khammam District) - NGOs, another Side of the Story[16]

Mr. Gandhibabu is the founder-director of the NGO, Agriculture and Social Development Society (in Telugu Vyavasaya mariyu Sanghika Abhivrudhi Samstha). In its brochure ASDS calls itself "a not-for-profit, rights-based development organisation (which) has been striving to empower the marginalised tribal communities in agency areas of Chintoor and VR Puram mandals of Khammam District, Andhra Pradesh...." Set up in 1985, ASDS works with a tribal federation of 112 grassroots women's groups and individual women under the umbrella of a 'Natwan Sangham' on a medley of issues that local NGOs in small viilages usually work on - agriculture, education, gender, health, natural resources management, etc. I came across this organisation in the course of my travels and initially found it on the forefront of opposing the Polavaram dam. It was also interesting that the registered office of ASDS was in Rekapally, in the submergence area, which actually meant that the organisation has a direct stake in the issue, since the staff (skeletal at the time I met them) and the organisation's founders are themselves Polavaram 'project-affected'! This fact, perhaps, made them stick to their stance which has been, at least till date[17]: the dam should not be built. Early in 2006 ASDS organised a public meeting under the umbrella of its 'natwan sangham' attended by representatives of some women's organisations in Hyderabad. Later the NGO held a meeting bringing together political party representatives and other activists opposing the dam. People who spoke at the meeting split into various groups (even among those opposed to the project, each not seeing eye to eye with the other). During the 2006 Godavari, ASDS gained prominence in local Telugu media, more so because most of the people, from the villages which Godavari visited, had gathered at Rekapally and helicopters and government relief supply reached affected people through this office.

International funding agencies, meanwhile (with whom ASDS had training and resource workshop partnership in the Disaster preparedness programme) apparently did not release any money for flood relief and sent an 'assessment team' months after the flood waters had receded and in its assessment one donor organisation in Hyderabad is said to have written in its report that ASDS had gained wide publicity and government assistance and hence did not need assistance, unlike other organisations). Gandhibabu says, "ASDS is in the open on the issue since 2005 August. At that point CPM was talking of reducing the height of the dam. We met Mr. Raghavulu (CPM State Secretary, AP) and informed him that height was not the real issue. At Kukunuru we held a meeting with activists and others. Later, CPM too changed its stance. In fact, at the local level even Congress party workers (living in the tribal areas) are not supporting the project..."

Speaking of what changed by being open about opposing Polavaram in terms of their activities, he says, "Today we are not getting any of the Government programmes that we used to be part of since we began in 1980s. Since last two years we have been talking against the Polavaram dam. Government institutions are not considering our applications today. They are contacting other NGOs and giving them the charge of implementing and assisting in Government programmes...." ASDS used to get most of its funding from Government projects, so this hit them hard. "From 1985 to 2004 we worked on many government schemes as facilitators. As an organisation, we see ourselves losing. And by that very same understanding we are opposing Polavaram dam – that it actually hurts at the tribal communities whose development was one of the Government's own agendas, which we facilitated so far."

About funding agencies he says, "Some are neutral. Since a year now some of them withdrew support to us. Some people working within these funding agencies in fact support Polavaram project openly since they have regional affiliations, and belong to delta districts. Some donors are afraid of our activist stance on Polavaram; they have expressed it. Our corpus is decreasing steadily since the last one year…We have eleven members (staff) and only three are able to get their salaries. We have got no funds since December last year...We are also facing problems from the government..."

While I was in the field area where they work, Haribabu (Gandhibabu's brother) was summoned by the police to give some information regarding their activities. Apparently, they are facing this questioning from local government authorities. About the general trend of changes in the tribal areas from the time the organisation started functioning, Gandhibabu says, "From 1986 onwards I notice greater awareness about immunisation, family planning, etc. There is development in the in means of communication and basic facilities. In terms of agriculture, initially, when I came here, tribal farmers used to cultivate jonna (sorghum), at least to an extent of 85 to 90 per cent, but they have now started cultivating commercial crops. Literacy has increased because of the tribal welfare ashram schools here. There are roads in more villages now than there were earlier. The bridge (between Kunavaram and VR Puram) was built a few years ago, which helped increase communication. But along with it came the menace of smuggling of timber (from the forests)...." Gandhibabu was uncertain about the future of ASDS. As of now, he says they still continue to participate in the government programmes as and when they are lucky to get some. They have to perhaps downplay their activism sometimes to gain support from funding agencies, as well. It will be interesting to watch, though, for how long.

***

Meanwhile, the Polavaram Project Progresses

As per the records that I managed from the officials in MRO office on the field in one of my visits in 2007 the statistics were as follows:-

Model Colonies – For Pochavaram village, R&R colony near Bhadrachalam, Satyanarayanapuram (in Cherla mandal), which is around 60 kms from Kothagudem. This was a local upper caste (Raju) land. He sold around 200 acres to the government for this purpose.

Velairpadu mandal (Khammam district) – Kotha Reddygudem R&R Colony is meant for the displaced from Puchirala village.

West Godavari District - Devaragondi – 8.45 acres of government land (at Polavaram); 2.79 acres of private land.

Singanapally – 9.78 acres of private land; Mamidigondi – 11.42 acres of private land; Thotagondi – 6.39 acres of private land; Pydipaka – 32 acres of government land; Pydipaka (Hukumpeta of Gopalapuram mandal) – 32 acres of government land.

'Land to Land' (for compensation) identified (as per Government records): –

Vinjaram (Polavaram mandal, West Godavari district) - 59.83 acres of Scheduled land; Pragadapalli (Polavaram mandal) – 263.63 acres Scheduled Land; Polavaram mandal (31.60 acres Scheduled land); LND Peta (Polavaram mandal) – 164.98 acres of Scheduled Land.

Total – 520.04 acres of Scheduled Area lands.

By the way, on 2nd April 2007, the Major Irrigation Minister, of AP, Ponnala Lakshmaiah announced (all over again) that AP would go ahead with construction of Polavaram dam by implementing "the second option" provided in the Bachawat Tribunal Award in respect of lands in Chhattisgarh and Orissa marked for submergence. The two options suggested by the tribunal were payment of compensation towards the lands lost to the FRL of the dam (150 ft) to prevent submersion. The tribunal incorporated various inter-state agreements signed by the riparian states of the Godavari into its award and recommended that "the Polavaram project shall be cleared by the Central Water Commission as expeditiously as possible for FRL of 150 ft." The Minster told the media that Chhattisgarh had sought and accepted Rs. 59 lakhs from AP to denotify submergence lands on its side through a joint survey as suggested by the Central authorities. Orissa had not responded. AP would raise embankments on its cost to prevent submersion of lands in both neighbouring sates. Their objections, if any, will be overruled, said the minister. He also informed that the project had gained 14 out of 18 clearances from CWC (Central Water Commission). The remaining clearances were expected soon.[18]

It has been a year now and there is relative silence compared to the opinion-deluge last year. "R&R" (Rehabilitation and Resettlement)/ "*packagi*" as terms have entered the everyday conversations in many of the villages under submergence zone and otherwise. And today the State government is using force to silence people affected by the dam by slapping cases against any of those who coming protest or question. While some groups and organisations in Hyderabad have been relatively silent on the issue, some others have been lobbying with the Orissa and Chhattisgarh governments to highlight the inter-state angle; some of them incidentally were identified as the 'think-tank' of the Telangana Rashtra Samiti (TRS), which has its own perspective on the Polavaram dam, being against the agrarian interests of Telangana people and demanding equitable access to Godavari waters for the Telangana region. The Telangana angle is firmly entrenched in this group's larger perspective on Polavaram dam. The district editions of the prominent Telugu dailies are especially interesting to observe as the local corruption of the officials working on the R&R comes to the fore in these editions; and at the same time, there are local stringers that have been co-opted (reasons not far to see) by the establishment. Telugu media-space then gives a fairly truthful representation of the things on the ground, both in the continuation of the project and the nature of localised corruption and discrepancies in the implementation of the same. National media has been relatively dismal in terms of representing/presenting the Polavaram issue, in the last one year. People from at least three tribal villages have already moved to the R&R colony. Some others have accepted the compensation.

***

In Search of Other Landscapes of the Godavari 2nd April to 7th April 2007

2nd April

My first visit to Kakinada! You find dominance of Kammas and Kapus[19] here, who are among the wealthiest. My first stay, in the journeys so far, in a (what they call in these parts) 'class' hotel (not my fault, it was 'booked' here by people who thought I come from a city). But it is located in a quiet residential area. Anyway, an entire day would be spent at the Coringa mangroves.

Towards Coringa Wildlife Sanctuary, the Mangroves of Godavari[20]

Coringa was declared a wildlife sanctuary (WLS) by the Government of AP (GO Ms. No. 484, 1978). Godavari in these parts is also referred to as Gautami, Akhanda Godavari, and Vasishtha Godavari. This is the largest surviving patch of mangrove forests in AP with more than 65 mangrove tree species. "According to the Forest Department, Government of AP, the total area under mangrove wetlands in the two estuaries is 2,363.32 ha under Godavari mangroves and 24,999.47 ha in the Krishna delta. Godavari mangroves are located between 16°30' – N' and 82°23' E in the East Godavari district."[21]

"As much as 70 per cent of the mangroves (in India) are available along the East coast…The Indian mangroves comprise of 59 species belonging to 41 genera and 29 families.... (of which) the east coast represents 48 species and 32 genera…[22]

Coringa in Colonial Records – "Coringa (vernacular Korangi) is nearly ten miles south of Cocanada. Population 4,258. It contains a travellers' bungalow, a native rest-house, a police station and the offices of a deputy tahsildar who is also a sub registrar. It was once one of the greatest ports and ship building centers on this coast; but, owing to the silting up of the channel which leads to it, it is now of no commercial importance...It appears that the present town of Coringa, which is on the east of the river was 'built' about 1759 by Mr. Westcot a resident of Injaram; while what is known as 'old Coringa' on the western bank, is older than this…Till quite recently…ships were repaired in mud docks at old Coringa…The river Coringa is said to have been brought to the sea by the sage Atri, and the bathing place is called the Atreya-sagara sangam. It is also believed that demon Maricha, who was sent by Ravana in the form of a golden deer to Rama, when he and Sita were at Parnasala, was killed by Rama at this place. Rama is supposed to have founded the Siva temple of Korangeswara..."[23]

The Coringa WLS has three Reserved Forests (RFs) – The Corangi RF, Corangi Extension and Bhairvapalem RF. Then there are other RFs in the non-sanctuary area – Rathikaluva, Masanitippa, Matlatippa, Balusutippa, Kothapalem and Kandikuppa. Among the species of flora found here, are *Avicenia officinalis* (*nalla mada* in Telugu), *Avicennia marina* (*tella mada*), *Avicennia alba* (*vilva mada*), *Rhizophora mucronata* (*uppu ponna*). The endangered Smooth Indian Otter, Jackal Monkeys and Fishing Cat are also found in this sanctuary. The WLS has an 18 km long sand pit in the north eastern side where Olive Ridley turtles nest between January and March every year. Over 120 bird species, including Little Egret, Cattle egret, Pied Kingfisher, Small blue kingfisher, Pond Heron, Reef Heron, Red Wattle Lapwing, Crow Pheasant, Brahminy Kite, Little Cormorant can be spotted during low tide when they manage to get their feed on elevated mud flats having small fish, shrimps and molluscs.

From Kakinada, with a team from MSSRF, we move towards a part of the Coringa Wildlife Sanctuary

Mangroves have their own distinct habitat and needs for sustenance. "Freshwater flows into the mangrove wetlands of the Godavari delta for a period of six months; the peak flow normally occurs during July to September, coinciding with the southwest monsoon season. During this period, the entire delta, including the mangrove wetland, is submerged under freshwater. A large bay called Kakinada bay is associated with the northern part of the Godavari estuary. It has long sand spit on the eastern side, which separates Kakinada Bay from the Bay of Bengal…"[24]

Eagerly Awaiting Godavari: People by the Coringa Mangroves

If there is one place where Godavari's coming is eagerly awaited, it is in the villages by the mangroves along the estuary of the Godavari. Not only by the people living in these villages, but also the trees, birds and animals that reside in its habitat.

"In Godavari…nearly 40,000 people are dependent on mangrove forest and collect juvenile shrimps, prawns, crabs, fishes and shells. The farming community, mainly Yadavas are mostly landless labourers and they possess only small area of land (1-2 acres) some of these villages are Chollangipeta, Ravimeraka, Polekurru, Mallavaram and Tallarevu and use the mangrove forest for grazing their feral cattle...."[25]

I meet the fisherman Mallada Narasimhamurthy, who says, "When Godavari comes, it is time for plenty of fish. We need Godavari to get the best fish. Our best fishing happens immediately in the months of the Godavari coming." He stays on the creeks on a boat along with his family, wife Narasamma and their twelve-year-old son Mahalakshmi. Mahalakshmi? Seeing the surprise written on my forehead, he says some people in these parts name their sons after the goddess; and this boy was born after years of being childless; hence they named him after her. They belong to the caste of Agnikula Kshatriyas (BC – A category) locally referred to as *vaddilu*. In the months preceding the monsoons (April through June), the family, like many others, set up their fishing boats on the creeks (*padava*, or *nava*) along the mangrove forests and almost live in their boats in these months. Mallada Narasimhamurthy told me, "We fish here from 6 am to sometimes late in the night." They earn somewhere between Rs. 30 to 100 a day depending on the fish catch. Narasimhamurthy says, "*mada adivi* (mangroves) keeps us safe from ravages of Godavari (when she comes)".

It is said elsewhere, too, that "Mangrove forests fulfill a number of well-documented and essential ecological functions in tropical and subtropical regions. They generate a variety of natural resources and ecosystem services that are vital to subsistence economies and sustain local and national economies...Mangroves provide breeding, spawning, hatching and nursery grounds for both coastal and offshore fish and shellfish stocks. They also serve as a physical buffer between marine and terrestrial communities..."[26]

Incidentally, the community-owned land in Narasimhamurthy's village, Gadimoga (Tallarevu mandal, East Godavari district), was sold to the Reliance group to set up their gas pipeline project. Around 80 acres (he says) were sold. Mallada Narasimhamurthy says, "I got Rs. 6,000 as my share from that sale. Though they were saying they would give Rs. 2.5 lakh per acre. They are doing some work every day. They say they will set up a factory there. And they will move us all very soon."

We visited the Metlapalayam *kaluva* (though the term is also used for channel or canal, here it is the local term for the creek, I was told) today. Tomorrow we would be visiting Rathipalayam *kaluva*.

At the MSSRF Office, Dr. Ramaswamy informs me that his organisation worked with three villages in socio-economic development and mangrove restoration and conservation – Kobbarichettupeta, Chollangipeta and Corangi (in two hamlets, Corangi and Dindu). They restored 10 hectares (ha.) of mangroves in Chollangipeta in 2006-07. But now they plan to work on other issues and not on mangroves, since there is no vacant land left for restoration. "We will take up mangrove restoration and conservation in Krishna. In this area, we will now focus on coastal non-mangrove development work. We have asked for permission from the Forest Department to work in Nagalanka in Krishna mangroves. Krishna mangroves were destroyed by construction of dams on the river, among other causes." He says for a thriving mangrove ecosystem, good rainfall, more than 2000 mm, is a must; a good supply of freshwater and sediment, and sheltered coastline with sand bars, like in Kakinada. Rainfall in the Kakinada belt on an average is 1200 mm. Livelihood is a major problem in the Godavari mangrove region. Untreated effluents from industry, mainly, paper mills and sugar mills, add to the organic load and kill fish species. Use of pesticides in the fields also leads to the growth of water hyacinth, which are difficult to remove in ponds and lakes. But these die naturally in the summer months. One could see many aquaculture fields in this area. Prawn culture has done its bit in destroying the mangroves, but I find that not many people, especially in the forest department, are keen on addressing this issue. Many cultivable lands that used to grow food crops in the past were converted into prawn culture ponds, mostly owned by rich, upper castes, mostly absentee landlords. Usually, dalits work on these fields on daily wages.

***

Meeting the Forester (Perceptions) - I met Mr. Anand Mohan, DFO (Divisional Forest Officer) Kakinada Division to get the department's side of the story about mangroves, Coringa, and the dam, of course. My interview is reproduced below:-

Which communities essentially dependent on the mangroves?

Anand Mohan (AM): Predominantly Angikula Kshatriyas and Golla (shepherd) communities (both Backward Caste) are dependent on the mangroves for livelihood.

What is the history of Coringa? Earlier records show it as RF (Reserved Forest) and later as WLS (Wildlife Sanctuary)?

AM: In 1882 the Madras Forest Act came into force. In 1883 Coringa was declared RF. After the Madras Forest Act, seventeen areas were declared RF, and Coringa was one of those. It was one of the earliest to be declared RF. In 1978 it was declared WLS. Right to fishing is granted here. There are three RFs here (in the estuary), Coringa RF, Bhairavapalem RF and Coringa RF extension.

Do you see any impact – direct or indirect – of the dams on Godavari on the estuary or mangroves?

AM: I don't know. They (the Government Irrigation department) are saying that only a small amount of water will be stored and it will not affect the flow of the river, by and large. We have been shown data and we are told it will not impact the discharge of water. Kakinada Forest Division is losing 149 ha of forest land[27]. Out of this (in Kakinada division) 14.13 ha is part of the Papikondalu sanctuary area. We are taking possession of forest land adjoining Godavari in east Godavari. But the Forest Department will get Rs. 200 crores to compensate for this loss.

[Incidentally, he also mentioned that the Reliance Petrochemicals gave the Forest Department Rs. 9 lakhs per acre for laying gas pipeline near Kakinada; which falls in the Kakinada forest division. The compensation is based on the NPV (Net Present Value) of timber, density of forest, etc.]

What is the significance of the Godavari floods on mangrove ecosystem?

AM: The 1996 cyclone affected some of the mangroves. But mangroves thrive on the principle of availability of fresh water and sea water. Floods are helpful in bringing the discharges. After floods, the mangroves flourish.[28]

He then informed me of the AP Forest Department and MSSRF joint implementation of the community based mangrove conservation programmes. Now the department is implementing an eco-tourism programme through trained EDCs (Ecotourism Development Committees). The Forest Department, incidentally, is also running this eco tourism project funded by the World Bank in Rampa and Maredumilli forest areas. Though the Forester did not see the direct linkage between construction of a dam and impact on mangroves, I found that the local communities see this connection, as do several studies that have documented changes over a period of time on the mangroves. Especially regarding the colonial anicut at Dowlaishwaram (which irrigates nearly 410,000 ha. in East and West Godavari district). "Two major shifts in the main course of the Godavari river and the formation of a sand spit have occurred since the construction of the Cotton Barrage at Dowlaiswaram in 1852. Until the 1930s, the Godavari flowed northwards, opening into Kakinada Bay. Between the 1930s and the 1970s, its course gradually shifted southwards…Since the 1970s the Godavari River flows eastwards.These shifts can be explained by a combination factors including the flatness of the alluvial zone, variations in river flow, and frequent cyclonal activity in the area…Fishermen unanimously reported that the catches have declined over the past 10 years…"[29]

Again there is a connection between flow of the river water and the mangrove eco-system, and it doesn't remain constant; construction of reservoirs can also affect the same: "Geomorphologically, the Gautami river has undergone changes after the construction of the Cotton Barrage at Dowleswaram in 1852. In 1893, Kothapalem mouth had deepened and widened considerably…By 1985 the Kothapalem mouth had gradually silted up and after the floods in 1986, the major outflow of fresh water started taking place through the Bhairavapalem mouth and only very little flow now takes place through the Kothapalem mouth. Due to change inflow pattern the Kothapalem RF, Masanitippa RF and Balusitippa RF have been affected due to the reduction in fresh water flow….Another notable feature after the construction of the barrage is that the peak flow of freshwater takes place only during four months starting from July to October. From November the flow dwindles very rapidly and this trend continues to the negligible flow during the summer months of April and May and sometimes to June. Hence this reduction in fresh water is also a significant cause for degradation..."[30]

Polavaram dam is also likely to cause similar impact in a future time. River communities (more than experts) usually understand the logic better.

***

Godavari Must Go to Her Sea!

Late evening, at Kobbarichettupeta hamlet (Chollangipeta Panchayat, Tallarevu mandal, East Godavari District) - This is a village of mainly fishing families. There are around 800 families in Chollangipeta panchayat. In this village, there are 260 families, most of whom are SC families, and the rest are BC. The fishing community here belongs to the Vadda Balija caste group (BC). Mydu Satya Rao, a former Sarpanch of this panchayat, is also present. Basically the meeting is organised for me to learn about MSSRF's work with them[31] but I manage to squeeze in questions about Polavaram dam in an effort to guage their thoughts on mangroves, the connection with Godavari, and Polavaram. Mydu Satya Rao immediately replies to my question on Polavaram, saying, poetically, "*Godavari samudraaniki vella-valasinde!* ('Godavari must go to the sea!'); "*adi prakruti dharmam*" ('that is the way/dharma of nature'). Cotton *dora* (Sir Arthur Cotton) built the anicut. But did it stop the river? No barrage can stop Godavari from going to the sea. People are saying Polavaram dam will stop Godavari wasting into sea ("*vrudhaga poye Godavarini aaputundi*"). But they cannot stop Godavari. Just how much water can you hold in barrages?"

***

For Whom Godavari Comes!

April'07

3rd April'07 - Enroute Bhairavalanka, I Polavaram Mandal, East Godavari district

We take a long jeep ride from Kakinada to Gadimoga via Yanam (administratively part of the Pondicherry). I take a look at the Reliance (Company) gas pipeline project at Gadimoga and speak to a few (Reliance) project-affected people, having their own share of problems from this major private operation. There is tight security at the work place. Security guards look suspiciously at a woman with her camera. It seems one has to take permission to take pictures! We take a boat from Gadimoga to Bhairavalanka (with Govindu, a dalit from that village as our guide), passing through Goggilanka, via Rathikaluva, Balusutippa, watching Gautami Godavari. On the way to Bhairavalanka, we come across Palipu Sesha Rao, a fisherman on one of the creeks of the Gautami Godavari on the way to Bhairavalanka. Fishing families like his, from Bhairavalanka, Chimitamapaka, are dependent on mangroves for their livelihood in the lean fishing season (as it was, when I met them). He makes about Rs. 900 by selling logs of wood from the mangrove forest to a brick kiln near Kakinada. This is their source of income, besides a lucky catch or two in the creeks. They hunt through the day and the boat becomes their virtual home on these creeks. The entire family can be seen on the boats, and you can spot quite a few such. However, there is a flip side - the Forest Department officials are always ready to slap cases, or leave them, for a price.

And this has been happening since the colonial times – Talk of Linear Power Narratives

From the correspondence files of the Godavari District Association, I discovered a letter, "To the Secretary to the Commissioner of Land Revenue (Forest Branch) Madras, dated Cocanada, 21st July 1913, No. 114."[32] I quote a few points from that letter, below:-

"1. It has been represented to the Association that during the last 15 years conflicting orders were passed by the Forest department for the levy of fees from fishermen of maritime villages and hamlets between Coringa and Yanam, fishing in the creeks running through the Coringa and Kandikupa Reserves. At one time, the fees appear to have been fixed at 12 annas per mensem, then reduced to 6 annas per mensem and afterwards not collected for some years together. The levy of fees appears to have been subsequently ordered to be continued at the rate of 6 annas per annum…

2. Under the existing rules, fishermen are exempted from payment of any fees for fishing in seas and backwaters. Rules framed under the Forest Act permit the public to remove free of charge, unclassified wood and fuel for bonafide domestic and agricultural purposes and unreserved lands. The fishermen of the villages in question were removing fuel free of charge from time immemorial, but in declaring the Coringa and Kandikuppa blocks as Forest reserves no land appears to have been left unreserved for the enjoyment of that longstanding right…The poor fishermen are brought under the constant observation and harassment of the Forest subordinates… In some villages situated within the Reserves, difficulty is being experienced even to get fuel for cremating dead bodies, for the above reason.

3. As these fishermen are very poor, and their position being analogous to that of hillmen who are allowed special privileges in the operation of forest rules, the Association submits the following for the favourable consideration of the Board:

i. that no fee be levied from these fishermen when the fuel is removed from the abovesaid Reserves for bonafide domestic purposes and for fish-curing;

ii. that sufficient land be ordered to be set apart in those reserves (a) for communal purposes and (b) for the exercise of privileges allowed to the public by the rules framed under Section 26 of the Forest Act."

Malladi Narasimhamurthy told me the Forest Officials 'caught' him for 'stealing' fuelwood from the reserves and he usually had to pay penalty to the forest guards. Other members of his community too face constant harassment by the forest guards in the Reserves by the creek. Couldn't help thinking – small time domestic needs of the fishermen seems to be a bigger crime when compared to the impact on the mangroves, however indirect, from the Reliance Company, in which case no conservation penalties are slapped.

At Bhairavalanka Village, Gogillanka Panchayat, SC Hamlet: Whom Godavari Feeds

We are now at the Bhairavalanka SC hamlet. Lanka Posarao, a fisherman, is the Sarpanch of the Goggilanka panchayat. There are around 150 Mala (SC) families here. The village has its own problems. The local government school has had no teacher since almost a year now. Golati Veeraraghavulu says, "We work as agriculture labourers in the aqua farms owned by landlords from Hyderabad and Khammam. Since last five – six years this place has become known for prawn culture. Earlier there used to be mangrove forests and small fields for growing food. We also work sometimes as daily wagers for the Forest Department. We go as far as Gutti sometimes in search of work."

Vijayalakshmi and Veeraraghavulu immediately catch the point I am coming to, regarding the Godavari and flooding patterns. They say, in turns, "When Godavari comes, she stays for two to three days (*Godavari vacchinappudu rendu-mudu rojulu untundi*). "We bless the Godavari and look forward to her eagerly. It is only in the three months when she comes that we plant our crops; our agriculture, on our small pieces of land depends on Godavari talli (mother Godavari). Without that we cannot cultivate our fields." I did not ask them about Polavaram dam, until that point, but suddenly Veeraraghavulu talks of it. "When Polavaram *barragi* comes up, we will only have this salt water, no fresh water. Only when Godavari comes we get good water. They are saying the dam will stop it. But then, there will also be no crops for us." Vijayalakshmi joins in, saying, "Rain and Godavari give us our crops. We grow pulses and some paddy. We ready the fields by monsoon time. Our cropping and farming happens in three months during and after Godavari comes. They say *'varadalu'* and all that but for people here, 'varadalu' (floods) means a lot of good. Godavari gives us water to drink, wash, and grow our crops."

This is the first time I hear this distinction made by the people themselves. She specified the term and qualified it. All this happened in a spontaneous flow. I had not asked them any question to that effect. I felt warm inside.

Personally, Intense Moments: Through the day, on the Gautami-Godavari, on the boat, through the mangroves, silently chugging past, unsuccesfully trying to capture the Cormorant not too far away yet elusive to the camera. Learnt a lot from a totally different vantage point, thanks to people who are not in the 'submergence zone' yet to be affected, and perhaps dislocated, in a future time if the dam happens?

***

Tourist Gaze, State, Consent and Dissent

4th April'07 - Back at Rajahmundry, and this time round the HPO was unavailable. In my diary I had noted: "From a stay in a 'star' hotel in the port town of Kakinada and the mind-boggling expenditure there on food (for the first time on a field visit so far) to a very calm, peaceful (barring one exception) stay in an absolutely unexpected discovery – the Regional Forest Research Centre at Rajahmundry in Lalacheruvu. With a very warm and kind attendant, Bhadram, a contract employee here who earns a monthly salary of Rs. 2,000, and has been working here for the last six years. And a considerate official, the State Silviculturist, who I have only spoken on phone so far, and never met (she is 'out of station'), who let me stay here for free, thus saving all my tension and dilemma, in an otherwise financially tight period. There are two 'suites' here, Charaka and Susruta[33] (!) on the terrace. Spotted the most amazing variety of butterflies. The only catch? This is far from town, on the highway. Unlike what they warned me about (which is why Bhadram said, "keep the AC on") there were no power cuts in the night all the three days I was there! Slept peacefully…And also got time to read in peace in many months, back from the visits. About each of the places I would visit; and I discovered a past in the context of the present; each page being a revelation. And would come 'home' to a warm cup of tea and Bhadram urging me to 'eat more'.

I met Mr. Satyanarayana, DFO Wildlife, at Rajahmundry and learnt of a fascinating aspect of the mangroves. He informed me that there are 34 species of flora in the mangroves, adapted to saline water conditions. He also said discharges from 'floods' are very important for the mangroves to thrive. What he found most inspiring about the mangroves (I loved the manner in which he described it, with excitement on his face, like it was happening right there for us to see) was: *"The 'masters' of this eco-system change. During high tide when the water is full, fish crabs, etc become the masters; during low tide, the bird and fishing cat come and eat these. They become the masters during low tide. It is a wonderful natural phenomenon."*

Late that afternoon, I met the advocate Trinadha Rao at his home quite far from where I stayed. This meeting would lead me on to some important people in the villages whose lives would reveal another story. He informed me about the tribal land alienation in the context of Polavaram R&R and how the system could manipulate records, to convert the term "continuous possession" (of the tribal people) to "enjoyment", and takes away those very lands for construction of the R&R colony. He has been a practising advocate here since many years and was also at some point, with the Human Rights Forum besides having published several reports on this very issue (irrespective of Polavaram). His suggested the route for the next day's trip which, thankfully, passes through Lalacheruvu and I can take a bus from Kambala Cheruvu not too far away. I would reach Gokavaram, where

Rambabu the legal rights coordinator from the organisation Laya would guide me. And he has a bike. Great!

I realise this guest house has nobody staying there besides me. The place is indeed quiet and by late evening even more so. Bhadram brought me food and tea in a flask. Sometimes I find these elements of a government guest houses disturbing. You are looked after, but you know they feel duty-bound and stuck in their subordinate positions to look after you. Well, it will not be for too long, the stay here. I decided to leave my luggage here in the room and return after my Gangavaram trip to the same place.

***

5th April'07 - Coming Back, Full Circle; but How!

This time I seek the Godavari as one among the tourists to see what it is like on this side. In the months of April through May and June, there isn't enough depth in the river to anchor bigger launches at the original starting point, Pattiseema, and hence the launch starts during this season from Polavaram, a few kilometers away. A bus takes us from Rajahmundry to Polavaram. We get onto the Punnami launch at around 9 am, but it starts only around 10.30 am. I have not informed anyone in Polavaram that I am here. I want to do this completely 'in' the sequence done for the tourist. Within an hour, before the launch motor is revved up, the riverscape has changed already. People are served breakfast of *idlis* and *vadas*, soft drinks. Godavari has become a dumping ground. And I go back and forth between my 3 am start, in 2006, and this. Now begins the whole story. Welcome, to the tourist Godavari.

On Travelling with the Flow: Godavari, Then and Now

"That was a different time", I would write of it later, "a different Godavari, a different flow and a different 'movement'."[34] It was langourous, as if time stopped by, for absorbing each moment of the flow of Godavari. It was in June last year, one's first trip across this stretch of the Godavari. It had poured the previous night...I sat on the deck amidst an assortment of groceries, vegetables, fruits, cardboard cartons...One was literally between the open sky above and the expansive lap of Godavari underneath. The timing was perfect, starting in pre-dawn darkness...Between Rajahmundry and Kondamodalu an entire range of feelings within... were matched only by the constant interplay of light and shade in the clouds above; the abundant silence broken now and then by Godavari waters hitting against the boat's underbelly, the sounds of the crude motor, and intermittent sounds of the women washing clothes along the banks. People were getting ready for another day's hard work. This was Godavari, at her magnanimous best...That was another time then... another *laanchi*...And in April this year, was another time altogether, on another launch named Punnami - a private tourist launch operating in the Rajahmundry-Papikonda Wildlife Sanctuary circuit. Here I was amongst some 60 odd tourists, a 'professional' guide named Raju (!)[35] and the sheer cacophony (in high volume) of the latest 'pop' Telugu film songs. In this April journey, the most significant companion, Godavari, was invisible; or she existed just incidentally, as a river that may well have been a tarred road! Each moment seemed like a well-rehearsed drama played out in that one long span of a day. Between these two strikingly contrasting images/experiences with the flow of the Godavari lies yet another significant dimension of the Polavaram dam and the Godavari and the people in the three districts of the submergence zone - destructions and exploitations, different in form, perceptible and imperceptible. The difference between the two journeys, last June and this April – twelve-and-a-half hours and little more than 8 hours, respectively – lay not merely in the pace of the journey or the nature of silence enveloping one, but also in understanding the local economy… between the villages as active agents of the State's socio-economic profile, and as objects of a curious tourist barely engaging with the communities, the sights and sounds of this vast beautiful riverscape that will soon be lost. Between last year's journey and this, also lies the difference between people's commerce and that of the State's and the private players' exploitative economy. In this entire journey from 10.30 am to nearly 9 pm, communities do not exist. Within one year the numbers of tourist launches plying on Godavari have increased phenomenally. People here tell me that on Sundays, sometimes there are as many as ten to fifteen launches between Rajahmundry and Papikondalu. The tourist launches move from Rajahmundry via Pattiseema (held sacred by people here) Posammagandi, Papikondalu and Perantapalli. There are two-day packages too, including all these places and a night's stay at one of the bamboo huts at a tribal village Kolluru (built by a non-tribal landlord) and a visit to the Pamuleru waterfalls. These tours are authorised by the AP Tourism Department, which gets a percentage as commission; some sanctions also seem to come from the Irrigation Department (AP) which also gets some commission, as one of the launch staff reveals. The tourist launch merely whizzes past the villages, oblivious to the peculiar handicaps of each. Tourists are from different places across AP, mostly from coastal districts. We are already on a massive consumption journey, I realise. After *idli-vada* comes *chai* in plastic cups. The latest Telugu film music continues through the journey. This includes the latest film, *Godavari,* which was shot on a launch specially created for the film and which gave a new spurt to the Papikondalu tourist circuit. Through the journey the guide announces names of different Telugu films shot in the pristine environs of the Godavari and the tribal villages…Nearly none of these films (from the 60s) ever had anything to do with the tribal people whose lands they utilised for the backdrop. In Telugu films, tribal communities (those dressed in 'tribal' attire) appeared almost always in token dance numbers, as a 'relief'. The Godavari was used excellently (in some of the films of the past) as a mood enhancer, or to highlight the situation of the characters (for instance, the old Telugu classic, Megha Sandesam[36]). But that is where it ended. Lives around Godavari have never been part of it. The tourist circuit today does not even 'enhance' the moods in synchrony with the river! Last year, along the way one spotted thatched hutments of fishing families... Most tourists seem least interested in these aspects. A few men play cards through the journey... Loud music makes sure they will only continue from where they last left – their homes and the TV sets. Only, the TV image here is replaced by that of a live young girl dancing. Yes, for there is Bhavya, a petite young girl aged 16, who dances everyday for a living on this launch. Her mother is an attendant in a hospital in Rajahmundry. She couldn't continue her studies after tenth. She earns Rs. 250 per day for dancing to the tune of Telugu film songs with her 'master', a 20-year-old Saikrishna, who gets other 'students' to dance here occasionally. When I ask her what she wishes to do in future, she says, without batting an eyelid, "*ila dance chestu undi povadame*" ('just keep dancing this way'). There is no choice, is there? There are other children working there too, such as Srinu, a 15-year-old illiterate from Vizianagaram. Srinu earns Rs. 1,500 a month. He has two sisters who have to be married off. He has been working on this launch for a year now. An acquaintance from his village got him this job. His parents are farmers, he says. All through the journey, the work of these children does not stop, including cleaning dishes, the deck, serving guests and of course, always eating last. These unorganised, unregulated tourist operations have all the scope for exploitation of ad-hoc workers and abuse. In a while, the Punnami launch stops at the Gandipochamma temple. Pilgrim tourism has become an appendage here. The temple caretakers ask for money for 'darshanam'[37] of the goddess. The narrow lanes by the temple are like any other (Hindu) pilgrim centre. I meet a few Padmashali (caste group) women (non-tribal settlers) on the way to the goddess temple. They say they dread moving out of the village when the barrage is built. The place has changed in character and seems discordant. The coconuts and incense sticks – quite alien to the culture of the Koya, Koya Doras and Kondareddis in the past – have come to stay. The inclusion in the circuit makes sure this will be the accepted universalised way of things hereon. Anyways, tll the dam is built. Then the temple will drown too. At that time nobody will remember there was a goddess here they had worshipped! About the temple, Chadala Venkatreddi, a Konda Reddi farmer from Nellakota had told me in a different context, "She was our goddess, but not the one you see inside the shrine. Our goddess still lives in the Godavari water, and comes during festivals and goes back. That is our belief. The present temple is located on a non-tribal landlord's property. Now they make money from the temple. Not a penny goes to the tribal people..."

We stop at Devipatnam next. Last year I had my lunch here (at an odd time, probably around 3.30 pm) at an eatery. This year I do not get off. The launch owners go up to the Police station to pay a toll tax - since they are now entering 'Agency'. The guide announces that Alluri Sitaramaraju, the revolutionary leader of the Rampa revolts blasted this police station. Even Alluri Sitaramarju is reduced to a tourist attraction, one realises. The conditions that prevailed in his time continue to haunt these parts today. Bhavya and her 'master' entertain trourists in the launch with their film dances. Her master, Sai, then calls up little children and aged women to show their talents, which they readily do. Everything remotely called entertainment is linked to Telugu film industry. The little children you see outside the launch, along the banks of the river, sometimes wave out to the tourists, and dance, too, to the same music. The tourist-launch passes Kondamodalu, Koraturu, Siruvaka, Tadivada on East and West Godavari ditricts. The guide points to the bamboo huts at Kolluru (Khammam district) ...Set up in the midst of the tribal hamlets, one dreads to think what this influx will bring in its wake...As we approach Perantapally, the ashram dedicated to the Perantapally svami, a mendicant from Kerala who initiated a system of procuring wages for the Konda Reddis through a cooperative society in the 1940s. This is the only place where silence is an 'order'. Here, Konda Reddis earn a supplementary income from selling bamboo handicrafts.

That Faraway Look, and What Lies Ahead[38]

It was late afternoon when I alighted at Perantapally for the first time though this was a place I had heard of, read about and desired to step foot on the 'next time', which did not happen, until this moment. After that loud blaring music, the guide told us to maintain silence in this ashram. As I climbed up the narrow pathway towards the ashram, right under the first tree, a *chinta chettu* (Tamarind tree), was this young, beautiful, dusky woman seated on its wide trunk, roots extending outwards, almost to the river. Her silent, faraway look even as I kept walking closer, oblivious to the whole world, would be a sight worth freezing in both mind's eye and the camera. As I walked closer, and even as I took her picture, she was not disturbed, lost in her thoughts. And on the ground, she had some beautiful handicrafts displayed. This Kondareddi woman, Vennela (I forgot her name, but would like to call her Vennela, means moonlight), like many of her tribe in this village (now part of this tourist circuit) makes lovely hand-crafted bamboo models of houses and trees. She sells each of these for Rs. 20 to Rs. 150 (the latter for the house model, which requires more labour, of course). Today, their sales figures are pathetically linked to the number of tourists visiting the place. Vennela comes here each day with her wares and bamboo reed. After waiting for the launches for nearly the entire day she manages to sell three to four pieces. Even the way the bamboo handicrafts are lined up is poetic: as you walk up hill towards the ashram, the colour of the sands and bamboo blends beautifully. Livelihood, then, is a double-edged sword – there are these tribal communities selling these wares they make with their hands, whose beauty belies the pathetic lives they otherwise lead in terms of access to a decent life; and there are the tourists who keep them in 'business'– the nature/form of tourism itself insensitive to their lives and their socio-cultural contexts. The image of Vennela then, was perhaps the most poignant image – a blank look into a future away from these lands and these homes, without the *chinta chettu*, if Polavaram dam would drown them all. On the return journey, the dance master Saikrishna points towards Korattur, Kachluru, and surrounding villages, saying, "In many of these villages Telugu films were shot – Godavari, Ramadasu...but of course these villages do not have electricity. They set up electric poles a year ago, but since Polavaram dam will anyway drown them, they did not give the connection!" Was he being smart or sarcastic? One can only guess. Then he points to cattle on the hilltops at a distance and says, "Look here, elephants with horns."

They now start playing the *tambola* (a sort of non-serious gamble with numbers). Incidentally, the driver tells me that the motor launch uses 180 litres of oil per day and they serve at least 20 litres of canned water to tourists per day. The workers eat last (includes the dancer, the children, the driver, his assistant). Sometimes, there is not enough left for them. But the driver brings food from home at times. Even the forests the hills and the Godavari are sights to be consumed. As the Punnami launch this year sets back on the return journey…Kondamodalu with smoke from its houses beckons one, but this is not the time for a tourist in their midst! That will be yet another time, for me, among them. And that will cost me only Rs. 20, on another such launch, like the first time. Not a steep Rs. 500, as it did now!

***

A 'non'-Submergence Zone: Encounters with Tribal Land Alienation and a Corrupt RDO

6th April'07 - I reach Gokavaram bus stand as planned in the morning. Rambabu has been waiting. His motorbike is a blessing. The ride happens through a vast beautiful stretch of land, with a large lake in its midst and several Egrets and Herons. Godavari does not flow past in these villages, by the way. Shall see the birds on the way back, if I can stop by. Now is not the time. It is a quiet, breezy ride, only broken by Rambabu's intermittent discussions on the situation we are about to see. We are on the way to the villages Indukur, Pedabhimpally, Chinabhimpally, home largely to Konda Kammara, few Koya Dora communities.

At Chinabhimpally, Indukuru Panchayat, Devipatnam Mandal, East Godavari District

Most of the beneficiaries of the Polavaram dam R&R package have been the non-tribal upper caste landlords. Kosi Ramalakshmi's was part of the tribal community-held land (in Pedabhimpally in East Godavari district). It was taken over by the Revenue officials for the Polavaram R&R colony. The compensation for the land, however, went to absentee landlords of the plains. Radha says, "We have been cultivating these lands.The RDO (Revenue Divisional Officer, Narsing Rao) came and obstructed us one fine day. He told us they (Revenue Department) have bought this land and we have no right to enter here. We have been growing these plantations (cashew, *jeedimamidi*) for twenty years. We sell these and make our living. They first dug a canal under the Musirimilli project and we lost that land. They said they would give us land in Ramaiahpeta but filed cases against us. And those who sold these lands are claiming the lands from us for cultivation! They did not give us pattas all these years. The RDO along with the SI (Sub Inspector) started digging and ploughing and did not give us compensation. When we questioned him, he sent us a notice. Even the ITDA P.O. is asking us not to enter our lands. Non-tribals from Tantikonda and Gokavaram are coming in (to take over the lands). They even beat up this old man the other day. A non-tribal, Pulapally Krishnamurthy and his sons (form Tantikonda) have been shown as owners of these lands." Pamula Veerasamy says, "They beat me up. I went to complain to the police station but the SI, Ankibabu did not register FIR. Ten days back, we saw the non-tribals taking away the cashew fruits. We phoned the SI immediately and waited until 7 pm and he did not arrive. He said the statement should have been given to the P.O. first."

Rambabu points out, "If tribals complain, the policemen say the MRO or PO, ITDA should recommend their cases, but if non-tribals complain, they immediately file FIRs. At least forty people here do not have pattas. We filed petitions on their behalf. The officials assured them pattas but did not give any. Even in Indukuru, they are taking away lands for R&R from tribals but compensation is given to non-tribals. Everybody has been charged with Criminal Tresspass cases. None of them own land, not even an acre, but have lived here for generations. Earlier they worked under landlords but in later years they were given the right of usufruct. Earlier, none of the non-tribals came here. Only when the RDO bought land in the name of some non-tribals they have started visiting personally."

Meanwhile, some more people assembled and Rambabu give me copies of petitions filed in the context of land acquisition for the Polavaram R&R. In a petition to the Collector, Kakinada (in Telugu[39], with their thumb impressions), dated 24th March 2006 (copied to the Human Rights Commission, Hyderabad, P.O., ITDA, RDO, Rampachodavaram, the MRO,

Devipatnam), Kosi Ramalakshmi, Komaram Satyavathy, Karu Ganga, Neram Sita and Madakam Kumari and others made the following plea (which I am inclined to share here below):-

"Our ancestors cultivated the banjar lands and wastelands in Chinnabhimpally, Pedabhimpally villages, growing cashew (*jeedi)*, mango (*mamidi*) plantations. Chinnabhimpally girijans were practising community farming on these *sadr* lands[40]. There is no other source of livelihood for us. Since the last 12 years, we have spent from our own pockets to bring to its present shape these *sadr* lands, placing boundary hedges (fencing), planting cashew, etc and have been making a livelihood out of that produce. But recently non-tribals, viz. Pasumarthy Mangaraju, of Gokavaram, Pulapally Krishna of Tantikonda, Miriyala Abbulu of Tantikonda, Satti Sur Reddy's grandsons Srinivasa Reddy of Rama(va)ram,Tiruvedhi Subramanyam, Nellore, Badireddy Narasayya Dora, Tantikonda, Baddireddy Nakku Dora, Gokavaram...forcibly entered our lands and tried to take away our produce. And the police are threatening us at their behest. We request you to please help us to continue to enjoy our lands. At present, the following lands (survey numbers) our being cultivated by us – 93, 177, 109, 110, 126/2, 175, 12, 89/1, 93, 96, 139/2, 143/1, 145/1, 146/1,143/4, 150/1, 187/2, 112/2. Around 80 girijans are cultivating on the sadr lands."

The sadr lands have been in possession of the tribal cultivators, and in tribal areas, continuous occupation of tribal community over land for several years should be recognised as legally owned under the provisions of the special laws in place for the Scheduled Tribes. Many of these petitions have thumb impressions of women. In fact it was through these women, many of whom are mentioned in these petitions, that I learnt of this angle to the Polavaram R&R story. I found them in the plantation lands they are seeking to protect, from where I was led to Pedabhimpally, M. Ravilanka, etc to hear the cases of police and RDO harassment.Few represent their cases in courts; in these cases, the organisation Laya was helping out with filing petitions free of cost, when the tribal communities (Kammaras in this case) approached them. But the follow-up, and all the expenses, running between offices, etc, are borne by the communities. Many a times, losing their daily earnings in the process. If they are charged with criminal cases, god save them, for until they clear their charges, an entire developmental earthquake has occurred in the meantime. So, the lands they fought for will hardly ever revert to them. And what irony that it is other tribal communities in whose names their lands are taken. Meanwhile, tribal communities from Bodigudem, D. Ravilanka, Paragasanapadu (Pudipally Panchayat) have started to come here (to the R&R colony). This was a case of the state inducing a conflict between tribal communities in the name of R&R of the Polavaram project. Karam Devudamma says, "Few girijans are already there in those RR houses. Though they are going back to their (original) villages now, later there will be conflict here. They have slapped cases against all the 60 families here."

Rayi Bhimaraju says, "Our lands in Agency Areas will give water to the plains. There is another canal downstream for Bhupatipalem project. In Surampalem project, the tribals displaced did not even have drinking water facilities. But some of the lands (in this case) are upstream of the project – how will they feed the agricultural lands, as they claim? Those tribals are all migrating as wage labourers all the way to Guntur and Bhimavaram. We cultivate lands together (there are no private lands here; these are community-held lands; around 54 acres will be give over to the Polavaram-displaced tribals). Some of the Koyadora men from here are in jail on charges of criminal trespass and sedition. The R&R colony is built on around 22 acres of land which is under dispute in the High Court. A Koya Dora woman, Padoji Nagamani, has set up a tea shack next to the R& R colony in Pedabhimpally. Ironically, it is her land that houses the R&R colony for tribal people from the submergence zone villages! Compensation for this land also went to non-tribal absentee (Kapu caste) landlords. She still awaits justice for her case pending with the High Court. About the tribal people now settled in this colony she says, with a sense of sympathy, "The government should give them daily wages (promised in the package) but they have not got a single day's wage as yet. They are here since six months now. They (Government) gave some people Rs. 2,000. People are hiring trucks on their own. They have given girjians' common lands to the girijans from the project (Polavaram-displaced). Those people came here in hordes, initially. Only few are left now...There is no work here; no wages...The RDO purchased 22.4 acres of tribal land from a non-tribal family."

In her case, she says, "The RDO changed the survey number on the sly and gave the compensation to non-tribals for the new survey number showing that it belonged to the non-tribal family of Kapus, from Tantikonda." She blames the tribal men (who moved into R&R colony here). "Men don't care – they thought they would enjoy the benefits from both places (their older villages and this new R&R colony)...They are now saying 40 acres of fields will be given to the Polavaram dam-affected people. When we asked for pattas they refused to give us pattas on some pretext or the other.They gave pattas to two villages on the same land – Indukur and Chinabhimpally."

There is a handwritten "Petition by Devipatnam, Addateegala mandal Girijans", "Buying lands from non-tribals for the Polavaram project. It records the hanky-panky (*avaka-tavakalu*) of Rampachodavaram RDO and his illegal activities", is dated 14.10.06 and signed (thumb impression against names) by Kunjam Chinabapanamma, Kosu Lakshmi, Karu Gangamma, Kosu Ramalakshmi and others (appended at the end of the book). One of its demands is to "Immediately transfer the Rampachodavaram RDO..."

In yet another petition (in Telugu) to the District Collector, East Godavari District, Kakinada, on the same date as above (from Kondalampalem), Bontu Chinnaramana, Midde Lakshmi, Padoji Nagamani, Pamu Potharaju and others submit that "We are Konda Kammaras, viz, Bontu Achhamma, Bontu Venkamma...(etc)... cultivating these lands near village Kondalampalem since round 30 years. Prior to that, our forefathers cultivated these lands, which then came down to us... A few days ago, the Jaggampalem Secretary, Jeeyampalem Secretary and the Gangavaram Revenue Inspector came here with a lawyer and told us that they would build homes for us, roads, schools and bring water facilities, etc, and...cheated us, and took our signatures on papers and left. After some days they came and said the lands we are cultivating are not ours; these belong to the non-tribal Devarapally Subba Rao's descendants Viravenkata Satya Prasad and Devarapally Butchi Ramayya. That these lands were bought from them for the Polavaram displaced tribals by the government. That we should at once vacate these lands, failing which they would take action...After a few days they came with some men and forcibly tried to uproot the trees and take away the produce which we obstructed. These atrocities are being committed against us girijans. We want protection for our lands from these unjust/unlawful (*aviniti*) activities. We hope you will do justice to us and protect us from this kind of harassment."

Seeking redress under the SC, ST (Prevention of Atrocities) Act, Madakam Kumari, Pamula Lakshmi and others submitted a "Petition *(darakhastu)* (in Telugu, with thumb impressions) to District Collector, East Godavari District, Kakinada By tribals residing in Chinabhimpally village, Indukur Panchayat, Devipatnam Mandal" dated 12.07.06 (Chinabhimpally). In this petition they charged the "Rampachodavaram Division RDO, Sri C.H. Narsing Rao" for having acted "in league with the non-tribal and other moneyed landlords, against the interest of the girijans." They also accused him of "showing false records of non-tribal ownership of lands under cultivation by the girijans in M.Ravilanka, Chinabhimpally and other villages." Seeking that he be punished under the Prevention of Atrocities against SCs, STs Act, the petitioners mention the language he used against the community (some of them being the petitioners) and his having threatened the Indukur Panchayat Secretary saying "*mee antu telistanu*" (lit. 'shall see your end') if he did not change the *adangal*[41] records showing non-tribal ownership of those lands. The petition reads, "RDO Narsing Rao came to Chinabhimpally with the police on 2.07.06 (July 2nd 2006) and rounding us all up at one place told us, "*girijana vedhavallaara ... meeru adavi jantuvulavanti vaaru, meeku panta bhoomulu kavalsocchinda?...*"

(I find it important to use the original Telugu words to show the level at which communities are treated. He was never penalised. It translates as – "you tribal wastrels...like the wild beasts of the forest, do you need farm lands? You are like wild animals, do you need cultivable lands?")

"(He) slapped criminal cases against us at Devipatnam Police Station and arrested us....Hence we request you to show mercy on us tribals and take action against the RDO...And to take back the wrongful cases against our tribal people and we request you, as a District Collector, to arrest and punish the RDO under the Prevention of Atrocities on SC, ST Act."

Incidentally, the conflict over tribal land alienation even predates the Polavaram project, but was compounded by the target-driven ruthless nature of acquisition imposed on the officials by the government and a lot of money was to be made by way of backdoor commissions and bribes in this process, something that local media (Telugu) had been reporting on, as well. By the time I had met the tribal communities of the above villages, in 2007, they were still doing the rounds of offices to gain access to their small pieces of land and to earn their meagre livelihoods. The entire land acquisition process for R&R of Polavaram project in this area is wrought with numerous episodes of corruption and violations of all constitutional rights to the tribal communities. An affidavit against the said RDO points out the flaws in the act of purchase of land for R&R. This Affidavit, "Before The Enquiry Officer/Joint Collector, E.G. District on Illegal Land Purchased by Revenue Divisional Officer, Rampachodavaram, E.G. District", Ref. A5/6068/06 Dated 13-12-06, Collectorate, Kakinada, Camp at Rampachodavaram" by an advocate from Rampachodavaram, Ch. Santosh Kumar, charges that "the RDO, Sri. Ch. Narsing Rao, purchased total lands of 572.91 acres situated in the Scheduled Areas of Addateegala, Gangavaram and Devipatnam mandal in East Godavari district from non-tribals in the name of Polavaram Project Rehabilitation programme, through 49 land transactions and paid more than Rs. 2 crores (Actual figure Rs. 20048350) in utter violation of laws, particularly Tribal Protective Land Transfer Regulations 1 of 70 and PESA Act 40 of 96 and Act 7/98..." And that the "tribals initiated legal proceedings questioning the legal title of non-tribal sellers before the Additional Agent to Government Court/PO (ITDA) Rampachodavaram. The connecting dispute is pending before the Project Officer, ITDA, as well as at High Court in Writ Appeal....(the) RDO... purchased lands an extent of Ac.37.50 cents...in Pothavaram village in Devipatnam mandal from a non-tribal namely B.L. Narsimhrao (appears to be a Retired Deputy Collector). The non-tribal landholder surrendered the said lands to the Government under Land Ceiling proceedings. The Government declared the land as surplus under the Land Ceiling proceedings. Against this legal provision the RDO purchased lands from said non-tribals and paid huge compensation amounts." The affidavit also states that the High Court of AP had admitted the Public Interest Litigation ("W.P.No. 18283/06") of tribal people "questioning...the land transactions held by RDO, Rampachodavaram and sought a declaration that all such transactions are to be as illegal." The matter is pending with the High Court at Hyderabad. In his long response, the RDO maintains that "...The then Collector, East Godavari... given a target of Ac. (acres) 1000 to Revenue Divisional Officer, Rampachodavaram to acquire the lands... by the end of May 2006 for the R&R Package of Polavaram....I have initiated the action for acquisition of patta lands of non tribals who are coming voluntarily to sell away their lands... On information and on the instruction of the Project Officer, ITDA myself visted the village along with Mandal Revenue Officer Devipatnam and conducted counselling to the innocent S.T. people who caused the disturbance[42]. The Revenue Divisional Officer has promised that the Revenue Divisional Officer and Project Officer are willing to allot cultivated lands for each landless poor ST family under IKP (Indira Kranthi Pathakam) but doing all these illegal activities are consequent for prosecution. For this...the Revenue Divisional Officer has issued a public notice in the village itself...and served on more than 50 persons...The Panchayati Raj act extension to the Scheduled Areas (PESA) Act No. 40/96 is not applicable to the lands acquired under Sec 3(I)(C) of APSALTR 1959 as amended in 1/70...The PESA Act is applicable only to the lands acquired under LA Act 1894 as amended in Act No. 68 /1984 in Scheduled Areas...So far the RDO acquired 595 acres and still we have to acquire nearly Ac. 4000 for the evacuated ST families of 24 submerging villages of Devipatnam Mandal for R&R Package..."

The said "counselling" was done by beating up Ekka Rajanna Dora, an old tribal peasant, and others in the process. Rajanna Dora, whom I met, had 'caused disturbance' by merely questioning the RDO when he with his men had started digging up his field. (Emphasis mine). Incidentally, few years following these allegations, the said RDO, Narsing Rao was transferred to another department with office at Rajahmundry where he found himself in yet another corruption controversy, 'outsourcing' his official work to a retired employee for a small fee , as reported by a Telugu TV Channel , TV9 where mention was also made of his 'disproportionate assets' during his tenure as RDO Rampachodavaram, which probably led to his transfer. One does not know what came of that allegation.

Incidentally, the High Court of Andhra Pradesh in a Writ Petition, "No 9232/06" directed the Project Officials of Indira Sagar Project (Polavaram) not to dispossess tribals from their 'D' form patta lands without following the due process of law. Revenue authorities and Project officials in Pothavaram village of Devipatnam Mandal, East Godavari District however only looked at tribal lands for R&R for the 'oustees' of Polavaram dam.

At the Polavaram R&R Colony at Pedabhimpally

I sought out some of the Polavaram "oustees" at the colony. Kangala

Naganna Dora, a Koya Dora from Pragasanipadu, in Pudipally Panchayat started off saying he is uncertain of his future here, as he too has come to know the land is in dispute. "We are 50 families and there are few non-tribals too, around 10 of them. It has been six months since we came here. It is not so good here. Godavari was close by back there. This land is is a bad one." They said they would give Rs. 64 per day as coolie (wages)." Naganna says, "Since six months we have not got a penny. It costs us Rs. 30 per day to go to our farm land. It is worse for those who are landless. They did not give us wages, which they had promised after we moved to the RR colony. We have to travel around 24 kilometers by bus to reach our farm land. We all own 2 to 4 acres of land there (in the original village). The P.O said 'you won't get any money, but we will give you work'. But there is no work here. Many of the families decided to stay back at the villages. They gave us only half of the transportation charges. Here, we cannot use the forests because other girijans are already using them. Half of our livestock is here and the rest there…After we came here we realised that the RDO purchased girijan lands from non-girijans (*girijanetrulu*). And girijans have not received compensation. We do not want to compete with fellow girijans here. We do not know what will be our fate…"

M. Ravilanka village has a mixed population of Koya Doras and Konda Kammara tribal groups. Around three decades ago, the excess/surplus land from non-tribals (owned illegally) was taken over by the Government and handed over to tribal people with 'D' Form pattas. But again the RDO has given the compensation for land to non-tribals, a large part going to a Retired Deputy Collector! A Writ Petition was filed in the High Court (9232 of 2006) – Villagers of Ravilanka vs. MRO and District Collector.

Ekka Rajanna Dora was in jail for four days under IPC 353 – "Assault or Use of Criminal Force to Deter Public Servant from Discharge of his Duties". That weak old man of 70 was beaten up; it was assault the other way round! Rajanna says, "We told them we were using these lands for which the government had given pattas. RDO and MRO filed cases against us. The ITDA P.O. sent a supervisor. They gave us pattas but are now showing us wrong survey numbers."

Rampa Nookarajanna says, "We have so many cases against us. We got to know that our lands are meant for the people of D. Ravilanka, which is under submergence. We used to grow brinjal, cotton on our fields (rain-fed)." Ekka Bhimanna Dora says, "The MRO came with the SI (Sub Inspector) and were digging up our fields (*dunnutunnaaru*). We went to the ITDA P.O. (Murali) who told us we had some other pattas. They showed us those lands, which were rocky and they were our lands in the government records. Why should we leave our own land here?"

All of them said they cannot afford lawyers to fight their cases forever. Nookaraju says, "only this village is ours but the land is non-tribals'! And so it goes on, same story, each one adds to it, Kosi Radha, Kosi Lankababu ...with an urge to share their woes. Kosi Radha says the "They built the RR colony on around 22 acres fields which we were cultivating. We used to grow brinjal, cotton, paddy, legumes, and millets (all rain-fed, *varshadhaaram*). In summer we would go for wage labour. Six months we work on our fields and the rest we manage with *coolie pani* (labour). Sometimes we pick cashew nuts (April-May) if someone calls out to us (*keikeste*)."

Incidentally, the term "*keikeste veltaamu*" (in these contexts) signifies the feudal past where farm workers would be called out aloud to work on the fields. In general, when it is not used for the landlord-labourer hierarchy, it signifies the manner in which people in these villages generally call (out to) each other across distances, homes, fields. It is a casual reference to 'calling'. But contextually placed, it makes a difference. I heard a strange argument in the midst of all these talks, from one of the Koya Doras here, perhaps it was Kosi Venkababu who said, Koya Doras believe that Konda Doras who are presently seen as tribals were not initially tribal people, but came to this region and changed their identity to Konda Dora to get benefits from the land here.

7th April'07 - I am at Rajahmundry Central Jail - "It was established in 1864...and will hold 1,089 criminal, and 20 civil prisoners. Cellular accommodation has been provided for 400 convicts, and the rest are kept in wards..."[43]

Mutchika Suramma, Kunjam Rama Rao and Jeedi Pentaiah from Chegondapally are housed here at the moment. The case on Suramma was filed in 2005, as with others. Charges include IPC 143, 341, 153 (A), 124 (A), the last one being Sedition, which is a non-bailable offence. I fail to get permission to meet Suramma, and the other two, at the Rajahmundry Women's Prison and Central Jail, respectively. The Jail superintendent (Central Jail) tells me media are not allowed to meet inmates. She even told me, if you mention "friend" or "relative" may be there was a chance! Did not know that earlier, now it was too late to change it. Nagaraju, the advocate from Rajahmundry who is fighting their case, tells me, at the time of bailing them out, they are framing newer charges and imprisoning them immediately. Their protest against Polavaram has brought them here. I am beginning to feel, it is a police state in the case of Polavaram project. If all that they claim was rightful, why the use of force?

At the Site of the IPC Charge 153 (A), viz. "promoting enmity between Classes in Places of Worship, Ramalayam, Chegondapally!

The Prajashakti stringer, Sitara 'master' (Sitaramachandramurthy) is also with me. I am recording a general discussion that is happening here, with more than 20 people assembled, as it always does; I came at an appropriate moment when they were having their own village meeting. Rama of the temple watches, as always. The following discussion was in progress:-

"Near Gunjavaram they showed the site (R&R colony). They have not laid the foundation yet. They have not given pattas but are asking us to take the cheques. They have not shown cultivable lands to us yet. We have not yet taken cheques. The lists are incomplete (47 names are missing) so we said we will not take the cheques until all our names are mentioned there. They surveyed lands based on old valuation (RDO never came to do the valuation of the lands, etc). How can they prepare cheques without valuation of our lands? Initially they told us we would get Rs. 40,000 for the house; and Rs. 82,000, or Rs. 76,000 per person as compensatory wages (for loss of wages). Now they are deducting Rs. 40,000 from this latter amount. They will build the houses with this remaining Rs. 80,000. Which means all we will be left with will be a cash amount of mere Rs. 50,000, which will finish in no time! Rythu Coolie Sangham has lands there in Gunjavaram (which they took for R&R colony). But they compensated the Chowdharies (upper caste non-tribals). If we say anything, they will arrest us too! See what happened to Suramma and others. Joint Collector Ramanjaneyulu had come to this very Ramalayam and assured us that the government will not ask us to leave until the entire compensation package was handed over to us!"

Yamana Srirangalakshmanasvami of Pydipaka says – "Now they will give the compensation amount to a committee (negotiation committee) and all transactions will be routed through these committees...We will have to give applications, and so on..."

"NGOs will also be involved in these committees" adds Sitara master. "The colony too will be built by the committee…"

One of them, not wanting to be named, says, "They are taking in (arresting) anyone who talks. So what should we say? So we think twice before talking these days."

I am amazed at the change enveloping the place since last year when these very people had assembled at the same Ramalayam and strongly opposed the project and argued forcefully with the BGR company staff for blasting dynamites; today they are afraid to talk. Midiam Singaraju says (and Ranga nods) – "We are so scared we are not attending the *santa* (fair) on Tuesdays at Polavaram. You never know who they will pick up the next time…Even my name is there (in police records). And I am not even part of the (Rythu Coolie) Sangham!"

Sunnam Raju says – "These cases are all related to the project (Polavaram). We met the RDO near Spillway site, and people were surrounding him that day (translated as "Wrongfully restraining persons…" case slapped against the three of Chegondapally). We never touched that engineer from Madhucon company" (Madhucon company staff filed a case against the people of Chegondapally in 2005; that they pushed the engineer and asked them to shift their machinery). Sunnam Raju says "We did tell them to please take away your machinery but we did not touch their staff..."

The government records show Ramiahpeta and Pydipaka were given Rs. 6.30 crores so far in compensation and Ramaiahpeta, Rs. 1, 85,000. Even for those who have accepted R&R package have not been given the entire amount. Even cheques will be deposited in the bank in the name of the committee to be set up for this purpose. People here say if the committee is corrupt, all the money will sink. By the way, the said committee is not democratically elected but nominated by the Revenue officials.

***

Jail Terms, Forced Consent, and Other Issues[44]

It is a question of building consensus by means of fear tactics and forced opinion-building... At least three people from Chegondapally, Tellavaram and Polavaram are housed in the Rajahmundry Central Jail, and the Women's Prison - Kunjam Rama Rao, Muchika Suramma and Geedi Pentaiah (the last two also members of the AP Rythu Coolie Sangham). They have been arrested under Sections 143, 341, 153, 153 (A), 124 (A), 341, 447, 434, 427 IPC (with variations for each of them). For the layperson some of these sections mean – being "member of unlawful assembly" (143), "wrongfully restraining any person/official on duty" (341), "promoting enmity between classes in places of worship" (153 (A), "Sedition" (124 A). Suramma and the other two (as many in their village) had some time last year, questioned the RDO and Joint Collector when the spillway work was under way. Chegondapally was among the villages that directly faced the effects of the same. Their requisitions on the same were of no avail. Now, every meeting in this village happens in a temple, 'Ramalayam' (and even officials meet the villagers in this temple). Hence the Section, "creating enmity between classes (read, pro-, anti-dam) in a religious place of worship"! Sedition, a legacy of the colonial times, holds the same intent today. In simple terms, opposing the powers; and that can be interpreted by those in power. In Polavaram, there are also others under different sections and liable to be arrested any time, including a CPM organiser named Banerji, and two others from Chegondapally, named Ravi and Sunnam Raju, among others. Things there seem to be at a political head at least for those openly critical of the dam. But (even in a place where) there is no such ostensible 'political' reason for slapping cases, (nor) question of submergence, the charges filed against (the adivasi peasants) are almost similar. As one has seen in the case of Rajanna Dora or Pamula Veerasamy. Chadala Venkatreddi, a Kondareddi from Nellakota remarks angrily, "If we took our bows and arrows like Alluri Sitaramaraju, will the P.O. (ITDA), the government still ignore us? We are all unarmed and helpless hence they are filing cases against us and doing what they want."

***

Godavari Feeds them Too and Incidentally, the Dalit Mahabharata and Gontulamma

Pydipaka SC Colony

When I met her last, she was busy overseeing the relief camps at Polavaram. It was during the 2006 Godavari and she had just returned from a tour with the former CM Chandrababu Naidu, who was surveying the damage. Kuncha Varalakshmi had another reason to be part of that survey – she is the Mandal President (MPP), Pydipaka Panchayat. She is a dalit (Mala) and won the elections from the constituency reserved for the SCs. She is one of the more articulate women of the area, and speaks her mind. She says, "for four months a year, Godavari feeds us; for three months, the agricultural fields (labouring on our fields or those of others), and the rest of the months these hills and forests feed us" (*naalugu nelalu Godavari mammalni penchutundi, mudu nellalu vyavasayam, vyavasaya coolie pani, migilna nellalu i kondalu, adavi mammalni poshistayi...*")

Godavari's coming is very crucial for the dalits in these villages (Pydipaka is the nearest village to Polavaram). They eke out their livelihood in the months when the Godavari swells up and envelops these villages. The Mala men gather the teak and bamboo logs that Godavari washes off in her rage, taking them along with her. The men go to the upper reaches in this time taking with them a wooden craft on which they lay their bodies, moving along with the river's rhythm, gathering the logs of wood, tied to a rope. The broken down trees and logs they thus collect, is sold to make some money. They sell these for anywhere between Rs. 100 to Rs. 5,000 depending on the amount they have managed to collect. Varalakshmi says, "We are asking for a colony specifically for the SCs (in the R&R package). We also need three acres of land; because the dalits, who will be displaced by Polavaram dam, will become landless. At least here some of us own small pieces of land. Where else will we get this kind of environment? We know we will lose out. But if we have to leave ultimately, at least the government can make sure the dalits do not suffer. We are sacrificing our lives for people in the delta. We are demanding (for dalits) free education. While all others (tribal communities) are getting Rs. 40,000 (for the houses) we are getting only Rs. 20,000. Our Joint Collector told us the other day if you do not take this package, how will you marry your daughters off? But I asked him, 'Where will you give me this house I constructed with my labour and money?' The CM is distributing land everywhere else to the dalits and weaker sections. He should do similar things for the dalits displaced by Polavaram project."

It was here that I heard a fascinating story of Gontulamma, which, to my excitement, I also found recorded in the colonial records (but in a different context). Subbanna, a Mala, told me this story (others also joined in with bits of information) which is specific to the Mala communities, when I was curious about a temple in Pydipaka, close to the river. There is no permanent deity there. He told me Gontulamma is actually the name of Kunti (from the Mahabharata). Her sons once came charging at her for having kept them in the dark over Karna's existence (her son from her 'experiment' with the Sun god[45]). She ran for protection; the Malas covered her up putting a basket over her, for a night. In the months of September-October, they make a clay form of Gontulamma, place it under a cane basket, to be immersed the following day in Godavari. Homes are cleaned and spruced up at this time and people wear new clothes. The Gazetteer records, "Scarcity of rain is dealt with in various ways...Malas tie a live frog to a mortar and put on the top of the latter a mud figure representing Gontiyalamma, the mother of the Pandava brothers. They then take these objects in procession, singing, "Mother frog, playing in water, pour rain by pots full." The villagers of other castes then come and pour water over the Malas."[46]

***

October 2007: A different start

2nd October'07 - First time, on the Road from Rajahmundry to V.R. Puram, through Gokavaram, Sitapally[47], Maredumilli

Lovely forest patch worth trekking all way through, if there was time. Will I come here again? Many beautiful sights enroute, apart from the trees. And interesting ones. For instance, just captured on my camera these clay idols of Rama, Sita and Lakshmana, resting their backs against a tree in a quiet corner; they were painted in lovely colours, though they seemed very old idols. No ado; no flowers. No worship. Had this been in Hyderabad, people would have built an entire edifice around it. How did these idols come here? Discordant in terms of tribal idea of godhood. Yet, I do notice the relatively large spread of the Rama legend in these parts, and people with names such as Rama and Seeta seem common, even among Koyas and Kondareddis. Years of living in a mixed social set up – with caste groups and their religiosity?

I have to reach Rekapally, from where I will decide as to when we will move on ahead. I plan to check out the interior forest areas of V.R. Puram mandal, probably close to Chintur. The cab driver, too, enjoying the drive as much as I am; and when I asked him to shut off the music in the car, he was surprised. This cab is a luxury, for Rs. 500, at the moment. But the moments are priceless, through this forest.

I Reach Rekapally, at ASDS office

Hear the same story of the omniscient "R&R". "No specific earmarked land has been shown to people", says Gandhibabu. "There are no colonies yet.

People are selling away their lands. Tribal people are cutting forests in the interiors. Then there are the tribals from Chhattisgarh who are settling down here, running away from Salwa Judum. The Joint Collector here said the other day that there will be no wages given to women (as promised in the R&R) since they will get married and leave. The person-days labour is only for men…Political parties have shifted priorities. CPM is into its "Bhu Poratam" (the land reclamation agitation begun not too long ago). There is harassment of those talking against Polavaram. No rallies are allowed. We cannot speak of Polavaram directly…"

3rd October'07 - Twice Displaced, Twice Hounded: The Gottikoyas[48] at Sunnamatka, Pedamatapally Panchayat, Matapally Reserved Forest Land (Khammam District)

Kurra Sannu, from Donnapalasa says, "I came here 15 years ago. We did not have land there (Chhattisgarh). We cleared some land here. Today if we have to take a bus from Dantewara to Kunta, it costs Rs. 25, Rs. 30 for an autorickshaw from Kunta to Chintur and Rs. 20 for a bus from Chintur to Sunnamatka (Rs. 75 in all). Sunnam Jogaiah used to live here; he brought the others (the later settlers) with him. We understand the Koyi language but do not marry into Koya families."

Vanjam Gujja Rao says, "There is a conflict with the girijans here. Since the Polavaram project is coming up, they now want this land where we have settled down. Earlier they used to fear clearing the forest patch. Because of the dam the local girijans want our lands, as this place is not under submergence. Pratipaka girijans will lose lands, so they are clearing these patches for *podu* and asking us to leave. We have voting cards and voting rights here. My father (Bhimaiah) was beaten up, taken to court. There is a case in the Court now. Forest Department has filed a case against him. It was adjourned. The Department people have their own interests in this. There is a Stay in this case now."

But the Gottikoyas are used by the Forest Department for labour: log carrying for Rs. 40 per day and plantations in June and July and sometimes the summer months. They get three months of labour from the department. Gujja Rao says the total extent of land here (which they have settled) "must be 20 acres approximately. Ten acres have been cleared. We grow paddy and *jonna* (sorghum), both rain-fed. Some have ration cards, too. Others living here since ten years got their ration cards cancelled, especially since the Polavaram project was announced. Since two months we are being asked to leave."

The nearest hospital for them is at Chintur, and the nearest school for their children is around 5 kilometers from here.

Diary notes

The village Sunnamatka (not mentioned as such as in the revenue records of V.R. Puram mandal, I am told), in Pedamatapally panchayat. The Gottikoya tribal community here settled since 15 years ago, or a little more. They came here from the erstwhile MP, which is today Chhattisgarh, in search of livelihood after one of them in the course of his migration found a patch in the reserved forest here in the Bhadrachalam division. Others followed. The issue today is that of old and new settlers; today there are reports of those running away from repression of Salwa Judum-Maoists crossfire. But to miss out these early settlers (may be others like them) would mean seeing the whole issue within a narrow purview of the recent Salwa Judum and Maoist context alone. The earlier settlers did not necessarily flee the Maoist-Salwa Judum crossfire. Because of the Polavaram dam now, these Gottikoyas are being asked to move out. There was constant conflict with forest officials (their homes being burnt down). The Gottikoyas have small huts and few livestock. There are around 15 families here. Today, they are caught between the forest department and the local Koyas because of the Polavaram dam and threatened spaces of tribal communities in the submergence area. They did not have any problems here earlier, they say, as they cultivated, together, around 10 acres of land; some of them have ration cards, and voters cards as well. Some voted in the recent Panchayat elections. But they do not have right to residence, which is funny.

***

Bhadrachalam – I met some people involved in the issue. Gilbert, from an organisation called Carmel Women's and Rural Development Society (I have heard of the organisation for the first time) says, "The forest is being destroyed (ever since Polavaram project was announced). ITC is growing saplings (Eucalyptus; *jaam oil* - as referred to locally) in reserved forests. They have a buy-back agreement with the Forest department, which grows plantations for them through VSS (Van Suraksha Samiti), JFM, CFM and so on. In Palavancha division ITC has linkage with the Indira Kranthi Pathakam. ITC calls it 'development of wasteland' and plantations and the yield would be used for village development. Now they are growing these on good soils. ITC has plantations in Kukunuru and Burgampadu mandal. The Forest department is employing its full-time staff for raising clones. ITC sends all waste water (effluents) in to Godavari river..."

Haribabu (ASDS) says, "Forest department has direct links with ITC, not through VSS, at the time of sales."

It is possible, as I glean, that ITC is actually growing its plantations (until the dam comes up) on tribal land. And that tribal people are being employed to grow these plantations as daily wage labourers.

'Packagi' and Bikes

Haribabu then casually mentioned that ever since the Polavaram package has entered people's lives, he has noticed a rise in the number of private vehicles purchased by them! So, the next thing to do was check out one instance. We went to the nearest automobile dealer iin Bhadrachalam. The manager, Venkataramana obliged. In 2006-07, he said the sale of vehicles (including tractors, three wheelers and two wheelers) was 276 (from his shop alone); while in 2005-06 it was 232. His customers were usually from Bhadrachalam, Cherla, V.R. Puram, Kothagudem, Kunavaram, Kukunuru, Manugur and Aswaraopeta. Eight to ten and even thirty vehicles are sold on full cash (down) payment. The ones who buy from him include contractors, heavy water plant contractors, and farmers. I heard even Sitara master telling me that the number of two wheelers (motorbikes) in and around Polavaram increased and in fact, you could be sure that the person had taken the package (*package teeskunnaru*) if you saw him on a brand new motorbike!

Reflecting – In this trip I have met a range of tribal groups: Koyas, Konda Kammaras, Koya Doras, Gottikoyas from the submergence zone. As I was ready to leave Bhadrachalam, to take a bus to Hyderabad, Lakhon Bhumia, another tribal man, from Malkangiri, Orissa, was on his way to Vijayawada and Hyderabad. He has been here since the last two months, he says. A "Bengali" brought him here and there are other Bengali contractors working in this area! He has not been paid any wages as yet, he says, for the labour he has put in. As I saw him and others with him in that bus stop that evening, an eerie premonition – will the decently-settled, relatively self-sustaining tribal communities of these parts be talking to a journalist someday elsewhere in a bus stop, unable to speak the local language, as migrant labourers like Lakhon whom I met? Will the State of AP join the ranks of Orissa and Chhattisgarh very soon in terms of the poverty that tribal communities suffer? I dread to think what the future holds. This journey has been symbolic – in meeting tribal communities from bordering states and reflecting on what it would mean to lose land and homes and settle in an alien land.

Meanwhile, some important news comes in - Responding to an appeal filed by Dr. R. Sreedhar of Academy for Mountain Environics (and represented by advocate Ritwick Dutta), the National Environmental Appellate Authority (NEAA), on 19th December, 2007 quashed the environmental clearance granted by the Ministry of Environment and Forests to the Polavaram Multipurpose project stating that the MoE&F had granted clearance without taking the opinion of the States who were affected by it (Chhattisgarh and Orissa). The NEAA order stated, "It is evident that no public hearing was conducted in the affected areas of Orissa and Chhattisgarh. Neither did the affected persons have any access to the executive summary of the project in the notified place, nor did they have any opportunity to participate in Public Hearing and express their view on the environment impact of the environment of the area…the question here is not one of majority or minority but it relates to the issue of enabling all the affected people to express their view/opinion/comment etc., whether they reside in one particular state or two or more states."

***

As the Journey Ends, Some Realities

What kind of value do communities attach to things? When you ask elderly men, women as to what a tamarind tree or a toddy tree growing around their homes means to them, they point out its antiquity in time, hence valuable; sometimes it is sacred (because it is old, or irrespective of its age but usually old and sacred coincide) and hence valuable. But governance systems post-globalisation (and even in preparation for the inevitability of globalisation, in late '80s and early '90s) have succeeded in altering this sense of 'value' in most cases. The new generation, the younger lot among both tribal communities and other castes, handling the R&R affairs, negotiating with the government authorities (usually men) and those who had accepted compensations for their land for the project, or were in the process of doing so, see the officials' failure to compensate the same 'old' trees in money terms. For the younger people, the tree is valuable in money terms because its fruits are sold to the Girijan cooperatives or its value is understood as the same value that the forest department attaches to the tree as a "productive" unit. There is no sacrality to it anymore. So the loss in its value for the community has happened successively, over time. It is amazing as to how the diversity within communities of value (of prestige, sacrality or emotional attachment) attached to trees, land, forests, rivers, wetlands, etc, has almost vanished to bring a uniform, common measure, or value of a monetary nature to almost everything in everyday life, including, at some point, relationships as well. The sheer diversity of the submergence area eco-system which I have seen in this one journey, is amazing. And there is not much thought spent on the nature of eco-system changes if all this comes under a singular, universal and centrally (state) controlled water regime. Officially, the agro-climatic zones of AP are as follows[49] (and I am taking only those which fall under submergence):-

Krishna-Godavari Zone – East, West Godavari, Khammam Districts: South West Monsoon (rainfall) zone, 800-1100 mm (rainfall)

Soil type - deltaic alluvium, red soils, with clay, black cotton soils, red loams, coastal sands, and saline soils.

Crops - Paddy, jowar, bajra, tobacco, chilli, sugarcane.

North Coastal Zone - Uplands of East Godavari: Southwest monsoon -

1000-1100 mm rainfall
Red Soils with clay base, pockets of acidic soils, laterite soils with pH 4-5
Crops - paddy, sesamum, jowar, bajra, blackgram

High altitude and tribal areas – East Godavari and Khammam Districts: southwest monsoon 1400 mm; hill slopes, undulating transported soils. Millers, pulses, chilli, turmeric

Just a small example of the government's own record of work in Devipatnam mandal should reveal the almost cavalier manner in which lands were being acquired for the project. A report, that I procured from one the MROs, gives the list of "submerged villages" in Devipatnam mandal, East Godavari district ('as per GO MS No. 111 dated 27.06.2005) as being "43". And then it goes on to talk of "Awareness Camps: Three (3) Awareness Sadassulu[50] on Implementation of R&R Packages were conducted. One at P. Gonduru on 14.06.2005 (villagers of 11 villages attended), second one at Devipatnam on 21.06.2005 (villagers of 16 villages attended) and third at Kondamodalu on 2.7.2005 (villagers of 12 villages attended). Again, panchayat wise Awareness Sadassulu was conducted from 31-5-2006 to 3-6-2006."

In stead of the mandatory gram sabhas with requisite quorum, in almost all cases the administration violated the PESA provisions in these Schedule V villages and cleverly used the term 'sadassulu'/meetings in their official records to show they did their job, to avoid being questioned on the same. But people did question them later, when they realised that these meetings were actually to be recorded as Gram Sabha consent to the project, when no such resolutions were passed in these meetings. Moreover, even before the project works (construction) began the government was already conducting these meetings, without highlighting all the aspects of the proposed dam or eliciting popular opinion. The above report also mentions "Household Surveys". "As per GO Ms. No.68 dated 8.4.2006 household survey was conducted in all the 43 villages and draft R&R package (Rs. 50.63 crores) was published in the villages by incorporating the MROs verification report."

The Total number of families is shown as 5,152, of which total number of STs (Scheduled Tribes) is 2,635, Total number of non-tribals is 2,517; landholders are 1,475; landless poor – 3,349; Rural Artisans – 221 and Others – 107. The "total extent of tribal land to be submerged tentatively (in acres)" is 4,549.14; "Total extent of non-tribal land to be submerged tentatively (in acres)" is 2,274.30; Total land (in acres) – 6,823.44. The funny thing is this is non-tribal landholdings in tribal area. An anamoly in itself which the Polavaram project has already become notorious for.

Under the section "R&R Package" the following is mentioned:

"A Model colony (165 families for 165 houses) is being constructed at Pedabheempalli for the villagers of Paragasanipadu (66 houses), Bodigudem

(21 houses) and D. Ravilanka (78 houses) villagers. For the construction of the colony so far an amount of Rs. 319.12 lakhs was incurred. It is nearing completion."

The expenditure incurred on this is shown below:-

Sl. No.	*Purpose*	*Expenditure (Rs. in Lakhs)*
1	For houses for ISL and Shopping Complex and Tree Guards	134.86 4.50 2.00 1.40
2	For electrification	5.90
3	For roads, community hall, school building etc, infrastructure	120.00
4	For drinking water	17.00
5	Wage compensations to landless labours	29.60
6	Other miscellaneous	3.86
	Total	**319.12**

Note: It must be remembered that by the time these activities had started, the project had not yet got all the clearances needed for a project of this magnitude.

Under "Land to Land" we have the following figures:

Sl. No	*So far Identified (Acres)*	*Expenditure (Rs. in lakhs)*	*Remarks*
1	1838.02.00 (Government land)	For the purpose of bush clearance, land levelling an amount of Rs. 20.00 lakhs was released to EE (TW) - Executive Engineer - R. C. Varam	Rampachodavaram (915.50 acres) Gangavaram (767.00 acres) Addateegala (113.00 acres) Devipatnam (42.52) So far bush clearance and land levelling has been completed in 1553.52 acres
2	571.81 acres (non-tribal patta land)[51]	So far Rs. 210.10 lakhs was released to RDO, R. C. Varam for purchase of lands	RDO, R.C.Varam incurred Rs. 210.10 lakhs by purchasing 571.81 acres of land under Sec.3 (1) (C) of APSALTR Act 1959 as amended by 1/70 with reference to rule 15 of APSALT of 1969. Possession shown for ryots is under progress.

"Out of 571.81 acres, so far 194.07 acres of land was distributed to 87 PAFs (Project affected families) of Bodigudem, D. Ravilanka and Paragasanipadu, who lost their lands in the project were given at Pedabhimpalli, Lothupalem, Indukuru, Pothavaram villages."[52]

The "expenditure particulars" are given as below:-

Purpose	*Expenditure incurred so far (in lakhs)*
A model housing colony at Pedabhimpalli for 165 houses -	319.12
Bush clearance and land levelling in identified government land - (1202.00 acres) for land to land	20.00
Purchase of non-tribal land for land to land-	210.10
Total -	**549.22**

Polavaram Mandal Land acquisition in 2007 (information gained from the MRO office) - Approximately 3000 acres acquired and for R&R approximately 900 acres acquired. Yet to be distributed. Only two housing sites shown. Erstwhile Hukumpet zamindar's descendants sold (rather, the government bought) 300 acres to the government, apparently for crores of rupees (obviously with the rates fixed for the lands in official records). One of the petitions also mentions this last point since the former landlord's family is from the non-tribal upper caste, and these were lands they held in Scheduled Areas. But the government had denied that some of the lands were in Scheduled Areas, and had initially said these were Plain area lands. However, the question arose again, as to why they gave R&R lands in 'Plain areas' (Non-Scheduled) villages to tribal communities? The Koyas of Devaragondi realised this in the case of their R&R lands.

***

Land Acquisition and R&R in Year 2007 – the Process Continues[53]

March 2007

1794.95 acres were acquired

June '07:

Survey done – 412.55

Handing over to Department (Land Acquired) - 453.64 acres

August '07

Survey done - 50.00

Handing over to Department (Land Acquired) - 166.37 acres

December '07

Survey done - 455.42 acres

Handing over to Department (Land Acquired) - 89.33

Cumulative progress ("Handing over Land to Dept"/Land acquired) for 2006-07 - 15,256.66 acres

Award = Rs. 29,211.35 lakhs (only!)

Cumulative progress ("Handing over Land to Dept"/Land acquired) for 2007-08 - 18968.60

*****

Notes

1. For a short while, I had started on a salaried editing job (with the flexibility of working half a month), thanks to a team of understanding colleagues at InDG, C-DAC, Hyderabad in late 2006. Mid-2007 came small editorial assignments from an organisation (DDS in Medak), a freak reporting assignment with UNESCO, and in the end of 2007, a small grant from PEACE coupled with a friend's generous help. All these have contributed their bit. Those were good days and Godavari flowed incessantly!
2. The idea was to start this time round with the dalit settlements. I thank Babji of APVVU for his support in giving me inputs on these villages and the dalit question in Polavaram.
3. The dalits, I noticed, refer to the tribal communities either as STs or girijans. They refer to themselves as SCs, as well; and to me this was the nature of their interaction with the state and tools of democracy for years. Even if they were not school-literate, terms like SC, ST, BC, Revenue, Forest, etc, are almost Telugu terms in these villages because of the constant undercurrent of conflict that usually defines their relationships with the government and Constitutional edifice dealing with their lives and livelihoods.
4. When I went to her office (she was Joint Collector, and supervising the land acquisition process), in course of time, she asked me if I wanted information (especially on the LTR cases in the submergence zone) "as a journalist, or under the RTI". This was a new phenomenon, which I noticed did not really help the cause; earlier government officials at least in the MRO offices, etc, shared information the moment you identified yourself as a freelance journalist. Now with this, officers like her fling a form at your face (as she did) and say, fill it up. And then you wait, and by the time the information reaches, the context and timing of the event or the story has ended. So, when I said "as a freelance journalist", Ms Yogita Rana ordered her subordinate to do the needful, send me the information via email. Last checked (early May'07), when contacted on phone, she continues to order her subordinate, who continues to dodge sending the 'soft-copy' version of the same, which they proudly claimed the department had.
5. A Telugu version of this article appeared in the Women's features page of Prajashakti Telugu daily; and the English version appeared in www.indiatogether.org
6. Unless otherwise mentioned, all the villages I write about fall in the submergence zone of the Polavaram project/dam.
7. *Madras Gazetteer, Godavari Districts,* 1915, p. 265. Incidentally, I came across several legends associated with Rama particularly in the tribal belt in Khammam district and met many named after Rama and Sita. For another time, I thought, it would be a fascinating journey to delve deep into the origins of these stories

in tribal areas, as against their own non-iconic representations of the divine. When did the Rama legend set base here? Was it with the coming of the upper caste landlords from the 'Plain areas'? Perhaps.

8. Babji, of APVVU had accompanied me here.
9. The clarity with which I wrote this later, was not available to my mind at the village on that day, for each one had some similar story to tell all of which I religiously noted down and recorded. I read more on the subject and later understood the situation better. When I had returned home, it gradually made sense. But I present it here, in this section, in the truth of my limited understanding of 4th February 2007.
10. Used as a generic term for jeeps.
11. At this juncture, taking over from Babji, who had left by then.
12. And this is based on past Godavari flood levels.
13. Incidentally the forms in Kukunuru (unlike in the Bhadrachalam division forms) read "Indira Sagar Project Affected familes agreement to give away land to government for the project" (in Telugu)
14. A Telugu translation of this appeared in the Andhra Jyothy daily. I had gone to the office accompanied by one of the CPI (M) party workers of Bhadrachalam
15. A version of the translated version of this article appeared in the Edit Page of Andhra Jyothy Telugu daily.
16. I mention this particular organisation, one because it has a direct stake in opposing the dam and two, in the relevance of this particular story to the politics of funding in a deeply political issue. I was trying to understand how subtle is the metamorphosis of organisations in such acute contexts - becoming part of the game they initially seek to change (if at all they did). I informed Gandhi Babu that I would publish this interaction with him.The day Godavari waters covered VR Puram mandal (a day after the 'floods') Gandhi babu had made a phone call to me asking if I could reach any member of this international donor agency or other groups for support. I sent an email and received the response that they were 'assessing the situation' and the moment the State government gives the green flag, they would be in a position to support with relief measures. Another funding organisation (also working with ASDS on a few projects) decided to send a team assessing the damage (a week after the floods). Two of them managed to reach the affected areas. It may be conjectured that the NGO taking a strong stance against a government project changes the dynamics of funding and donor-recipient paradigm of international agencies and local NGOs. Considering recent state interventions in cancelling FCRA of NGOs involved in any form of protest, it seems to be exactly the case.
17. As on date (of this journey). Though the very confrontational anti-state stance changed by later years and this organisation too has expanded its base, and has itself entered into negotiation for 'adequate compensation' and R&R. But it supports activists and others, including researchers, who seek its base. It has been active in the Telangana statehood movement, as well. It is also involved in administering government programmes in Gottikoya villages of displaced tribal people from Chhattisgarh settled in Khammam district.
18. The Hindu dated 3rd April 2007.
19. Kamma, Kapu, Reddy, Raju, Chowdhary are titles of erstwhile landed aristocratic groups some of whom also happened to be rulers of smaller

principalities and kingdoms in the South. In the context of Andhra, Rayalaseema and Telangana regions, these castes have continued to be the most dominant and most wealthy caste groups. They have also been groups with interests in entrepreneurship and industry, besides larger landholdings.

20. I thank the M.S. Swaminathan Research Foundation (MSSRF hereafter) Kakinada, Dr. Ramaswamy Ramasubramanian and his team - Swarnaraj, Prasad, Praveena and Nagaraju - at the office for guiding me through part of Coringa and other mangrove forests of the region, with their expert inputs, when sought. And for sharing their material on mangroves, not all of which I could utilise here. I specially thank them for allowing me to talk to the people at my pace, in my way all through, and not prompting questions nor answers. My contacting MSSRF was by itself no less providential. It was an indirect benefit of having worked shortly at C-DAC Hyderabad, where one of my colleagues had incidentally worked with MSSRF! I thank him, Srinivas, for his insights on the mangroves and connecting me to his former colleagues.
21. T. Ravishankar, R. Ramasubramanian, D. Sridhar, N. Srinivas Rao, M. Maqbool and D. Ramakrishna, "Community Participation in Joint Mangrove Forest Restoration and Management", in S.K. Patnaik, H.N. Thatoi, eds., *Mangrove Conservation and Restoration: Proceeds of the National Workshop on Mangrove Conservation,* Bhubaneswar, 2001, p. 110.
22. H.M. Kasim, "Mangrove Ecosystem and its Relation to Fisheries", Discussion Papers, National Workshop on Restoration and Conservation of Mangroves through Participaory Mangrove Management, Rajahmundry, 12th February 2002. p. 47.
23. *Madras Gazetteer, Godavari District*, pp. 210-212.
24. *Joint Mangrove Management in Tamilnadu*, MSSRF, Chennai, undated. Courtesy, MSSRF office, Kakinada.
25. T. Ravishankar, et al, op.cit, p. 110.
26. Dahdouh-Guebas, S Collin, T Ravishankar, et.al, Analysing ethnobotanical and fishery-related importance of mangroves of the East-Godavari Delta (Andhra Pradesh, India) for conservation and management purposes, *Journal of Ethnobiology and Ethnomedicine* 2006, 2:24, Published: 08 May 2006, http://www.ethnobiomed.com/content/2/1/24 (Open Access). Authors request page numbers (within article), not to be cited.
27. However, in the AP Government's report on the project (by I&CAD office, Hyderabad) the extent is shown as 112.31 ha.
28. Emphasis mine.
29. Dahdouh-Guebas, et.al, op.cit.
30. T. Ravishankar, et.al, op.cit, p. 110*ff.*
31. Some obligations you cannot refuse. But thanks to this meeting, I would hear one of the more philosophical statements of my journeys.
32. *Selections from the Proceedings of the Godavery District Association in 1912 and 1913*, From the India Office Records Collections, British Library, London. Though a confession needs to be made here, namely, I have jumped two years ahead in pointing to this reference, which I discovered only in 2008. But the context here seems far too appropriate to ignore.
33. The 'traditional medicine'/flora and fauna 'specialists' of ancient India famed for the texts, *Caraka Samhita* and *Susruta Samhita*.

34. An edited version of this experience found a very nice space in The Hindu, Sunday Magazine with a more poetic title, "Tale of Two Rivers", dated Sunday, 24th June 2007. Pity the use of an unconnected picture taken by someone else, of Godavari, and not mine!
35. The other Raju guide I knew existed in writer R.K. Narayan's famed novel, *The Guide*.
36. Which I discuss later on in this book.
37. Literally, 'to see' the deity in the shrine or temple. Money is charged depending on how 'close' physically you wish to get to the deity in the sanctum sanctorum.
38. Edited version of this was published as a short photo feature in Prajasakti, the Telugu daily, last week of May 2007. I needed to place her name, so named her Vennela (since it means moonlight).
39. Ms. Uma (Lakshmi) and later Ms. Sharavani (Prajashakti Publications, Hyderabad) read out for me Telugu petitions. Translations mine. I thank both of them.
40. Assigned land.
41. Records of the land administration: the Register showing the area, classification value and assessment of a holding with the name of the land-holder.
42. Questioning the officials, that is.
43. *Madras District Gazetteer: Godavari Districts*, 1915, p. 195
44. A telugu version of this appeared in the edit pages of Telugu daily Andhra Jyothy and an edited English version appeared in the *EPW*, "Polavaram: Rehabilitation via Forced Consensus", June 23,2007, pp. 2385-2387.
45. Karna was born to Kunti – mother of the Pandavas in Mahabharata – and the Sun god; she left the baby (in a basket) on the river to avoid the stigma of being an unwed mother. She realises his truth only by the time of the great battle of Kurukshetra, when she comes face to face with him.
46. Gazetteer, Godavari Districts, 1915, p. 47.
47. About Sitapally, I was to realise much, much later, that the Godavari District Association in a letter of 1913, mentions, "a proposal known as the Sitapalli project to divert certain Agency streams towards the Rajahmundry Taluk by means of dams and by forming a reservoir which it was said would irrigate 70,000 acres. The Deputation requested the appointment of a special office to investigate the proposal adding that the local Overseers were not willing to make a careful examination of it owing to the locality being malarious." Talk of the greed of water; thank the mosquitoes!
48. I was writing this story after my visit and, as usual, made a list of places that might print it. But finally it remained in my notebook. This event of my meeting the Gottikoyas remained a sad memory. However, not so sad, since their story did not remain hidden to the larger world (which was more important than *who* wrote). Local Telugu media had covered this problem. Funnily enough, I went to Sunnamatka and surrounding villages much later, in 2012, for an assignment solely about the Muriya/Gottikoyas and displacement. I met one of the men who I had first seen in 2007! Their numbers have increased and the conflict with local population continues. Some – the poorest among them – live a wretched live on tenterhooks, clearing small forest patches to stay since they do not wish to return to Chhattisgarh and their homes burnt down by the Forest Department often. Some others have learnt the ropes and become small

time contractors in the process.

49. State Agricultural Profile, Commissionerate of Agriculture, 1999.
50. Telugu word for meetings.
51. Again the pattas in name of non-tribals has been contentious in these villages where land was purchased from them, which is sought to be explained through the convoluted amendments, etc. of the Land Transfer regulations in Scheduled areas (under 1/70), which again, have been disputed. Many of the cases were still in dispute in the courts when they were purchased by the RDO, one of whom we have come across in this journey.
52. I have mentioned these villages as well. Here the case of uprooting one group of tribal communities to settle another was going on, creating its own set of contradictions and conflicts.
53. Published on the official website of AP Government Irrigation and Command Area Development Authority Source:http://ppms.cgg.gov.in/DetYearProjLA (Centre for Good Governance). Website open to public access without subscription.

January-February 2007

A CPM Protest Meeting at Bhadrachalam Town

Dalits of Polipaka SC Colony

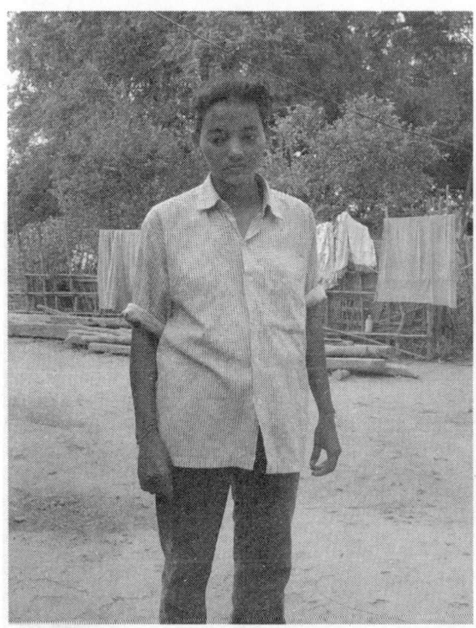

3rd February-Seetamahalakshmi of Sriramagiri

Sriramagiri Rama Temple

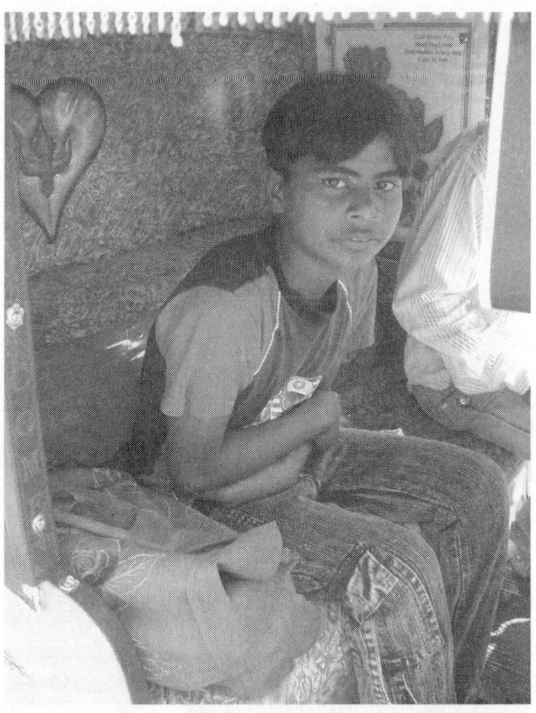

This Premchand Loves English Literature

Dalits of Rudramakota

Board showing R&R Colony-Rudramakota

R&R Colony under construction, Rudramakota

A Surreal Sight

Chitturi Subba Rao collects Lotus leaves

5th February, at Palavancha MRO office, upper-caste landlord, with the dalits to sell lands

In Search of Other Landscapes: The Mangroves - April'07

A section of the Mangroves (Coringa), birds flying away

Fishermen on the creek

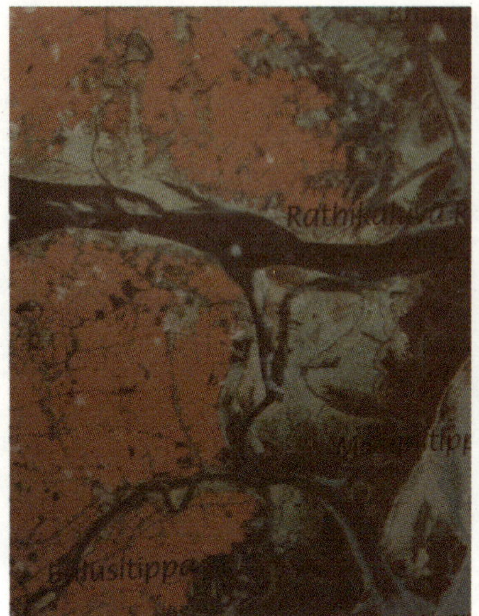

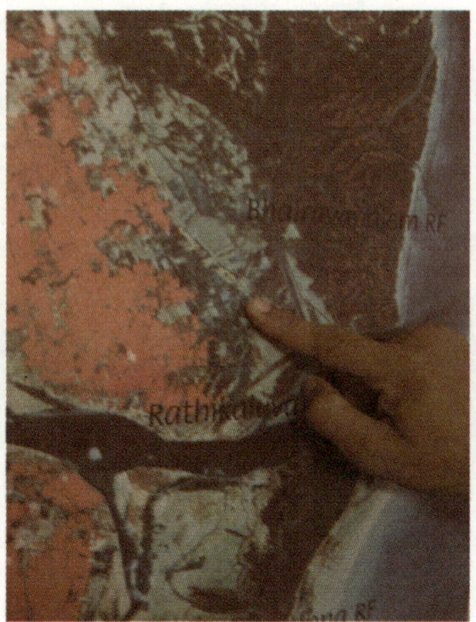

Satellite Imagery of the Villages and the Reserved Forests of Coringa (pic Courtesy: MSSRF, Kakinada office)

For Whom Godavari Comes

Mallada Narasimhamurthy and Family

Palipu Sesha Rao and family at Kandikuppa, by the mangroves

Inside Coringa Wildlife Sanctuary – fed by fresh and saline waters

Fish being sold at Kobbarichettupeta (Kakinada district)

Dalit lands fed by Godavari (I.Polavaram, Bhairavapalem)

Aqua farms of the rich landlords, destructive exploitation

Tourist Gaze (April'07)

The Punnami Launch at Polavaram

Bhavya dancing inside

Readying for the second round, wearing her anklets

Continuous Entertainment inside the boat

Tribal villages that are passed by

Enroute Gandipochamma temple, the Padmashali women (setting wafers of sago flour for sun-drying)

Children with a tutor at a home near Gandipochamma temple

That Faraway Look: Vennela at Perantapally

Encounters with non-Submergence Zone Tribal Land Alienation

Kosi Ramalakshmi

Pamula Veeraswamy

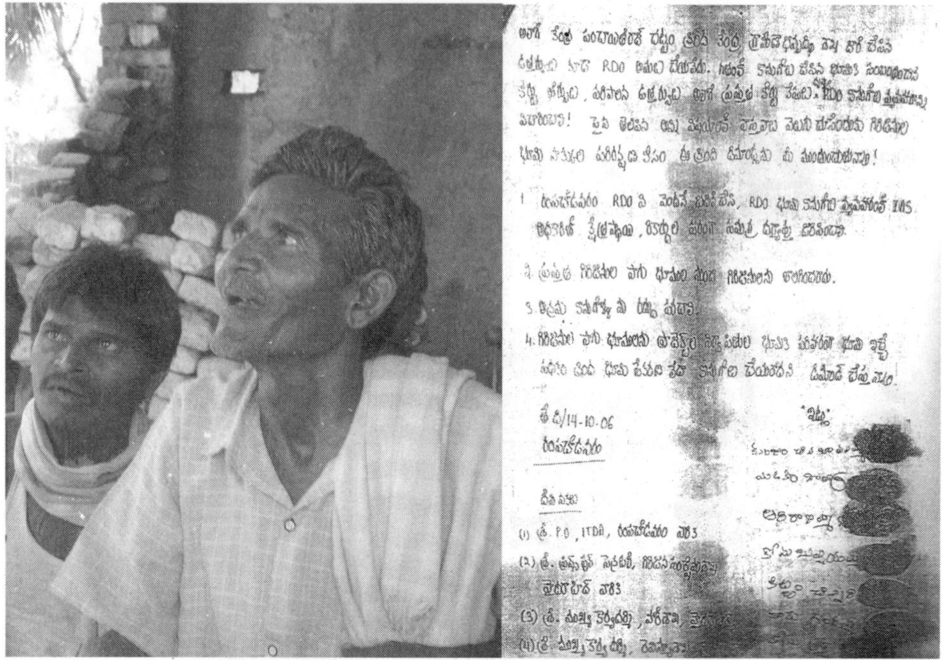

Ekka Rajanna Dora

One of the Petitions with thumb impressions

R&R Colony on disputed tribal land, Pedabhimpally

At Site of IPC Charge 153 (A)-Ramalayam,Chegondapally

Kuncha Varalakshmi at Pydipaka

Strange livelihoods, the log dalits float on to collect logs washed away by Godavari

A calm Godavari; where Gontulamma is immersed

Where I heard the dalit Mahabharata, Gontulamma story

A Different Start - October'07

Through Maredumilli forests

Gottikoya Children and Women at Sunnamatka

Gottikoya men at Sunnamatka

Godavari Journey 2008[1]
Some Grand Designs of History and Linearity: Tribal Assertions, 'RR' Games, the Colonial Route and Other Accounts

The mainstream images in films and in politics reflect the two major 'themes' of the nature of discourse on River Godavari. On the fringes were the tribal communities of Godavari who had understood that the essence of freedom was to retain hold over land and forests. Post-Independence and formation of States, the hegemony of landed classes, with direct links to water, land and sea continued. In tribal areas, the hold of non-tribal landlords also continued unabated. Godavari seems to have connected these interests (landlords, industrialists, politicians, and at one end, also the film industry) and also pitched the marginalised (with no say on Godavari) against the rest. Amazing, it is only the rich and the landed (and, of course, sections in bureaucracy; it can be counter-checked) who have been speaking of a river wasting into sea, while for others, the river-to-sea idea is one long, unbroken journey and so long as it fulfills your basic needs, its seaward movement didn't cause high levels of resentment.

About the 2008 Journey[2]

I couldn't pull a stop over with Godavari once all the supporting resources of 2007 had ended[3]. Are there ways – to continue the trips – that do not bind one to conditionalities? Then, an advertisement came with a sign of hope: the NFI media fellowship[4]. It ultimately came through in the month of March[5], with enough freedom (sans conditions/agendas, except to publish '10 articles') to travel back to the 'source'. It was great to continue my journey as a freelance journalist, considering the nature of work and circumstances of the work, as was becoming clearer to me by the day. This journey was difficult to connect in a linear sense – between one visit and the other. Many things about it were disjointed; personally, too; it was like an obstacle race, at some levels. There were more silences in this journey than in the earlier ones, within and outside, as things on the ground seemed to have moved on to another plane altogether. Of all my journeys, the years 2008-2009 were the toughest, and threw far more challenges of every form and size in my path than I could ever have imagined. After a point, Godavari became the singular meaningful point of anchor in those years, consuming and life-giving, at the same time. And, outside of oneself, there were people who became part of one's own extended relationships never revolving around how much of their words I wrote made it to print. This was a journey on Godavari as well as outside of it. More than the number of visits, I got insights into crucial developments around the dam. I began to observe Rajahmundry, too, at close quarters and was led to some unconnected areas, such as a 19th century library. This time round, I did not always stay at the Rajahmundry HPO, but also tried out a different vantage point altogether, helping me understand the other impacts of the dam-building on Rajahmundry town/city. There was no mad rush to report every new development on the ground. The 'stories' from the field significantly made me question aspects of democracy; its contradictions in terms of certain guarantees to the marginalised turned upside down; history and idea of a past remembered by some of the people one met. Godavari this time was relatively silent as a phrase and metaphor, but in the background a strong presence and the reason for aspects to be pondered over; and crossed over, several times. The year 2008 was strange: earlier that year, the roads led me to the point where Godavari enters the State, Andhra Pradesh (in the Telangana region)[6], and by the end of the year, to a completely different landscape and mindscape altogether, where too, river Godavari had been discussed often in the 19th-early 20th century. I would seek Godavari[7], but through a crucial Godavari-colonial connect, outside India.

January 2008[8]

10th January'08 - The Ritualistic Godavari, Adilabad

Just as many of the natural phenomena having close links to human civilisation (in the sense of being crucial to it), Godavari too, finds herself worshipped in a ritualistic sense at certain points of her journey. At Rajahmundry I had seen glimpses of it, where people visit the Godavari banks and offer prayers to the river and at the temple by the river, as a 'must-do' ritual without which the visit to Rajahmundry (at least for outsiders) is incomplete. As for people in Rajahmundry itself, many of them begin their days here, starting with worship offerings to Godavari. Importantly, I observed that the people who do not offer a ritual worship of this kind are the fishermen, whose days actually begin with Godavari at 3 am and end with her at dusk; their worship is of a different kind, compared to the brahminical. It is devoid of riff-raff; no coconuts broken before starting on the *nava* with the fishing nets; just a prayerful mind which understands when to leave the river alone, to the extent possible (extent in an economic sense, where necessity can take over at times, but almost always does not) – before throwing the net the first time that day into Godavari's still, dawn-time/tide waters. I am in Adilabad district, at the Basara Saraswati temple and happen to see the ritual on Godavari. People, mainly, and mostly, from Maharashtra are offering prayers to the river after worshipping at the temple of Bhairava. Earlier, on the road towards Basara (with an old shrine dedicated to Saraswati, said to be one of the very few), passing through Ramayampet, I stopped for tea and breakfast at the New Kaveri (it is owned by an Udupi[9] person, hence the name) Hotel! People here hail Godavari as *Gangamma talli* (mother Ganga). The hotel owner tells me that not so very long ago, 75 per cent of pilgrims here were people from Marathwada (from around Nanded, and villages bordering Adilabad and Medak districts) and barely 25 per cent from AP; today it is 80 per cent from AP and the rest from Marathwada! There is increased commercialisation of the place. Others at the hotel, who join in, say the farmers from Marathwada, most of them poor, can no longer afford to visit this place often, as everything, including the worship in the temple precincts, is expensive. While they do come, most of them travel by passenger trains, bring their own food packed from home for the journey (chapatis, usually) and offerings, too, take a dip in Gangamma (Godavari) and leave. They do not even light the lamp (on a leaf basket with flowers) which people from AP do. With farming hit hard, it has also affected this pilgrimage of the poorer farmers from Maharashtra. Later, I noticed myself that when I came here four years ago, this was a small and quieter place and a quieter temple, too. There were no serpentine barricades (in the style of Tirupati temple) and one could go up to the inner recesses as well as the sanctum sanctorum of the goddess (Saraswati) directly to offer prayers. It has radically changed. Probably this change started with the whole brouhaha over bringing IIT to Basar (there was an entire discourse on it in the years 2001-03 in Hyderabad among some circles where it was seen as a 'prestige' issue[10])? Also, the nature of the temple itself as a small, quiet shrine as it was then, has altered. The goddess, of course, looks beautiful in turmeric yellow with a smiling face, and large eyes. Incidentally, I hear the place is being readied for the visit of the Chief Minister (Rajasekhara Reddy) in two days from now; which means there will be longer barricades and it will take as much longer for poorer farmers to meet their goddess, and from a longer distance, to be able to barely spot her eyes, and pushed aside (like they do at Tirumala Tirupati temple). But people do not complain. Their faith is humbling.

11th January

I watch the Godavari banks buzzing with activity. Families here leave earthen lamps on palmyra (?) leaves to float, along with offerings of coconut, flowers, etc. And bathe in the river. This is the Telugu side of it. The Maharashtra farmers offer hay, wheat balls made into lamps, flowers and balls made of gram flour. Their offerings are essentially prayers for a good harvest. This tradition seems to exist outside of, and irrespective of, the Saraswati temple – the Saraswati temple is not the reason for them to offer worship at Godavari river. This distinction is important. For, not all of those who come here, I noted, visit the Saraswati temple, (at some distance from the ghat) as well; sometimes they just make these offerings, and leave. Many, especially the Maharashtra agricultural labourers /workers, and small farmers, have come in hired jeeps, and are returning to their homes after the ritual offerings to the river. But at some point, both these traditions seem to have converged – offering worship here and going to the Saraswati temple. The temple is not situated at the banks of the river. In fact there is a different temple right on the ghats, of Bhairava, I am told, by the priest within, an old rock-cut shrine again; where it is mandatory to offer worship before sending the offerings down the river on a leaf-boat.

And then...Who would have ever thought, so soon? Just three days ago I was wishing aloud to a friend that I might see the spot where Godavari enters the (political) State of AP, leaving (political) Maharashtra. And the point where she rises (Triambak, near Nasik). I am yet to see its origin, but today I am at the point of her entry to AP! Absolutely unplanned and sudden; happened over a cup of tea, literally! I asked someone about the place where Godavari enters Adilabad. They said I might as well go there (not too far). So I reached Kandukurti (Navipet mandal, Adilabad district, Dharmabad, on AP-Maharashtra border). It is called Triveni sangam in these parts; someone said Godavari, Manjira and Indravati (?) meet here. Apparently there is a temple at the meeting point, where you can only enter at the onset of summer, sometime in March, when the river is lying low.

So, wish fulfilled, as I stood on the bridge where, as a form, the river has entered a different pace and course of life.

***

June 2008

12th June - Rajahmundry

This time, I could not find rooms in the decent lodges in Rajahmundry; the HPO room wasn't free either but they suggested the Telecom Sanchar Bhavan office guest house (at Kambalacheruvu) which is at a great distance from Kotipally bus stand and railway station. Got the room – "Suite I" (what is it with Suites)! But there were problems – in an otherwise decent room - with the bathroom literally falling apart from all sides. However, thanks to this Telecom office I chanced upon one of the oldest libraries in India, the Gautami Grandhalayam, remarkably close by. After the *upma-pesarattu* breakfast (a delightful, quintessentially Rajahmundry breakfast), I decided to pay a visit. Started as a private library by Nalam Krishna Rao in 1898, it was taken over by the State government in 1979. Prasada Rao, the Assistant Librarian (since 1971!) told me they have a stock of 45,000 old books and 75,000 updated collections, mainly of English and Telugu literature (the classics). The present structure is new but the complex has been here since 1898, including the tree with the older building right behind the new one. I learnt from Prasada Rao's friend that the Vishakhapatnam Collectorate has a stock of "22 volumes" on Alluri Sitaramaraju. I have not confirmed the information so far. He also told me of a new library by Kotipally which has a park of statues of women freedom fighters from this region. I must return to this library. Just take in a bit of the place. Let us see when. Wonder why I feel this particular journey is taking me to history on its own, or there are people referring back to history.

13th June

I was back this morning, at the HPO room after paying Rs. 150 for the Telecom office 'suite'. I started off soon after a much-needed bath to Polavaram and beyond. This time, I had to hire a cab. It was a good decision, as I later found out; though costs have risen (with too many officials too hiring local cabs to visit Polavaram), especially when one is on a shoe-string budget. Yet, compared to Hyderabad and other cities, Rs. 500, or Rs. 700 (then) didn't seem so bad, especially on some of those roads (up to Sivagiri). I called Saikrishna (the Prajashakti newspaper agent) at Polavaram. He knows a bit about the Polavaram project 'whodunit'. But most times, I do not even see him as he seems rather busy with his Party work and his household. Incidentally, his wife too is a member of the AIDWA, the women's wing of CPM.

Gunjavaram R&R Colony, Pragadapally Panchayat, Polavaram Mandal (West Godavari)

The cab would take us (Saikrishna and me) to the R&R Colony at Gunjavaram and later to Sivagiri, from where we would cross Godavari to reach Kondamodalu and return to the cab, and get back to Polavaram. I see the colony under construction and a few workers on the site. A young boy is supervising work. He is Boragam Buchiraju, perhaps in his 20s, an Engineering college drop-out (had to give up his studies owing to the financial situation and this new 'responsibility') building his house in this colony. He says, "We had 12 acres land. Mamidigondi panchayat has been given land and house sites of 3 cents each. For house sites, they have given us Rs. 80,000 in the first installment and will pay the remaining Rs. 40,000 later...We have to pay daily wages of Rs. 80 per person per day and Rs. 150 for *maistry* (mason). The money they have given us is not enough to build the house. They will evaluate our original houses and give compensation for them. We did not wish to come here; we had to. For the landed, they gave 200 acres. We have not yet started the work on those fields. We cannot do that before building the house. But in our own fields they have dumped soil and debris. We have just about laid the foundation and a base wall for the house; once it starts raining; we will begin work on the fields (R&R)."

Saikrishna says, "Iron cost Rs. 50,000 to 60,000 per tonne at least four months ago. Now it may be much more. Even Rs. 1,50,000 won't suffice to build these houses."

As for expenditure on an agricultural field, "For an acre you need to invest at least Rs. 15,000 for seeds, fertilizers, etc; labour costs in Gunjavaram are going up to Rs. 150 per person per day...Gunjavaram lands (bought for R&R colony and fields) were owned by non-tribals and there were a few tribal lands, as well. The RDO says the LTR cases (Land Transfer Regulation) have been cleared and the non-tribals have records to prove their ownership on these lands which the tribals do not even know of, if they were cultivating these lands."

We reached Sivagiri. The cab driver was to wait for us here. He was quite a character, though, and started fuming and fretting about "these roads", and at a point it was very irritating. I wished I had taken the bus. Many of the cabs from Rajahmundry regularly ply between the town and Polavaram, bringing visitors, officials, etc. But they never get into the interior villages. The road to Sivagiri was indeed challenging at a few points. Now he looked at us even more annoyed when we asked him to wait!

I observed that the Koya people were building their 'individual' homes, on their own, each with the compensation money they had received from the state. The labour was hired (the ones labouring on these homes were the SCs). Something was very different from what I had seen earlier, in all the three districts in tribal hamlets. I had observed neighbours pitching in with the building of a home; especially the roof, or bringing in material. It seemed to be a community building exercise. Here, the community was missing. So, even if they were a cohesive unit for R&R purposes, the act of building homes, and the aspects of social life, which made for that cohesiveness, is lost. May be it was lost even as their lands were regularised as individually owned land?

Incidentally, the Haimendörfs had written years ago about building of houses in a Kondareddi village. "All cooperate in the work of building the houses, and as a rule, though not necessarily, the village founder's house is erected first. When it is complete, the first fire is kindled in his hearth by friction of two bamboos; this rite is in many villages repeated once a year, when the pujari, who is generally a descendant of the village founder, clears out his hearth and rekindles the fire by means of a fire-saw. When building a new house, and even when rebuilding it in later years on the old site, Reddis put a little uncooked millet, usually sama, into the hole which is to take the central post, so that the house shall stand firm."[11]

***

At Kondamodalu, Madi Mutyem of Kokkarigudem, (Kondamodalu Panchayat, East Godavari) and History of Resistance

Madi Muttem (of Madi Mutyemma, officially), the young member of the MPTC (Mandal Parishad Territorial Constituency) of Kondamodalu Panchayat has come over to meet me this time at Kumari's, though I wish I had gone to Kokkarigudem to meet her, instead. But word got around here, even before I arrived. She speaks with a slow, sing-song intonation, typically Kondareddi. She represents, and relives, parts of the resistance history of the place (and strangely enough, I had not asked her about history; she just decided to speak of it). She recounts what she has heard from her parents, of the times when non-tribals from the plain areas entered into their region afresh post-1960s. "Non-tribals came here to set up small businesses. They were from Tuni and Samalkota. They set up shops selling oil, salt, general provisions, etc, bringing all those things initially as itinerant merchants. They took over lands in return for loans they cleared for our people. They took the lands to give them Rs. 20. Nagayya, *pedda sahukar*, Sivaya Patrudu, etc, called tribal people over, fed them, and overnight, on the sly, took possession of over 15 acres! In these kinds of surreptitious ways they took over tribal lands and became land owners. Nagayya (his family is now in Mettugudem) has a patta and he may have sold those records, we do not know. We struggled (to get the lands back) – people like Kunjam Rajulanna (Peddagudem) and Madi Lakshmayya (Kokkarigudem, his mother is alive, alone) died in police firing in 1969. The Collector then went to our *gudem* (hamlet) and announced, after all this, that these are Koya and Kondareddi lands. Our naïve ancestors had given away lands which we have (Agency Girijan Sangham) recovered through a tenuous struggle. It was based on one of our demands to the ITDA P.O. then, following the firing which killed our people, they lifted the police station from here. We used to say, '*tella cheera erupaavotthi.*" I asked her what did that mean, 'white saree' and the colour 'red'? She explained, "unless our blood falls on this soil, we won't get our lands back. Two people died, didn't they? That is what it means." [It means a white saree must be stained with the revolutionary blood.]

It was the earliest Kondareddis of the region who first broadcast seeds and grew maize, pulses, and later, paddy (of the 'wild' variety), entirely rain-fed. To supplement this basic need-based (in what can be today called 'eco-friendly') agriculture, they had the forests. But life was always about negotiating, seeking concessions for, or when it came to it, fighting for their place on the earth. Forest policies in the colonial period and thereafter and the question of land grab by non-tribals were some of the reasons why this place lent itself to resistance movements and radical Left politics. The original inspiration for this peasant agitation of 1969 (which Muttem speaks of) came from the Telangana Armed Struggle led by the united CPI, the last phase of which movement saw some of the cadres move base to the Godavari forest region, especially East and West Godavari, in late 1960s. The region around Srikakulam became a mass support base for the naxal movement. But these parts of East and West Godavari developed a separate line of thinking (especially among the Kondareddis, and a few Koyas) different from the hardliners who had led the Srikakulam movement. In course of time, the leaders (initially from the non-tribal leaders who were earlier part of CPI and then fell out with it, under Tarimella Nagi Reddy) built a tribal mass support base, which gradually gave way to leaders emerging from among the young Kondareddis and Koyas of the region. In course of time the Agency Girijan Sangham and Rythu Coolie Sangham were formed. But I have observed closely that the tribal leaders here largely call the shots, especially in the context of Polavaram rather than meekly follow the command of non-tribal members of the Rythu Coolie Sangham. The sense of politics, egalitarian ethos of their own cultural contexts, celebrating its tribal festivals, observing the rules of the community (respecting the elders' decisions, for instance on matters important to the tribe), equality of the sexes (pronounced in the case of Kondareddis here) cohesiveness, etc, seem to blend with ideas of how to educate their children, etc. It is fascinating to find Karl Marx sometimes in some of their homes. What is important in a larger historical, political sense is that conditions of the tribal peasantry gave rise to distinct political contexts. Today, Kondamodalu is one the last panchayats that is strongly opposed to the present form of the Polavaram dam and R&R package.

Muttemma says, "They have been conducting surveys for long. They will build barrages in five or ten years. We will be asked to leave. For those ten years, we will not be allowed to touch our own lands which will be enjoyed by the government. They are giving lands on lease elswhere. In Koyda the government gave Rs. 1,30,000 to the adivasis, who have spent all that money. What will they live on when they go to those (R&R) colonies (not fields) without any money? We do not want to go if the barrage will come ten years from now! We are sending away the officials who come for surveys...The government is now asking us to show pattas for the lands that we reclaimed in 1969. Non-tribals have pattas for those lands. Since no survey was done here the non-tribals have not taken the compensation. Wherever Agency Girjan Sangham is strong, nobody has taken compensation. The government is not ready to hand over the lands reclaimed by the Sangham (in spite of several petitions in the Courts) even though the tribal people originally owned these lands. We cannot survive in plain areas where everything has to be bought: fruits, vegetables, tamarind, etc. Here we get all these for free, except salt and few things that we buy at our *santa*s (weekly fairs)...Nobody is recognising the rights of tribal people...One person called Mangaram, a Kondareddi, went and saw the R&R lands. She took thirty people with her. But as yet none of them has taken compensation..."

Speaking of other developments in the Panchayat, she said "Remember that road[12]? They have laid the road till Koyyalagudem as of now. The ITDA PO came here and said 'the road is okay; if there are 'floods' this time they will send us provisions by this road'. Our own people built the road. We haven't seen the road yet! But we hear it is good!"

Saikrishna started reminding me it was late; we have to get the Sivagiri boat, lest we get some more barbs from the cab driver all the way till Polavaram!

14th June - Diary Notes: "Again, boatman Punnam Mutyam and his smile and his boat and his blessings! Again, Kondamodalu, which envelops me. Again, Madi Muttem of Kokkarigudem and again, Kumari and the quiet courtyard! Mutyam says he belongs to the Vadda Balji community (Balijas), the Dasipoyina Vada balji ("SC", he says). The man who saw Mahatma Gandhi during the freedom struggle (a story I never got time to hear). He says, he started from Rs. 1.50 (passenger fare) long back and has been making making Rs. 550 (for a month) from his ferrying. "I am happy, though today even 5,000 rupees do not suffice!"

The Last Crop and Life, in Between[13] (Mamidigondi, Mamidigondi Panchayat, Polavaram Mandal, West Godavari)

The day Boragam Rama Rao saw the fresh stocks of corn crop crushed by large excavators and earth movers parked on his 2 acre (0.8 ha) land, he knew he had made that transition - from tribal farmer to tribal 'beneficiary'. At least, until such time as he starts reconstructing his life all over again: with Rs. 1,20,000 and a piece of land as yet uncultivable. The Koyas of Mamidigondi present just one instance of a larger problem for the tribals to be displaced, being in that '*trisanku*'[14] state – between the promise of a heaven above and dispossessed of a piece of their land below. Rama Rao lost his land to the ongoing saddle dam work of the dam. "They placed orders of arrest against our people (on charges of 'criminal trespass on government land') even when we went to harvest the last crop. At least they could have allowed us to enjoy that last yield from our land. Earlier they had assured us until the *barragi* (dam) was built they would let us cultivate on our fields here." Rama Rao bears the unfortunate burden of representing a tribal panchayat in this mandal which was among the first to 'accept' the 'RR package'. Mamidigondi's is just a case in point exemplifying the irony of the R&R package hurriedly dispensed with. Lush fields of rain-fed crops cultivated by tribal people have turned into vast stretches of potholes (water logged from the initial monsoon showers) and dumpyards of rock and stones. Thousands of acres of prime agricultural land in Pydipaka, Ramayappeta, Devaragundi and Thotagundi panchayats were rendered useless for cultivation within months of disbursal of R&R cheques. Lands once owned and tilled by farmers are now on leased out to private contractors engaged in the Polavaram dam construction. When officials in charge of land acquisition (including the Revenue officials of the mandal as well as the Collector of the West Godavari district at that time) made several rounds to these panchayats, their constant reassurance to tribal land owners that since the dam would take at least 5 to 10 years to be completed they would not even be evacuated unless the waters came in. In the meantime, they could continue to cultivate their lands and also accept the cash and land compensation. I had been introduced to Boragam Rama Rao as the Sarpanch of Mamidigondi at the MRO office in Polavaram sometime last year. Today I am meeting him again, by his home, in a different time. I meet him now, accompanied by Saikrishna. He says, "It is very difficult to cultivate those R&R fields. We have to spend a daily bus fare of Rs. 20 per day. We have not started cultivating. We had already lost our crops by November last when they gave us the compensation. The lands there (in Gunjavaram) need to be prepared. By then we lost a lot. We have lost all our agricultural income. We used to cultivate paddy and corn. We used to get 40 bags paddy per acre; and for the second crop, 30 to 40 bags of corn. Everything is gone. They have not given any compensation for crop loss. My entire field has been dug up. We (the village) lost 200 acres under the project; entirely cultivated land. We did not even need too much water, it was entirely rainfed and we got excellent yield here. They said they will not compensate crop loss, since they gave us lands."

Boragam Rama Rao laughs as he says, "Gunjavaram RR land was bought by the government from non-tribals, especially from Rajampalem. Each one of those had 55 acres and 100 acres of land each in the Agency Area! Look how 'just' that is – the non-tribals of Gunjavaram earned Rs. 1,85,000 per acre from the government for selling them for the R&R colony! Here they had not even conducted gram sabha about the project at the time of clearance. We questioned the government officials then. They said there were orders from above. They told us we will be moved from here and so they will be compensating us. We had not passed a resolution in favour of the project. They just told us we have to leave. It's going to be hard, starting from the scratch. We have to spend Rs. 10,000 there, in stead of Rs. 5,000 here. The field here absorbs water very well (*baaga gunjukuntundi*). The soil is so rich. There you will have to use (pump) water every three days. The soil is drier. And then there will be labour costs (if we let our lands on lease) – for a year, about Rs. 10,000 or 7,500. Here it used to be Rs. 3,000. Those working on our lands on lease are crying and have lost everything. The SCs are losing all they had. They have got only work days wages for one year and nothing else to bank on."

I couldn't help asking him what happened to their earlier resolve (when I had met him previously) to refuse consent to the project. But he went on with other issues – "When one of us took the compensation, and we were told if you do not take it you will lose all, we had to take it. What could we have done? People from Devaragondi have been pushed to the plain area (non-Scheduled) and lost all their rights as tribal people, since the R&R colony is in non-Agency area (in Polavaram). When they moved there, they were unaware of this. Today they are in great trouble having lost all their rights which they enjoyed in an Agency (Scheduled) Area…Landless tribals got Rs. 1,90,000 but no lands. We asked them what about landless among us, the officials said it was the government's orders; if they wish, they can take loans from the banks but they will not be given lands. When we demanded work under NREG (National Rural Employment Guarantee Programme) they told us since they are giving us the R&R compensation, they will release some amount from that package even for NREG! We have to manage with Rs. 1,20,000 – building our homes, preparing the fields, starting agriculture, and generally re-starting our lives. They promised they will give compensation for the houses here and the toddy trees – Rs. 1,500 per tree – as well. We cannot be sure until they do. They only evaluated coconut and toddy trees and not the others. They will also have to give us money to shift, which is part of the package. Until then, we remain here."

As we walk towards Rama Rao's field-turned dumpyard, Rama Rao continued (talking like his thoughts were gushing out from dam gates just opened up[15]) – "Now they have this new *adavi chattam* (Forest Rights Act) to give us rights over forest and identify *konda podu* within the forest. They have set up committees for Devaragondi and Mamidigondi, but according to the MRO, Sarpanch should not be part of the committee…" And he cracked the joke about the duality of the forest department, saying, "Now the forest department is telling us we have no rights over the trees we grew and the forests we conserved. They tell us since those trees were planted under Joint Forest Management, we have no right over them and since we are anyways leaving now to the R&R colonies, we cannot touch those trees anymore. They have been logging them off and selling timber. Anyway, we are hoping the Forest Rights Act will bring them in line."

Back in his village, Boragam Pandamma (landless Koya) says, "They have dug up all our fields. We have no agricultural work; no livelihood source. We are very troubled. What about the landless who make a living from the forest? What will they do? None of the women wanted to leave… We are getting some work under the 100-days (NREG) work. We are managing that way. They gave us Rs. 1,70,000 (as one-time settlement). We are bulding a house. It is not enough. Cement, brick and sand cost like gold…"

The plight of the Koyas of Mamidigondi is a forewarning of what lies ahead between now and life in an unpredictable future (with the elections to a new government next year) whether or not the Polavaram dam is built. The government thinks their responsibility for the tribal communities (and the dalits) ends with disbursing the 'RR" cheques. A disturbing, tiring day. Back in Polavaram. Saikrishna's owns his home which is still under construction. Adjoining is his brother's house. There are two tiny rooms and a kitchen. There is a narrow verandah with a pyol-like seating arrangement. The house is on the road leading to the Panchayat office. The toilets and bathing space (as is the custom here) are outside the house, in the backyard which overlooks Godavari at a distance; and we remember where Godavari 'was', when I last came here. I know I will sleep like a log.

***

15th June - Meeting the political 'Convicts' and Seditionists: Chegondapally Again

Since I could not meet them last year at the jail in Rajahmundry, I decided to look them up, having heard they were out on bail. So, after a breakfast of *upma* that Saikrishna's wife lovingly made, both of us moved on to Chegondapally, to meet up with two of the 'criminals', namely, Sunnam Raju and Muchika Suramma. Posi Babu, too, joined in the conversation. Sunnam Raju says, "Since last 18 years we have been fighting against non-tribal landlordism and reclaimed nearly 600 acres from the non-tribals, which we have been cultivating, without pattas. The non-tribal land owners said they have D-Form pattas[16] and received RR cheques for those. Our boys, Ravi and Bujji Dora had questioned this at the Ramalayam the other day. How can they give compensation for lands we are cultivating? The officials kept saying they will give us pattas and didn't. They slapped cases against the two of them for questioning. They said we have no rights since there are no records..."

Suramma joins in, "They won't give land for Sangham lands. We will not let go of the Sangham land (reclaimed by the Agency Girijana Sangham). Government should give us government lands, not lands taken forcibly from tribal cultivators. We also need forest land..."

Posi Babu adds, "The officials are trying to settle us on disputed lands (in Venktareddygudem, Managopalam, Kunkala–Jillelagudem panchayats, Pragadapally and Vinjaram panchayat, Polavaram mandal). Tribal people there are cultivating on landholdings as small as 50 cents and 30 cents. The officials are showing those kinds of lands to us. We do not want to enter into conflict with our own tribal people. The government has not yet shown us gram sabha decisions, land surveys, house site surveys till date. We raised these questions even at the *praja padam* (government programme)... We had ourselves surveyed 680 acres (that should ideally be given in compensation for the entire Chegondapally Panchayat) but the officials are saying it is only 250 acres. The newspapers reported 300 acres."

Sunnam Raju, too, says, "The cases of 2005 and 2007 against me are still pending, and they keep adjourning hearing. We have to keep running back and forth to the court. We have to take a bus ticket of Rs. 100 to go to Jangareddigudem and back. There are ten of us fighting cases slapped on us...We have spent a lot on lawyers, at least three to six thousand rupees each time."

Posi Babu continues, "Baddam Krishnamurthy was charged with cases of conflicting with police. They want Krishnamurthy to convince people here to leave, and they tried to bribe him; said if he convinces people they will give him Rs. 5 lakhs. He refused. Some people told him he has a golden chance if he leaves Agency Girijan Sangham. They told him he should take people of Chegondapally people in a lorry to show the R&R (disputed) lands. Krishnamurthy has stood up against that. They now allege that he slapped a policeman with his associates…If we do not take the cheques, they will deposit them in banks and then we will have to go to the Courts to fight and get the money released…"

***

July '08

17th July - Rajahmundry[17]

I came back to do the Kondamodalu story of resistance and history, and the library story (came for one, but got two as a bonus!). I hope Kondamodalu gives unto itself its own history of the historic struggle…

Diary notes – Am at River Bay.[18] From a different vantage point – at an expensive, touristy place for the first time in Rajahmundry. HPO and the State Siliviculturist's guest rooms were unavailable…The place is right on the banks of Godavari; and who knows, it is likely built in violation of some rules. The approach road to the place does not seem too safe for women, as I experienced it first-hand. But it is interesting to see Godavari from this point, as well, and to observe the kind of people who come to stay here. A high-end inn at Rajahmundry. It is interesting to watch the number of government officials either lunching or dining here and businessmen staying here. It is meant for them, obviously…the price is…relatively high by Rajahmundry standards, which have also increased in the last two years, ever since Polavaram project has commenced. Got the library stories which I wrote here and some others. But my mind is restless. This is not my place, though Godavari is so close, and can see it from my window. Just looking at the distance between people and me from here, which staying here exemplifies. Yes, visited Draksharamam, after promptly getting into the HPO in a few days, after leaving this place. It is an ancient Siva temple. This visit brought home the point about the network of canals and irrigation and the access to these resources, especially Godavari waters, for the upper castes, Brahmins and others. Perhaps there is a longer line of linkages which precedes the colonial period in this region, where brahminism itself seems to have smoothly transited into the colonial system, with people so used to getting water for their fields and tanks for their temples, continued to plead with the colonial rulers for continuing with the past concessions that they, as a community, enjoyed.

***

18th July – The town, Rajahmundry, incidentally, seems to seek its modern identity through the Cotton Barrage at Dowlaishwaram built in the 19th century and the irrigation system created thereby; for whenever one is here, there is someone asking you if you have been to Dowlaishwaram. It almost has a reverential and religious aura to it, that Godavari anicut, which Arthur Cotton built. Rajahmundry also got its Government College and the Schools built within a colonial set-up but it has had a long established tradition of Telugu classical language and literature which of course pre-dates the colonial period. I could not, much as I would have liked to, make my trips down this route of history, considering that the other 'route' that I was already on, then, had most of my head and heart in it. But I managed to return to the old library in town, for starters... Sri Gowthami Regional Library. Libraries such as these are links to the fascinating past of Rajahmundry and the freedom movement. Even if you weren't treading on that path, or past, you still end up intrigued enough to start on a path based on what you hear. Or read there. In my case, it led to a meeting with the family of late Padala Rama Rao, Communist freedom fighter from Rajahmundry who has apparently written a comprehensive and authentic biography (in Telugu) on Alluri Sitaramaraju. Even as I was flipping pages from records in this library there was also a fascinatingly coincidental link to a chapter from the past that resonates in the present. The story of the Koya people arrested for 'seditious' activities in Rajahmundry Jail (last year) came back to my mind as I read through this reference from the colonial period titled, "Report on Seditious Preachings in Rajahmundry Taluk, 30th July"[19] (which was a "confidential report, No. 5) submitted by C. Rukmangadha ("Your most obedient servant"[20]), Inspector of Police, CID, to R.W.D. Ashe, Esq., I.C.S, Collector and District Magistrate, Godavari. The letter states that "there is more swadeshi feeling in the Rajahmundry town than in any of the places visited by me in the delta taluks. I have collected some reliable information from some students that the students of the Rajahmundry college who were prohibited from taking part in political discussions are privately meeting whenever necessity arises such as for the raising of a relief fund for the defence of Aravindo Ghosh (or for the starting of the national volunteers for the pushkaram[21] festival, etc, they have given up periodical meetings which used to be regularly held first year). A number of students go to the Vireshalingam Library and there decide as to what they should do and then send out subscription list, etc and carry on business. They have decided to stone to death any student who will give information to the college authorities and thus turn a traitor....The merchants of Rajahmundry are unlike many others of their class in other taluks, all swadeshi at heart. For the past 4 months they closed their shops on Amavasya[22] days and met in the Santha *chatram* to discuss about matters of general interest. When my sub-Inspector attended this meeting, there were about 200 men present. On Monday the 27th which was an Amavasya day met in the Santha Chatram from 4 to 7 pm under the presidentship of Nyapathi Subba Rao Pantulu (Hon'ble Mr. Subba Rao Pantulu)[23] and discussed about the amount they should subscribe to the National High School, Rajahmundry and about the necessity of national education...."[24]

Another interesting point noted by this official is that he "could find one hundred and one swadeshi shops called "Indian Stores" "National Warehouse", "National Emporium"...etc so much so that one cigar merchant had a board "Swadeshi shop" another merchant calls his shop "Swadeshi dyeing shop" and curiously enough a Muhammadan tinker has a board "Swadeshi tin shop"...One Dharani Praggada Suryanarayana of Rajahmundry tried to have the Vande Mataram Natakam enacted at Rajahmundry but hearing that the Kothapalli people who are the actors were prosecuted, has abandoned the idea..."[25]

I am discovering Rajahmundry anew, for the first time. It feels like something has opened up just now, a small door, to a long passage with several doors yet to open up, seeking my discovery, as if.

***

21st July – An Alluri Sitaramaraju Rewind

Today I met the late Padala Rama Rao, a Communist freedom fighter from Rajahmundry, through his son Veerabhadra Rao. They stay at this place called Gorakshanapeta, which seems to be an old locality. And for the first time in these last four days, had sumptuous, home-made Andhra (incredibly spicy) food. Hogged, indeed! He spoke a lot about Alluri Sri-ramaraju, better known as Sita-ramaraju, whose biography his father had written. Padala Rama Rao was also closely associated with Mallu Dora, the first tribal MP from Bhadrachalam constituency at a time when Bhadrachalam was part of the Godavari districts (or East Godavari). Mallu Dora had closely assisted Alluri Sitaramaraju. I saw the photograph of Mallu Dora, too, from a magazine, besides a short documentary made by Veerabhadra Rao. On the way to his house, I spotted a large life-size statue of Padala Rama Rao and Alluri Sitaramaraju. Finally, Veerabhadra Rao gave me a portrait, created from an original photograph apparently, of Alluri Sitaramaraju as I took leave.

***

22nd July - Thoughts in my Diary – It rained all night yesterday. Morning felt so nice. But it's been hot again. Just sat by the river in the evening; decided to consciously not think, of anything. Couldn't help thinking though of Alluri Sitaramaraju, my Godavari journeys – where did it all begin and will it end? I focussed my eyes on Godavari and the last fishermen as the sun set; looking, for a while, without seeing.

***

I left part of my luggage at the reception for safekeeping – will get my train from Rajahmundry on the way back. I took the bus to Polavaram.

An 'Un-settling' Issue: at Devaragondi, Devaragondi Panchayat, West Godavari District – 22nd July

Met Boragam Kannayya/Kannaparaj twice, in 2007 and now again in 2008, both times quite by accident. Both times, he had accosted me as I was walking past, or talking to others and had insisted on my recording his points of view. The first time he was totally opposed to the dam coming up and was representing the voice of the Koyas in his village saying they would not accept cheques. However, the first time, there were many people who had gathered around and there was unanimous voice against the project. The second time I met him, he had resigned to the fate of many like him, and consented to giving his land. And this time round, the village wore a deserted look, as quite a few were working as wage labourers outside the village, since many of them did not own their fields anymore. This was one of the first few villages where the digging had commenced for the project within the forest area, and the forest had been barred for the Koyas since. Kannayya seemed to have realised, in an almost eerie-looking village that things are not so easy from the moment of his signing his consent. Getting the money is not too simple. He virtually rushed to me as I was talking to a few others and showed some papers (including his letter to the MRO, etc), saying, angrily, "RI, MRO, everyone is taking their *vaata* ('commission') from our compensation; if the government has sanctioned Rs. 3,80,000, all these people's *'vaata'* process will begin: the Collector gets the first, a big man, isn't he (*pedda ayina, kadaa?)*? So, he will take full *vaata*, half (*sagam*) will go to the RDO, MRO gets 3 per cent, R.I (Revenue Inspector), 1 per cent and then the *gumasta* (clerk)'s *vaata*, and so on. So you see, there are so many mouths to feed – Collector, MRO, RDO, their family and children…" Did he know this village will be resettled in the 'plain area' (non-Scheduled)? He said, "When we came to know it would be a plain area, we questioned the authorities; they got us beaten up."

Kannayya realises what a grave mistake it was to have given up his lands for the money promised, which did not come to him in the manner he thought it would – as a hefty cheque of Rs. 1, 40,000 which was promised. The distance between the R&R housing site and the fields (shown to few, not all, people of Devaragondi) is about five kilometres. Kannayya said, in jest as in anger "how can we go so far? Cultivating those fields would be like leaving one's first wife back home and going to a second, far away. We cannot do that." Kannayya was still living in the belief that he would be allowed to till two fields at the same time, but pattas had not yet been disbursed. But he did see the futility of the resettlement of his people in another kind of land altogether, as he showing me the centuries' old tamarind tree, "look at this tree here; it has been here even before I was born. If this were to be grown there, in a new place, it would take years before our grandchildren (and not even our children) can benefit from it, sin't it? In our Agency areas, we do not have to buy *taatakulu* (leaves that are used to cover the roofs), or *kunkudukaaya* (soapnuts), nor custard apples or mangoes. We only buy some oil and onions from the bazar. There (R&R colony) we will have to say, 'hey give me a few leaves, give me a dozen custard apples, give me some mangoes, for money. And we are not people who keep savings like the non-tribals do; we use money for the day and do not worry about tomorrow...Just what are proving by going there when nobody is giving us any importance?…"

Madakam Kondaiah, a Koya woman, about 60 years of age, looked totally lost to the world in her makeshift shelter: a hut with all the household things scattered around, as if she was to move out any moment. In fact, she was to. But nobody, least of all her, knew when. Her home in the R&R colony was not yet built and her life seemed stuck. I had seen another home of an older woman here last year, which her son was building for the family, a cemented one, relatively better than the rest. It remained incomplete even this year, because the land on which the house was being constructed was just beside the road leased out to the BGR Company (then) as part of the spillway works of Polavaram. It was right where the forest starts. Now you could see rubble and soil dumped, and the old woman, sitting all alone, like she sat last year. Last year, it was a home being constructed, and this year, it is a home abandoned: just the old woman and a few hens loitering around, for a morsel of food from her plate. She was a symbol of a desolate village or what it would look like, once people are forced to move out and before (if) the dam comes up. Madakam Kondaiah said, "They gave us some money (from the package) which we used to lay the foundation of the house (in the R&R colony) for Rs. 30,000. We told the officials our money was exhausted; they gave us Rs. 40,000 more, which was used to build some more. We got the money through the committee – with Balraju (present Congress MLA from Polavaram, who is also part of the negotiation committee for R&R package in Devaragondi panchayat), Midiyam Venkayalakshmi, Virabhadram, and others. The officials made that committee gave this compensation. We do not know whether the committee will give us any more money." She did not know the R&R colony was not in the 'Agency area'. "After the committee was set up and people built some houses, they said this is not Agency Area...It is all chaos (*ayomayam*) for us now. We have our fields here but we cannot cultivate because they have dug up all our fields. We have not built those houses yet, because there is no money left..." As for the NREG works, "They cannot show us work because there are no fields here anymore; all dug up and out of bounds for us."

How are they surviving? They go to the forests (at that point when I met them, they had some access to minor forest produce) and collect some nuts and fruits and leaves. Kannayya's family lost "12 acres and 38 cents". He has five brothers and his share is 2 acres and 38 cents. "And I cannot get even half of that in compensation on the disputed land they showed us. Total land of tribal people in Devaragondi must be approximately 370 acres, including common and private lands. They did not show 370 acres in Gunjavaram (R&R). Kangala Veerasamy and few others have not even got alternate lands. I was shown land but not 12 acres and 38 cents. We do not have our fields here anymore to cultivate, no work either. They told us we would get wages for 365 days (in compensation package); they gave me a cheque for Rs. 60,000, but are yet to give land."

***

Pydipaka and Chegondapally Again - Marginalised, for 'Public Good' and the NREG Scam in Polavaram Mandal[26]

Mutchika Rajamma and her daughter-in-law Lakchamma, landless Koyas of Chegondapally (Polavaram mandal, West Godavari) went to this new employment programme in their village with great enthusiasm. A new 'hundred-day' work on a defunct check dam in their village was to begin. But, within ten days, their enthusiasm turned into sheer frustration. Rajamma says, "We did what we were told to do – dig up the soil. We went for ten days and at the end of the tenth day, got only Rs. 300." It didn't seem worth the effort. After a few days, however, the officials decided that works here should be stopped. All this happened within the month of June this year. Gradually people gathered at the Ramalayam where this discussion had commenced, since the *upadhi haami pathakam/vanda dinaala pani* (local name for NREG) was bothering them at that moment. The Rama within the temple was witness, yet again. "I got Rs. 200. I worked for 7 days", says Midiyam Saramma. "The gumasta (field assistant) told us they ordered to stop the works; there is not enough money for the works in this village. They told us not to come anymore for the work." They were fortunate that their village got at least ten days worth of work. But Pydipaka in the same mandal is a different story altogether. The computerised list of NREG programme in the MDO office at Polavaram marks "No work" against the village Pydipaka. Suresh, operating the computer informs me, "it is going to be submerged; it was difficult to convince the Collector to get work done here."

According to records for Pydipaka, 349 job cards were issued for 729 wage seekers (384 Male; 345 Female). Of these SCs have 154 cards; ST-2; BC – 159 and; others 54. I am speaking to the dalits (Mala and Madiga) in the SC hamlet of Pydipaka, some of whom I kept seeing during other visits. Andru Peddaraju says, "The Assistant Project Officer (Y. Srinivas) told us, "you are all going to leave; your village is going to drown; why do you need work?" They gave us job cards but no work. We staged a dharna at the MDO office in Polavaram. The RDO and the Project Officer assured us they will solve our problems. But till date nobody has come. We have no land and no jobs. No SC here owns land. We used to work as labourers on fields. We are just idle. Some are going to Jangareddygudem and Buttaigudem for work. Actually we wonder just what our life is all about." Even an Act which is supposed to guarantee livelihood to the poor, especially the socially excluded, stands no chance before the concept of 'public good' it seems. In the case of Chegondapally and Pydipaka, construction of the Polavaram dam takes priority over guarantee of livelihoods before the dam is even built. Polavaram mandal has shown the most pathetic record of NREG programme by far. East and West Godavari districts were among the newer districts to get the NREG underway, this April. NREG in Polavaram mandal started in April this year, on paper. B. Peramma (SC, Pydipaka) has a child to look after and her husband died few years ago. Her home is in shambles, which she cannot repair. The work she did as agricultural labourer on the non-tribal farmers' fields is no longer there for her. Even the NREG scheme did not take off. She cannot work too far from the village on account of her child. She says, "They keep saying we will be moved from here. There is no work, no money."

Pydipaka exemplifies all the paradoxes that social hierarchy, oppression and land tenure history has done in the Godavari districts since ages. The Polavaram project, in fact, only accentuated this economic and social divide. Today you see a *crorepati* (millionaire) –mostly upper caste non-tribal in the same village as an absolutely on-the-brink Mala or Madiga. All the compensation for land (at the rate of around 50,000 to Rs. 1,50,000 per acre) has gone to the non-tribal landowners of Pydipaka. People joke about this package that has made some (such as "Kuncha Venkatratnam", they say, in a chorus, he is also referred to as "Dora Babu") millionaires almost overnight. The dalits here have slipped back to where they originally were in the socio-economic table pre-Independence. When they received compensation for their lands, the non-tribal landlords of Pydipaka (small, medium and large farm-owners) stopped cultivation. This rendered the dalits, who worked as agricultural labourers/workers here, unemployed. The R&R package was supposed to compensate such people in terms of wages for 100 work days. But that has not reached a single one of them. The only hope for them was the *'vanda rojula pani'* (NREG works) which was not 'given' to this village because in effect this is a "submerged area." So they did not wish to "waste" money on this village. Jadla Mary of his village says "We are dying of hunger; and these bombs (dynamitc blastings from the spillway site next to their village) have made it a living hell for us. Nobody cares for us. People like you come here and go, but our situation remains the same – no food, no work. The other day that CI (Circle Inspector) Rajkumar said they will put us in jail if we question too much, when we went to the MDO (Mandal Development Officer). As SCs we have nothing from the Polavaram RR package either; now they are denying us the 100–days' (NREG) work as well."

The officials cited technical problem for this. Since all the land has now been acquired by the government, which has in turn leased out portions to contractors working on the Polavaram dam, how could they get works initiated on government land, virtually 'owned' by contractors? The Assistant Project Officer, Y. Srinivas, when contacted, said, "Pydipaka comes in immediately submerged area. We have to show asset creation in the NREG works. So how can we do that here? It will all go waste. We have shown works for other submerged areas but Pydipaka's case is different. All the farmers have given away their lands to the government and we cannot create assets there anymore..."

Forget about 'showing' work, many in Pydipaka, like Chittumalu Sitamma and her husband Chinnaseeni do not even have job cards. Peddaraju is distraught. He says, "We are all in deep distress. For generations we lived here, and worked on the fields of big rythus (non-tribal farmers). Godavari also fed us." But this time round, says Peddaraju, "even Godavari had no pity on us." It didn't rain enough until late July in these parts for *'pedda Godavari'* to come for them to collect the logs washed by the river. "The landlords have sold their lands. They have put their money in banks. They are not interested in the fields. We are losing out...We are already displaced" says Peddaraju. He also points out that dynamite blasts carried out at the spillway site everyday without warning poses another problem in their village, forcing them to believe "the government truly wishes the dalits would disappear into another horizon altogether."

At Chegondapally, Koyesi Ramasalaimma had said, "The MDO came here. She said the kind of work we have done (at the check dam) won't suffice. There was always a problem – they would say, the way we dug up was not good enough; or that we did not remove enough soil; that it did not add up to a meter in depth and so they would not pay. I worked 9 days to get Rs. 400! They say we are not interested in working, can you believe that landless agricultural labourers (*cooli*) like us would shirk work?"

Srinivas (Assistant Programme Officer) says, "We paid them according to some measurements. For a meter of soil dug (height, depth-wise) we give Rs. 54. Some people think like other food for work programmes they can just give attendance and get Rs. 80. When they do not fulfill measurements, we have instances where we have paid even Rs. 30 and 40, and at the same time some are getting Rs. 100 per day."

There were other discrepancies on the NREG work done here too. For instance, only in Polavaram mandal they gave wages through the Village Officer and not through the Post Offices. They did not create accounts for the job card holders here. The APO says it was because the works started in April and they felt opening an account in the Post Office would mean delays in payments, so they gave the money through the VO (Village Officer). For Chegondapally, he says, "we gave cheques for 6 days of work, which they encashed from the banks. We still have to give amounts in other villages of this mandal. After verification (unclear what those verifications mean), amounts up to Rs. 2,15,000 have to be given."

DWCRA leader Andra Suryakantam of Pydipaka has a point when she says. "The *pedda rythu* (landlords, as they are referred to) have got their money, but where will we go? ...The NREG field assistant is getting her salary. When no works are happening, why are they paying the field assistant?"

NREG programme in Polavaram mandal (14th April to 14th June) has been pathetic. For instance, only 2 households in the entire mandal got 100 ("completed") days of work! For 19,419 wage seekers, in 15 panchayats, 10,415 job cards were issued. Value of work (148 sanctioned) was Rs. 3,14,76,824. Works completed were ten. Average employment generated per household was 12.82 per cent. The field assistant says they were out of station so she did not give them cards. There isn't a single notice showing details of the work in this mandal. The APO says, "We gave a proposal to set up walls later in the next season."

In Chegondapally, for 828 wage seekers, 387 job cards were issued (the APO said, *"vaallu godava manushulu"*! – they are trouble makers). But the problem comes in the following statistics, which raises several questions: No of person days – payments made Rs. 1,48,641. Per day average wage for Male was Rs. 55.18; Female – Rs. 50.85. Male work days – Rs. 1,186; Female work days – Rs. 1,186; Total wages given – Rs. 83,195.

Statistics – NREG in Polavaram mandal (West Godavari district) in the "totally submerged" (if dam is built)/"project-affected villages"[27]

In Polavaram payments were made on 14th June and there were no funds thereafter. NREG started in April all over Polavaram mandal, except in Mamidigondi, where only one payment was made in June of Rs. 2,040 for one week of work, which was "jungle clearance." "Total wages" shown for Male were Rs. 1,836 for 19 days of work and Female were Rs. 204 for 2 days of work. This, while the number of "job cards" issued for 313 "wage seekers" (145 male and 168 female) were only 151. The average wage (both male and female workers) was Rs. 102.

***

23rd July - To Sivagiri, Across Godavari

I called up an old acquaintance, Srinivas Rao, the autorickshaw driver in Polavaram whose vehicle I had hired on a few occasions before. He would take me to Sivagiri (he came all the way up to the main road near the Panchayat office cleanliness worker, Jyothi's house. He will charge me Rs. 200, he said, since he has to return, even if not 'empty', but probably empty half-way down. We pass through the spillway site which has transformed some more and the hillocks look barren, with more trees chopped off and the fields are of course filled with rubble; there is dust in the air. But as we move ahead, uphill, on the road to Sivagiri, everything looks beautiful and green at this time of the year. Srinivas obligingly stops by for me to take photographs, including one of the old Tamarind tree. Strangely, this is the first time I take photographs on this stretch though I have passed this by many times before, mostly in a rush. The silent ride is blissful. Though the roads as are bad as they were the previous times and it is quite a bumpy ride. Reaching Sivagiri, I find that Mutyam is not around this time (he is attending a wedding in the family at Bhadrachalam and won't be back for few days; yet again I shall not get that story of his seeing Mahatma Gandhi). But someone from his family is ready with the boat and waiting for the bus (did not know there was an afternoon bus, which I could have taken, had I known) to arrive so the boat can ferry all of us across.

At Kathanapally, Somarlapadu

Kondla Bullebbayi will accompany me to Kathanapally up to Kondla Gangaraju's house. Nobody knows I am arriving today. I surprised Kumari by reaching her place. She wasn't around when I reached. Her husband was. This was the first time I was seeing her husband. In a while, she came home, seemingly happy to see me. Her sons haven't come home in a while now, she tells me. She gives me their phone numbers (in Hyderabad) and asks me to let them know they should come home. Kumari's husband offered to accompany me half way through the fields (since I had the usual bags and it was a hot afternoon); one of the farm-workers, Bullebbayi (a Kondareddi), would take over from there. It seems like a never-ending walk.

Bullebbaayi is a short man and chatters away non-stop (on different subjects) pausing only for some questions which I manage to 'insert' even as I walk, panting. I miss out on quite a lot of things he says. Am thankful for the voice recorder, which is 'on' throughout and helps me relive the moment, later. He has never been to school, and runs errands for everyone. He is one of the four sons of Kondla Chinna Reddi. Two survive. Kondla Gangaraju is his brother (was surprised to know) but they don't seem alike. He has his own tiny piece of land (that is, reclaimed by the Sangham) but works on other people's fields on daily wages. He would begin the *vanda rojula pani* whenever it starts. "We need work. Usually there is no work in our villages. Not all are into *konda podu* (shifting cultivation). We grow lentils, castor (*amdalu*), etc. and some tubers. In the past our agriculture was far better. Now we have paddy, *mirchi* (chilli), *pesari* (different varities of gram). In *Uttara karthi* time, (winter season), we grow lentils/pulses. In our *konda podu* lands crops grow without any fertilizers (*purugu mandulu)*. On those lands we grow *jonna* (sorghum), *kandi*, *bobbara* (cow pea?), *amdalu* (castor), *gummadikayalu* (gourds), all in one place. Here you have to grow the same kind of crop, unlike in our konda podu. Those who have settled lands have stopped doing konda podu…In the past we used to depend on konda podu, mainly. We used to drink our gruel, especially *java*[28], and go to the fields." Bullebbayi remembered people from his community who participated in the freedom movement (*swatantra poraatam*) and mentioned in that context Subba Reddi, Kanthi Raju. When we came across the 'martyrs' memorial', a tall stone structure in red made in memory of the people who died in the police firing of the 1969 agitation, he tried to read a few words written on the memorial. Kondla Gangaraju took over mid-way – the bag and the talk. We walk some more before we reach his place. Questioning me on my route this time, he informs me that "They discontinued the service launch from Polavaram to Kondamodalu. Today there is just the 7.30 am launch to Polavaram which reaches there around 12 noon. The same one leaves at 2 pm form Polavaram to reach here at around 6 or 6.30 pm. They charge us Rs. 25 (*paatiki rupaylu*). It is a five-hour journey by launch and takes two hours by bus from Sivagiri. There is a bus at 6 or 7 pm from Polavaram which reaches Sivagiri at 9 or 9.30 pm. The boatmen there will ferry us across. And if we take the 2.30 pm bus from Polavaram we will reach here here around 4 pm."

Routu (route), a term used by them to denote road, is a very serious affair for people here, and if you pay attention, the timings and frequency of the boat and the connecting bus can tell you a lot about the 'connect' of the everyday people here to all the politics in the area, on a river. The outside world then seems (from these villages) the virtual illusion created for an unreal reason.

At Kathanapally - Listening in to New Developments and the Glorious Tribal Past

I want to start off, immediately. So I take out my notebook and recorder, and camera (amazed as to what has become of me?) It has been a long journey, under a piercing Sun from the banks of Godavari upto here, with my bag (gage). Gangaraju could see that. He ordered me to "sleep first". Thought I would simply rest my limbs in a brief nap on the coir *mancham* (bed). But that warmth of that quiet – of the hens and their chicks and the faraway voices of women attending household chores – put me into deep slumber. Heavenly! Later, after lunch of lovely (but spicy) *avakkai* (mango pickle), rice, *pulusu* (boiled tamarind water with spices) and *perugu* (curd), washed down with tea, feeling sated physically and spiritually, Gangaraju allowed me to pick up 'work'. His caring, the space of his home and its courtyard (matched by space in their heart) has touched my soul. It had been a while since someone said to me, 'eat, rest first'. I was happy I would be staying here tonight. His daughter (her little child is now a year older and can actually walk around) is very pretty, with curly locks of hair, unkempt. She came by to greet me. At intervals someone or the other from my last journeys would come along and greet *baavunnaara*? (Are you well?). And suddenly my thoughts rewound to my first visit here and the same old tamarind tree where the women had danced. Somethings had changed, yet many had not. At least there was a place I could simply reach and hit the bed at and take in its own pace – gentle, slow – without the teeming rush to record, click, write simultaneously. Saroja, Anuradha (trained ANM who lives in Kothagudem, belongs to Kokkarigudem), Reva, and suddenly, from out of nowhere, Madi Muttem, come by. We speak of latest developments here. They didn't expect me here. Kothagudem Anuradha tells me that the school in Kondamodalu was shifted this June and 'merged'[29] in a school at Rampachodavaram, for the boys, and at Devanapally for the girls. They are going to teach English there and appoint new teachers.

Suddenly someone added, "The Road (Kondamodalu) is not yet completed!"

Other developments around the place add to the problems here. Anuradha adds, "If a child is unwell now they ask us to go to Swatantra (hospital) in Rajahmundry which charges anywhere between Rs. 12,000 to Rs. 15,000 for treatment. There is a Girijan cell at this hospital; it is located in Rajanagaram....Though we are given Aroygasri Policy Card, only tests and bed are free; medicines are very expensive." As for the PHC at Kondamodalu, same story; a doctor comes now only on Sundays for an hour. People from Khammam and East Godavari tribal villages come to this PHC. But for deliveries you still have to go to Devipatnam, and they charge you Rs. 1,000. Doctors travel between Devipatnam to Rampachodavaram and if you miss them, there is no way out. Gangaraju's daughter died mid-way (this is news, I did not know that) when she was to deliver a baby (as the launch takes three hours to reach Devipatnam from here)."

The other news was that RR colonies for Kondamodalu people have been 'shown' in Rampachodavaram, Fasulabad and Korukonda. This includes 600 acres of land of Harishchandra Reddy (non-tribal) in Rampachodavaram (who hails from Anaparthy, East Godavari). They also showed lands belonging to Hukumpeta zamindar's family in Venkatreddigudem, Kunkala and Pendrala. Not a single person has taken (R&R) cheques here. That was a collective decision. And I hear the shocking news that last November (2007) a young woman was found raped and killed and her body had been thrown on the banks of Godavari around here; she was apparently brought here by two male tourists from outside n the tourism launch. Apparently the news appeared in the local editions of newspapers here. But the case was hushed up later.

The Last Post of Resistance[30]

Gangaraju, always the kind elder here, gets talking about history as he sees it. But once he starts, he can go on forever, and at times into several unrelated issues. This time round, there is no stopping his flow of thoughts as he remembers the stories he had heard as a child and youngster before he participated in the 1969 agitation. He recounts, "In those times (*aa rojullo*), people like Alluri Sitaramarju, Bhagat Singh would say, 'why should foreigners take away all our wealth? Can't we govern ourselves? Have our own government?'... Alluri Sitaramaraju started living with girijans, Kondareddis, and others, trained them, worked among them...I have heard from our grandfather that he used to write a letter (and send it with girijans who acted as courier boys), saying at this time and date we will blast a particular police station. Many Kondareddis (but from forests in the uplands) cooperated with him. It is indeed a wonder that he blasted all these police stations on the same day (in the context of Rampa revolt)! He blasted (*kotyaadu)* Chodavaram, Addateegala, Y. Ramavaram, three in a day! He also blasted the Devipatnam police station. Our elders used to say he used to have herbal portions, which he took from the forests, which gave him that kind of strength. When there was a *tirugubatu* (revolt) going on, our people knew. *Pituru* (*fituri*), they used to call it in my childhood; when our elders used to talk of it. But all the direct action happened in the hills on the upper reaches and forests of Chodavaram, mainly...In the past there used to be a Reddi Polavaram, near the present Panchayat office, and Kondareddis were the heads, such as Nadupuru Rami Reddi and Lachi Reddi; that has become Polavaram as it is today after the Hukumpeta zamindar took all those lands...In 1960s we struggled between the oppression of the landlords and the government (forest department) which started creating restrictions on our grazing and use of forests. *Kudipi-lo elaka laaga kottu-mittadutunnaamu appudu* (our situation was like that of a mouse struggling inside a bowl of gruel[31])."

Late that evening, around 7 pm, Gangaraju, and I walked to Somarlapadu, followed close at heels by *the* dog, (a new addition, I noticed; extremely friendly with no name, except that he responds to *'chey'*. He has clipped ears. Gangaraju says they cut off part of the dog's ears in these parts when it is still a pup, as they believe it will make the dog more obedient). He has a small torch, for my sake (which I feel isn't necessary; it takes just a while to adjust the eye to the dark, and it takes the joy off walking under the stars). We will meet the elusive Illa Rami Reddi at his home. I am recording the sounds of the night in my voice recorder as I walk along. And feeling the moments – a beautiful village, hills and a river, lush green fields (cliché, but it is the truth here) a Lapwing that has been hovering above. I love its call – always gives the sense that it comes from some great distance, while it is quite close by; if the concept of nostalgia were to be created as a sound, this bird's call would be it, especially at dusk or dawn. It is almost leading our way. And suddenly vanishes as we reach Rami Reddi's home! Seeing Rami Reddi seemed a different persona – of a family man, who has been running around for his daughter's college admission; and worried over a second daughter's medical condition (she gets epileptic seizures). Life is also about these things 'in between' all his political activities for the Agency Girijana Sangham and this Polavaram issue, now on fast track.

Kondamodalu, meanwhile, has been a region of tribal *tirugubaatu* (revolts/'fituris' of the colonial documents) against the British. And today, in independent India, it seems like a last post of tribal resistance to the dam. The Kondareddis here have been standing their ground against moving or accepting compensation. Officials have been unable to coax or coerce the Kondareddis to yield…Administration in the Agency Areas – a term coined by the British, with an Agent managing affairs in these tribal-dominated pockets –meant shrinking access to the forests from colonial times. Later, regulation of their land – meant to 'protect' their interests – intensified their problems. The Kondareddis may not boast of high literacy rate today, yet have accessed the judicial system for the longest legal battles over land and forests, some fought by different generations of a single family. Since 1960s Kondamodalu locked horns with the administration in independent India. Through a long drawn agrarian movement between 1960s and 1982, the Kondareddis reclaimed around 3000 acres of land, grabbed by non-tribal landlords from the plain areas, in Devipatnam mandal (East Godavari) alone. Two young tribals fell to police bullets in 1982 during that struggle. Since then, the Kondareddis have sought legal rights and pattas to that land, but in vain. Rami Reddi recounts, "In British times the government was always in conflict with girijans on podu. They brought in the reserved forest system. Our grandfathers told us they would demarcate a boundary line we were not to cross. Our people opposed it even then. But Nedupuru Lachi Reddi supported the British. A long fight ensued for six-seven months. Many Kondareddis stood up against them, too, such as Karukuntla Venkata Subba Reddi. We heard once there were revolts for seven long months (in the 19th century). Subba Reddi stopped supplies to the British forces coming in boats between Cheduru, Thutigunta, and so on. Finally, they caught him and executed him at Polavaram and another one of us at Buttaigudem. Vetla Subba Reddi was also hanged in Polavaram. In the past Kondareddis were hill chieftains of Reddi Polavaram until the zamindari estates were set up and Polavaram passed into non-tribal hands."

Subba Reddi in Colonial Records: I had found Subba Reddi in the colonial Gazetteer, which notes, "Polavaram village contains some tombs which are locally stated to be those of European soldiers who fell in the *fituri* of Mangapati Devu at the end of the eighteenth century.They bear no inscriptions. Another grim relic of the old disorders in these parts which existed here till recently was the gallows on which Subba Reddi and Kommi Reddi, the ringleaders of the *fituri* of 1858 were hanged. This was carried away by the floods of 1900."[32]

Rami Reddi says Subba Reddi's descendants are in Koraturu. He says, "Even after independence our government continued the reserve system. They used to charge us Re. 1 for kondapodu (2 annas before that); we had to give *pannu* (tax to the amount of sometimes Rs. 5 per acre for konda podu) to the forest department and they would give us receipts on that. We were agricultural labourers on our own land and if we did konda podu we got into trouble with the forest department…And we paid *beda* to muttadars. Later we paid *dastu* of Rs. 2. In 1969 Tarimella Nagi Reddy (Communist revolutionary, who had broken with the CPI on issues arising out of the Telangana Armed Struggle) came here and gradually we felt inspired by what he was telling us about revolutionary path which should emerge from among the masses. All our lands were with non-tribals from the plains so we felt all the more the need to fight and get back those lands. I went to jail many times for 5 to 6 years in those days. Some of us still continue to make rounds of Courts!"

The paradox is, having occupied the lands reclaimed from non-tribals, the Sangham has been trying to get a legal sanction for the ownership of these lands in the form of pattas, which has not happened in so many decades. In this sense, they have been trying to negotiate with the instruments of democracy, which till date seems to be elusive or non-responsive to their demands. The revenue department and even the forest departments have been in the know of the extent of land cultivated by the

Kondareddis and other groups. It is through their assistance that the pattas for many of these lands are with the non-tribal settlers, who may not be residing in the area. In 1917 the Godavari districts came under the provisions of the Agency Tracts Interest and Land Transfer Act. This rendered all transfer of immovable property (except that among the tribals) to a non-tribal, except under the authority of the Agent or concerned official invalid. Civil justice was administered through the Agency Rules of 1924. In 1959 the Andhra Pradesh Scheduled Areas Land Transfer Regulation (APLTR) was passed under the Fifth Schedule incorporating part of the 1917 Act. In 1976 it was amended to prohibit transfer of immovable property in the scheduled area by a non-tribal to another non-tribal. This was made to regulate the transfer of lands in the Scheduled Areas of East Godavari, West Godavari, Visakhapatnam and Srikakulam. Later Adilabad, Warangal, Khammam and Mahabubnagar came under its purview. As per the provisions of Regulation 1 of 1959 any transfer of immovable property situated in the Agency Tracts by a member of a Scheduled Tribe, shall be deemed null and void unless made in favour of any other member of a Scheduled Tribe or a registered society composed solely of members of the Scheduled Tribes, or with the previous sanction of the State Government, or subject to rules made in this behalf, with the previous consent in writing of the Agent or of any prescribed officer. The provisions of LTR 1/59 were amended by Regulation 1 of 1970, making null and void any transfer of immovable property situated in the Agency tracts by a person (whether or not a Scheduled Tribe), unless such transfer is in favour of a Scheduled Tribe or a society registered or deemed to be registered under the Andhra Pradesh Co-operative Societies Act, 1964 (Act 7 of 1964) composed solely of members of the Scheduled Tribes. In 1978 all offences under ther LTR were declared cognizable with a one-year imprisonment and fine up to Rs. 2,000. The non-tribals took the issue up to the Supreme Court which dismissed their appeals in 1988. In case of a land transfer, the Agent or Agency Divisional Officer under this act was authorised to restore any such land transferred, back to the rightful tribal owners or their descendants, based on a application by any one interested, or on information given in writing by a public servant or *suo motu* decree ejectment against a person in possession under a void transfer. Post-1986 the tribal areas of Devipatnam Mandal in East Godavari district witnessed a fresh land occupation movement by tribals with the support of A.P.Rythu Coolie Sangham.

Rami Reddi points out, "Even on mango trees there were restrictions on consumption and there were restrictive rules and charges; on certain kind of timber, fees were collected for use. There was much revenue accruing from timber, minor forest produce from the Agency Areas in erstwhile Madras Presidency. Some of us still have those receipts! At the time of AP state formation, many non-tribals continued to hold lands here. We used to collect bamboo and earn daily wages. We used to be paid paltry sums, Rs. 9 or 10 per month. We did konda podu at Kontikonda, Burugulakonda hillocks..."

"We do not need R&R colonies" says Rami Reddi. "We need lands first. We seek pattas on lands we have been cultivating. What use are colonies without land to feed us? Our homes have a place for everything, cattle, goats, hens... All R&R houses that I have seen so far seem to be made for the slum dwellers in cities like Hyderabad. Rich people in apartments in Hyderabad may have lots of money in hand, yet they buy food from markets. We grow our food. How can they equate us with Hyderabad and Vizag city people? Have you seen our grain storage (*gadi* – a raised foundation, thatched bamboo storage space)? We store grains for years that way. Their RR colonies do not even have that concept. From farmers they want to make us pavement dwellers. Displaced tribals from the Bhupatipalem, Surampalem, Musirimilli and Kovvada projects are yet to get land. NGOs had thronged like ants over a lump of sugar then, but finally the girjians are left fighting court battles...We have fought a long battle without arms. We seek implementation of the Acts meant for tribal people. And this project will take twenty or more years to be completed. Don't we need roads until that time? We do not even have buses till Koraturu!"

As per reports of Department of Tribal Welfare, Government of Andhra Pradesh non-tribals are now holding more than 48 per cent of land in Scheduled Areas. The Supreme Court of India declared in the year 1988 that the provisions of Land Transfer Regulations are constitutionally valid. Rami Reddi says, "There are so many Acts for tribal people after Independence, including 1/70 and PESA, yet there are violations...The Reserve system (Forests) did not allow us to collect fuelwood or wood for building our homes. We would run between officials and the Courts for months to clear the cases, which we continue to do today."

Gangaraju adds, "We believe that since the earth was born, we girijans have cultivated these lands; if we cannot make use of these who will?"

I stayed that night in the peaceful home of Gangaraju, and after a simple meal of *pappu annam* (rice and lentil soup), slept outside on the cot as *the* dog settled down by the cot too. Gangaraju's wife gave me a sheet to cover myself. A good night's sleep to come to me, at long last! Sleep, here, in Kathanapally (or it could be a village around here with any other name) just has to be deep; there is something in the air here. The quiet of the night is not exactly quiet, as night crickets and a night bird or two fill the air with their mutual chatter, but above all, this is an enveloping silence which lulls you into deep sleep, no matter how strenuous the day had been. Nights are nights here, in their characteristic darkness and dawns move into daylights of different hues. Here at Gangaraju's home, the night is not broken by light. For there is no electricity most times, and no TV set either. Nor is it broken by a phone call, for cell phones do not work here. Each season has its own corresponding night and day-light, of distinct hues and fragrances. I understood the essence of *payiru gaali* for the first time in these fields - the breeze, or the whiff, *gaali.* That there can also be different kinds and fragrances of breeze in months, is a poetic idea, rooted in the agrarian seasons. Much poetry in Telugu is written using the term *payiru gaali* – a Telugu term I love – which is how I first learnt that term (especially the songs in the old films).

"Light north-easterly breezes in January and February, the driest months of the year, are followed in March and April by light south and south-east winds which blow during the day but die down at sunset. This south breeze is called by the natives *payiru gáli,* or the 'crop wind'...In December the wind blows from the east during the day and from the north during the night. The latter is called the hill (*konda*) wind."[33]

I also used to try locate the ancient Tamil concept of the *tinais* (eco-zones as evocatively represented in Sangam poetry) in the villages of the submergence zone (in all the three districts), the *marutam, kurinci, neytal,* here, mainly (the agrarian, riverine, forest-hill, littoral) – the beauty of it in the fact that these are contiguous geo-physically; one eco-type slips and slides into another almost! To think that all this rich bio-diversity would give way to a grant modern design of a dam does feel painful.

24th July - Oh, what a dawn!

But cannot stay on. Mists on the hills and over the fields, and smell of the moist earth. Am happy to walk this time, a long walk till the banks of Godavari down below, to get my boat. Incidentally the young boy who walks me to the river is one of those who walked with me back in 2006 on *that/<u>the</u>* road.

Afternoon at Polavaram

For a while now, I have not been able to spend much time with Sitara mastaaru, as he was busy covering sundry assignments. So, it was fixed I would visit his family before I take the evening bus back to Rajahmundry. When I reached there, his wife was already deep into preparations for a huge lunch – eggs, rice, *pulusu, majjiga* (buttermilk)...! Much as I would question them as to why they should take all that trouble, they would not listen but give a convincing reply that the lunch was for them, I was merely joining in. So, over two cups of tea, as the lunch was on its way, I got quick updates on Polavaram from the person who knows so much. But this time, I asked him to read out some of his stories to me, as he had been covering the issue since 2005 now. And there were the regular laments about money and the problems with it. And his contemplating quitting his stringer position with Prajashakti which was not sufficient to take care of his familial responsibilities. Life poses many questions. The major handicaps in his attaining a 'staffer' status in mainstream Telugu dailies were the fact that he has been writing for a Communist newspaper which runs purely on subscription and that he lives in this small town of Polavaram. Otherwise, his reports would document with rare ingenuity the local political dynamics and issues in and around Polavaram, sometimes missed by the Telugu mainstream media. The reports he sometimes showed me had an amazing sense of timing. As regards the dam, he wrote a phenomenal number of reports, especially with regard to the R&R aspects.

***

Late Evening, at Rajahmundry

I meet B.D. Sharma (activist, former bureaucrat and one of the petitioners in a case pending with the Supreme Court against the Polavaram project on R&R) who has incidentally come on a visit, to address a meeting on the Polavaram issue. I meet him in the evening before I get my late night train back to Hyderabad. He says, speaking of the cases slapped on communities protesting the dam, "All Cr.PCs (Criminal Procedure Codes) , etc are archaic; they were made for slaves of the empire and are incongruous in a democratic set-up…There are the four Ds – displace, decimate, disorganise and demise – that are being carried out (by the AP Government). In the Tennessee River Valley project the entire command area was resettled and space was created for those who were displaced. In Maharashtra and Madhya Pradesh, rights were created (in policy) for beneficiaries in the command area to share the benefits with the displaced. Here (Polavaram) if 7 lakh acres are to be benefitted, can't they give 1 lakh acre (to the tribal communities displaced)?"

***

25th July - Back Home, and a Different Lens: Two Films and Godavari

I watched the B&W Telugu classic, *Muga Manasulu* (Akkineni Nageswara Rao-Savitri-Jamuna starrer) on a CD at home, something I had long wanted to do. Earlier, I had watched the more recent *Godavari*. I reflect on the time and context and the change in the nature of 'consumption' of Godavari. The 'gaze' of the camera, of a film crew on Godavari is related to the larger developments in the economy and the way Papikondalu, and river Godavari have developed into a tourist spot. The film Muga Manasulu (Silent Hearts), directed by Adurthi Subba Rao[34], was released in 1963 and Godavari, made by Sekhar Kammula, in 2006. Muga Manasulu was entirely shot in the backdrop of Godavari; and one of the rare films of the time to have shot outdoors. Every scene, literally, has the Godavari (river). If this were theatre, Godavari would have been the only prop. The moods of characters correspond with the moods and flow of Godavari, so poetically shot in monochrome. Emotional upheavals of the three main characters in the film correspond with Godavari's flow. Each song and each dialogue (almost every other dialogue) has reference to Godavari, including the very popular song of those days, *Godaari gattundi,* picturised on the actor Jamuna, around a tree by the banks of Godavari. Godavari in this film, perhaps the fourth character of the film, has been shot with the luxury of time. And time, as well as the camera, dwells, absolutely, on the river, with the act of crossing the river on a *padava* (boat) being central to developing the story of three characters who indulge in the crossing, their lives were linked to her flow – Gopi (played by Nageswara Rao), Radha, or *ammaayi gaaru* (Savitri) and Gowri (Jamuna). All of them live on the banks of the river, and are representatives of different classes (and caste, too, of course, which is unstated but revealed in the dialects used). *Muga Manasulu* is a story of unexpressed feelings, of love between the Gopi, Radha, and Gowri and the fourth, Godavari, who are interconnected emotionally, physically and metaphorically. *Godavaari talli* (mother Godavari) is the not-so-mute witness to the play of emotions between the three characters and through them the extended, related, characters in the film. She is as 'mute' (but dense with emotion, like the river in the time immediately preceding the coming) as the three characters, developed so beautifully in the film. Love is expressed in many ways other than the word 'love' (or *prema*), per se. Unrequited love of Gowri for Gopi; Gopi's exalted love for (or worship of) Radha (*amaayi gaaru*, as he calls her, signifying the class status), or Radha's unexpressed love for Gopi. Gopi and Gowri seem to belong to the community of boatmen, or fishermen (the occupations are not necessary to the director here). All their emotions are non-verbal expression (except for Gowri pestering Gopi to marry her, which is seen as being childlike). Radha hails from a landlord's (upper caste) family and the river is the route to her education (perhaps in Rajahmundry, where her her college is). Each day, in this act of crossing over with Gopi (and most times Gowri), happen other expressions of these emotions. For instance, Gopi offers Radha a *banti puvvu* (marigold) everyday, without fail, which she puts in her hair ever so sensuously. In that act of the day, day after day, Gopi's expression finds meaning. Their friendship is shown as being of an evolved nature (hence the names, suggesting the love of Krishna and Radha never consummated in marriage). In time, a classmate in her college (actor Jaggaiah) falls in love with and marries Radha. But their happiness is short-lived, as he dies very soon. The changed dimensions of their lives have been portrayed against an overcast sky when Radha is arriving home as a widow, and there is silence enveloping the three characters, Gopi, Gowri and Radha. The pain is shared among the three. All the most beautiful songs in the film (*paadana teeyaga, mudda banti puvvulu..*) happen in the film after this sad event. Ultimately, a series of other sad events overtake the characters, which includes rape of Gowri by a relative of Radha's, the demise of her father. The film ends with Godavari (the river) ultimately consuming Gopi and Radha in its rare whirlpool (*sudigundam*), after Gopi and Radha's separate lives, as also that of Gowri's, have gone though their emotional 'rites of passage'. In the end, caught in the *sudigundam* the heroine Radha finally tells Gopi that this was the moment of their expressing their bonds to the world, by being consumed together, hands held tight. That is the only moment of their expression of love. Godavari begins and ends the film and the story, first frame to the last. The director of *Muga Manasulu*, Adurthi Subba Rao, also made other memorable, remarkably sensitive films such as *Sudigundalu*, *Maro prapancham*, among others, which were also films that were discussed and debated in their times and later. That he was from Rajahmundry perhaps made Godavari a natural choice for a film like this. No other river (certainly not Krishna, for instance) could have given the film its understated, yet intense, drama.

Switch to today's time. Year 2006, Summer and the film *Godavari*. Here, only the title of the film gives the river a larger than life sense, but once you are into the film, she is incidental and a mere scenic backdrop that may well have been a lake in Switzerland. Extraneous to the characters (except that the central characters meet on a launch to Bhadrachalam). Thereafter, it is their lives, very urbane and city-centred; and to an extent (as the story goes) self-consumed of their desires and ambitions. The characters here are also three in number, played by the actors Sumanth, Kamalini and a third actor whose name I do not know. The film was released in 2006 (strangely enough, the real-life Godavari 'came' in August 2006) and you see the changes that have happened over, around Godavari in these many years. The characters in this film (again two women, one man and love, this time expressed) are urbane, and have no living bond with the river, per se. In this film, neither the time (within the story), nor the camera, dwell on Godavari, nor do they 'stay'; the river is there, of course, important, yet in a limited sense. The man is an idealist who wishes to enter politics to change the world, the woman he loves does not relate to his world-view. The third woman enters by chance into the scene – a feisty, career woman who wishes to open a design boutique of her own. They are forced together in the circumstance of a marriage 'procession' on a steamer from Rajahmundry to Bhadrachalam. Yes, every scene is on the river, but not so much connected emotionally, intrinsically to it; rather, it is like being on the tourist boat (here garish, which was specifically made for the film), which in a real sense today, is the mainstay of tourism industry in Rajahmundry. Every scene evocatively depicts people, and the pace and their excessive consumption – eating, playing, the untamed rush to consume, the cooking on the boat, the journey to Bhadrachalam, and so forth (not critically, but as a happy coming together of family and friends in a typically depicted Telugu film where marriages are always happy family events). Nobody, not even the characters, dwell on the river, either on its beauty, or on it simply being there. The dialogues do not mention the river. The river is not important to the development of the characters or the story. For, naturally, the hero and his love interest are city-dwellers who shall return to Hyderabad once the consumption of the Godavari on the boat is complete.

Muga Manasulu may have been one of the inspirations for this film but that does not make a difference to it. Just coincidentally, though, actor Sumanth is part of the Akkineni Nageswara Rao family. But if this is the portrayal of Godavari in a film titled *Godavari*, one may not fault the director Sekhar Kammula (who made a name directing 'off-beat' films like Dollar Dreams and Anand) really. In fact it is a very true and contemporary reflection of the larger canvas: the way the nature of Godavari and movement on the river changed in all these years. As also the gaze. If Godavari, and crossing her over, was a necessity of those times, for the people living beside and along the river, today it is a tourist attraction for people who just arrive and leave. There is no need for the dramatic 'crossing over', unless, of course, you were among those who continue to live in villages all along its banks in east, west Godavari, Khammam districts, etc. In the context of the audience for which *Godavari* (2006) was made, today you can just enjoy the feeling of being on a river, without the related crossing over, and it does not ever touch the moods or upheavals (there are no upheavals, incidentally in *Godavari*, in the sense that you see in *Muga Manasulu*) ....

Godavari by Shekhar Kammula (considered to be among the more sensitive directors in the present crop of young Telugu filmmakers) depicts the Godavari post-commencement of the Polavaram project, incidental to the making of the film, but perhaps indirectly connected. For, tourist consumption of Godavari (especially the special launches that started making round trips from Rajahmundry to Papikondalu, or prior to that, to Bhadrachalam) had begun on a scale unprecedented; and the film made it even more prolific – both the tourist traffic and thereby consumption, of Godavari. The river now exists as mere scenery, with the Papikondalu hills as a backdrop as each day scores of private tourist launches take off at Polavaram or Rajahmundry (depending on the depth of the river and the season) towards Papikondalu or Bhadrachalam. The garish boat in Godavari (the film) is the main scene of action. Here the main characters (protagonist Raja and the heroine, played by Kamalini) meet. The journey is important for them to get to know each other. The boat may be the scene of drama throughout but on a 'by-the-way' Godavari river. As incidental as she is to the tourists who cruise past every day in these parts. Her presence does not add to or correspond with the emotional play between the three main characters. Their lives are untouched by her presence. Unlike in Muga Manasulu where, without Godavari, the boatman Gopi wouldn't be and without his presence, nor would there be a woman travelling to and fro; not even the play of their lives in that boat. And no open blunt expression of love or romance as it is conceived in the film Godavari. Later, there was another film, titled *Gopi, Gopika, Godavari* (2008). Intriguingly though, they pick up for 'prop' a mobile health clinic on a boat. Incidentally, there was one such boat run by a local NGO some years ago, giving ad-hoc services such as medicines, etc (I remember seeing one of their boats when I first journeyed along Godavari in 2006 June) but those services ended in the years of the tourism launches taking over. So some observation has gone into conceiving heroine hailing from these parts, an interesting, independent woman, a doctor in this mobile health clinic that she manages, for local villagers. But it ends there: her radical character. Once the hero appears – from Hyderabad – it is yet another tame love story, and has nothing to do with the clinic, nor the river. The strange new thing that one noted, however, this was the first film with Godavari for its backdrop that uses the word Polavaram (that is where the heroine is supposed to be living, though by the looks of it, they seem to have shot the film somewhere near Singanapally). Of course, the characters are the non-tribal upper castes who have settled in the tribal belt. It is important that the boat, the heroine and her mother are upper caste non-tribals.

What is the point of these comparisons? In it somewhere lies the increasing spate of Telugu film crews turning up along the banks of the river Godavari in the recent past for shots of song sequences, or more. None of the films show the characters of the place. They are as removed from the river as could be. Yet, if you see the sub-text of the films in question, there is a very strong caste and class character to the entire story, which reflects another historical truth. *Muga Manasulu* builds on, imperceptibly, perhaps, or unconsciously, the nature of crossing-over allowed for the characters in the story. Of Radha's, to an education which was the privilege of the Kapus, Kammas, Brahmins of the time, with their landed property; and Gopi's is a life that is tied to the river and the boat and cannot cross over, either to express love for the woman he loves nor to the college he has merely heard stories of from Radha. At a third angle, Gowri cannot belong in the high status that Gopi's heart has reserved for Radha – in love – and she is usually referred to as a *picchi daana* (mad girl) whenever she expresses her desire for him. Their characters were not allowed to go beyond their class or caste. The other recent films, too, have been about class, set against the urban backdrop. But essentially these are films made by filmmakers from that very background, in caste, class terms, though Adurthi Subba Rao, is known to have experimented with Left leaning thinking in some of his work, especially *Maro Prapancham.* Today there are local contractors in the submergence zone who are contacted by the film crews who wish to shoot in these environs and the mainstream films shot here are almost always about the picture-postcard scenery around the place and either part of a dream sequence or the backdrop of an agrarian landlord's family living in the place. Between 2006 and 2008, I was told several film crews set up their units for a few days of shoot and left. Unconcerned about the larger goings-on in the area they seem to love for its external beauty. [35]

***

Seeking Godavari's Colonial Connections/Constructs November 2008 – January 2009

In late 2008, early 2009, I seek Godavari in London, the capital city to the 'Empire' of the past. Not by design, but by a chance-offering (call it destiny) of some distance. It helped make sense of all that I had so far seen in my Godavari journeys. I did not go to London seeking Godavari, really, initially, but it so turned out that I stuck to "Godavari" in my library and archival searches and happened to find a few linkages in historical time that make my London trip, too, part of my Godavari Journey 2008. From November 2008 until January 2009 (browsing through the India Office Collections and other sources at British Library) was not a planned itinerary, but one of those detours that lead you ultimately to the road you initially started off on. In my case, it also led me further into a thought-process that had begun by 2006, as to where the Godavari anicut fit into the larger scheme of things. Here I was, between a contemporary and a colonial design on the river Godavari – the Polavaram project and the Cotton Barrage/Godavari anicut. The latter having its own imprints on an idea of Godavari on the former. This physical detour to London led me also to a quaint citizens' Association formed in late 19th century. But it gave one insights into their responses to the irrigation and public works edifice built by the colonial rulers. And there were Arthur Cotton and his many lectures, as well. I must confess I wonder, in amazement, sometimes, as to the way in which this journey has panned out in terms of places I ended up at so far in the course of the Godavari journeys and this book, a long road with several connections to the past. There was the Godavari river, Arthur Cotton's anicut, a trip through Manchester[36] where Arthur Cotton would have been lecturing endlessly between 1840s and 50s, convincing entrepreneurs (including the members of the Manchester Cotton Supply Association) about the economic 'sense' of his grand project on Godavari and his views on the usefulness of riverine navigation as against the Railways. The Government of Andhra Pradesh, meanwhile, was doing the same kind of convincing activity back home, seeking investments in Godavari and the State's coastal corridor from private entrepreneurs in the present time and context! It was also funny, in a sense, to think as to how the erstwhile colonial power continued to retain most of India's written records of history (although in a democratic and transparent edifice of the contemporary times), at times seeming like a symbol of power that comes from being the repository of so much 'knowledge' which colonialism constructed around it. Had I not been to London, I might have missed the Godavari District Association and an important chapter in the history of the Godavari districts, which to me signified the nature of state control over the lives of the farmers accessing the Godavari through the channels built by the empire, something that continues today's times, especially the idiom control, perhaps with a few alterations. Strangely, ironically, but fortunately, for once in my life, my doctoral research (rather post-doctoral, in essence) this time round subsidised my journalistic (even if not merely journalistic) engagement with Godavari and not otherwise. Ironically, again, there was little to be found regarding the Tamil Jainas in the British library which was my primary interest in this case (barring just a few nuggets of information) when compared to the wealth of information around 'public works' and Godavari and Sir Arthur Cotton. Later in the book I share some of these archival discoveries with the view to comprehend the legacy that politics in Andhra Pradesh has been built upon, to a substantial extent. This became clearer to me through these papers and records that I discovered in 2008, though I had already written about (by then) the connections between the Godavari anicut[37] and the idea of 'control' and dominance. The anicut was not simply an engineering enterprise, but a symbol of an economic perception of India as being part of the structure that built an imperialist power-language. And the economic dimensions of the Godavari delta were important for the self-perpetuating colonial empire. Most of the political contestations have wrested on Godavari and Krishna delta in Andhra Pradesh; controlling and managing these have meant controlling and managing political power over the entire State, post-Independence. But the idea of building political power over rivers had taken root in 19th century under colonial rule (although I would say that control over water resource – through irrigation channels and tanks with the creation of an entire mythological schemata – happened within a Brahminical religious paradigm many years prior to the British entry into the region) through myths around the divinity of the tanks, around which you found Brahmin settlements and land grants given to Brahmins by kings and chieftains in various periods in history. However, strangely enough, as far my research in the Tamil country goes, though Buddhism and Jainism were also patronised by many kings and chieftains here, there were relatively lesser number of temples with tanks or associated water bodies in their case, when compared to the number of well-endowed (in terms of natural resource distribution) Brahminical ones. The myths too helped in giving divine sanction to these water bodies and resources by virtue of some divinity having visited these. The Kaveri delta region has been historically shown to have had one of the largest numbers of Brahminical settlements patronised by royalty in different times. In the case of Godavari delta, too, inscriptions from temples in these parts have also revealed the existence of several such settlements from the early centuries A.D. Some more work on this needs to be done but the idea is to show that the regions which have had a longer history of dominance under the Brahminical temple worship paradigm smoothly transited into as well as submitted to the colonial politics and ideas of modernity, where (this is important) it helped them maintain their religious and caste status.When I talk of the Brahminical (as a system), I include the upper caste categories of the Reddys, Kapus, Kammas, and the whole lot, coming within the purview of what has come to called 'Hindu' religion. In places where these clans (which became caste categories over a period of time?) were the royalty, we again have historical records showing land grants and other patronage extended to the Brahmins for a ritual sanction of their powers.

There is also the idea of a river and its surrounding areas which has developed on a British model of land tenure, revenue, forest administration. I would like to believe that the same segment of people, who in the past comprised the influential landed segment, where agriculture was their mainstay, over the years transformed into an influential landed segment from the delta region into industries, so the nature of demands now have also smoothly metamorphosed into demands for land and water, again, for different purposes. Continuous exploitation of water and land, for them, has been the constant through centuries. In this case, if there is anything permanent, it is un-change.

***

Land Acquisition Progress[38], Polavaram Project 2008-09

April'08
Survey Done & Sub-division record prepared (acres)
453.00
Award passed
453.00
Handing over land to dept.
812.42

June'08
Survey 50.00
Handing over land to dept. 166.37

September'08
121.50

Award passed 21.00
824.99

November '08
180.40
181.69
6.00

March'09
367.65
178.80
178.80

Cumulative Progress
24162.71
11089.42
22802.12

***

Notes

1. Most of this journey happened alone (in the sense that no activist was accompanying me in the field, etc and the routes were chalked out alone; except in one case, partly because I had wished it that way). Until 2010, it remained that way, much to my satisfaction.
2. This journey was almost entirely possible because of the NFI Media Fellowship, in addition to a friend's help in times of need.
3. Including smaller assignments and the flexible job - the latter by personal choice. If one found an offer for a salary, one wasn't sure if that would allow one the freedom to travel to the villages as and when one wished. Or there were workplaces that wouldn't indulge me on my Godavari affair. Or spaces that weren't politically compatible. And media (alternative, or mainstream) did not assure regular spaces. On hindsight, perhaps Godavari journeys and a salaried job would not gel.
4. Though officially the fellowship came through in March, it was a while before I received the first instalment amount; and properly resumed my travels in June'08. March/April also brought in news, of another fellowship, as a Nehru visiting scholar at the Victoria and Albert Museum, London.
5. March was a strange month, with overwhelming events – good and bad – including an indulgence I brought home, in addition. I was to realise much later the important role this new companion of mine – a pup - was to play in her own unobtrusive ways in all my journeys undertaken for this work. And she still does.
6. Which was then a region; it has become a State now. I retain the state names of that year, to retain the historicity of my journeys, or the truth of that time.
7. While nearly all my time, journalism subsidised my PhD, this was time for my

PhD to support my ostensibly journalistic engagement with Godavari. As Nehru Scholar at the V&A museum, London, I did not just revisit the Tamil Jaina history, but also to chanced upon the Godavari in several papers of India Office Collections in the British Library. I found more material on Godavari in the library than I did on the Tamil Jainas!

8. This visit was not done with the NFI fellowship, and was more of a personal trip, which somehow led me the ghats and the first sighting of rituals over the river. I now realise this was also part of the larger Godavari journeys.
9. Originally from the Udupi region in Karnataka.
10. For Telangana. Finally IIT did come, but to Hyderabad; not Hyderabad physically, but it is called IIT-H, in an area that has forcibly become part of the construct called Greater Hyderabad.
11. Christoph and Elizabeth von Fürer-Haimendorf, *The Reddis of Bison Hills: A Study in Acculturation,* Macmillan and Co Ltd, London, 1945, p. 53. Incidentally, he mentions *sama,* which is a generic term for millets. There are different varieties of *sama* or millets grown. Incidentally, in the Preface to the same book, the Haimendorfs had written (and one can perfectly understand why), that in Peninsular India "the aboriginal populations are among the disinherited of fate and to identify oneself with their interests means a depressing struggle against oppression and exploitation, a heart-breaking sense of frustration in the face of the tribesmen's loss of land, economic freedom and self-respect." (p. ix) This was in the early 1940s.
12. To Kondamodalu.
13. An article on my meeting with Ramarao and the issue, with the title, "The Last Crop" was published in *Down to Earth* magazine, November 16-30, 2008. By the way, it was sent in June 2008, right after I returned back to Hyderabad). I am not reproducing the abridged published version entirely.
14. The story goes this way, I think. Sage Viswamitra planned to use the strength of his penance to send Trisanku to heaven but the king of gods, Indra, unhappy with the proposition, scuttled the plan; ultimately, Trisanku stayed in a land which was neither heaven nor the earth nor the netherworld. The term is used in everyday parlance (in Hindi) suggesting a 'neither here nor there' fate - between two impossible possibilities.
15. Not in the manner of Godavari's coming.
16. Assigned land. That the pattas will be non-alienable, and be brought under cultivation within three years of their being assigned.
17. Yes, this time I have a B&W film roll given to me by a friend who doesn't need these anymore.
18. Not my type, in terms of being expensive. But 'elite' sense in Rajahmundry is not the 'elite' sense of Hyderabad or any other big city. For the class character in these places is not the same as in bigger cities or does not express itself similarly, though it is present. The workers are from local places and nearby villages, and there is no air of speaking in English and behaving a certain way. It is too expensive - Rs. 850 - 1,000 per day. Am here thanks to a small assignment I did and a friend chipping in. And the strangest part is to be writing on Kondamodalu, sitting in this place. In all my journeys, this is the first time I have stayed in this expensive a place - feels guilty. But it helps me look at some other aspects that I did not know existed. And the river from my window

temporarily sets to rest other self-doubts.

19. M. Venkatarangaiya, ed, *The Freedom Struggle in Andhra Pradesh (Andhra), Vo.II (1906-1920 AD)*, The Andhra Pradesh State Committee Appointed for the Compilation of a History of the Freedom Struggle in Andhra Pradesh, 1969, pp. 271*ff*.
20. Ibid, p. 277.
21. Pushkaram is the Godavari puskharam festival held once in twelve years, when the river is supposed to cleanse people of all their sins, which they do with a ritual dip or bath in the river at various points; it is one of the most important festivals in the region, held as highly sacred.
22. Last day of the dark fortnight (of the waning moon), considered auspicious or inauspicious, depending on cultural contexts. Some mercantile and occupational groups do not work on these days.
23. He happened to be one of the co-founders of The Hindu, the English newspaper founded in the 19th century.
24. M. Venkatarangaiya, op.cit, pp. 271-72.
25. Ibid, p. 276.
26. Sent to the EPW in early August 2008, an edited version of this appeared as "Accentuating the Social Divide", *Economic and Political Weekly*, January 31-February 6, 2009, pp. 17-18.
27. Source: MDO office, Polavaram; and APO. Y. Srinivas, M. Jyothi, DWMA (District Water Management Agency), Eluru.
28. *Java* is a gruel made from either millets or sometimes rice. In their case, it is prepared with millets.
29. A state government policy that had drawn flak from many, but was continued; where two government schools were merged in order to 'address' the problem of teachers and lack of infrastructure.
30. An edited version of this story (written and sent in August 2008) was published in The Hindu Sunday Magazine as The Last Resistance, on 29th March 2009. I use parts of that version, too, here.
31. *Kudipi* literally refers to a preparation made of grains meant as cattle feed, of thick consistency. The mouse feels trapped in it, once it has fallen in, to taste it, and cannot get out of it. The idiom suggests being trapped in a situation, struggling to get out.
32. *Madras District Gazetteer: Godavari*, 1915, p. 283.
33. Hemingway, ibid, p. 12.
34. He was born in Rajahmundry in 1912 and directed his first film, Amara Sandesam in 1953. Maro Prapancham, his controversial film hailed as revolutionary, came out in 1970.
35. There was another film recently with Godavari as a backdrop – Gundello Godaari, with the 'floods' as a central context of a love story of a couple from a fishing community.
36. Passed through Manchester on my way to Lake District invited by a friend to share some thoughts on the Godavari-Polavaram with his neighbours and friends, and activists, with their own concerns on ecology in the UK. It was a revelation to them, the numbers of displaced (as they had not heard too much being talked about Polavaram, unlike the Narmada long ago).

37. I discovered the *Gazetteer of Madras* at the MIDS Library in Chennai, in late 2004, and got myself a copy of the same, which later became a point of reference for the colonial context during the Godavari journeys.
38. Source:http://ppms.cgg.gov.in/DetYearProjLA (Centre for Good Governance). Only few months taken here, to see the general progress of land acquisition.

The Ritualistic Godavari (January'08)

A Farmer Family preparing a straw boat with offerings

Filling it up with an earthern lamp

Setting it to float, thanksgiving to Godavari for a good harvest

Gunjavaram R&R Colony, Of Homes and Homelessness, Fields Crushed (June'08)

Kondareddis building a home at Kondamodalu—a community affair, in year 2006

Koya homes being built with hired labour at the R&R colony today

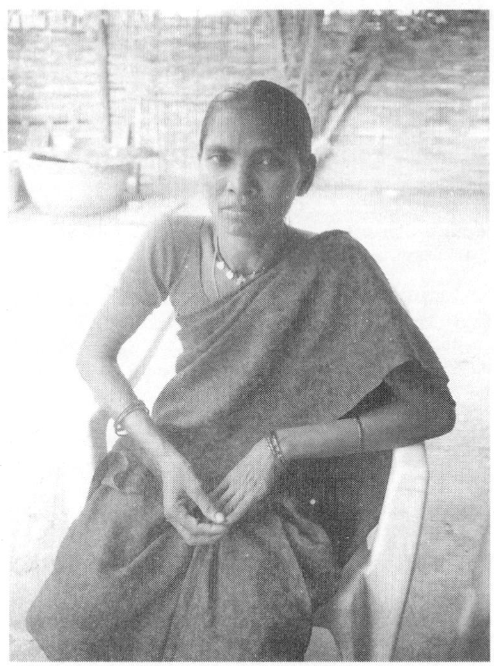

Madi Muttemma of Kokkarigudem

Last Crop, Life in Between, NREG Scams (June'08)

Fields Broken through (Mamidigondi)

Twin Tunnels of Dam through the villages (Polavaram mandal)

Boragam Rama Rao on his destroyed field (Mamidigondi)

Meeting the Political 'criminals' - Chegondapally Sunnam Raju, M. Suramma

Boragam Kannayya - Devaragondi

Old woman at her home in year 2007 (Devaragondi)

Same unfinished home, abandoned, in year 2008

Madakam Kondayya in between homes (Devaragondi)

Women at Chegondapally Ramalayam

Landless, unemployed dalits of Pydipaka (the NREG Scam)

Pydipaka Jadla Mary outside her home, with cracks from dynamite blasting at spillway site

Chittumalu Sitamma (Pydipaka)

Pydipaka Andru Peddaraju

Ancient Tamarind Tree enroute Sivagiri

Kondla Bullebbayi at the Martyrs Memorial, Kondamodalu

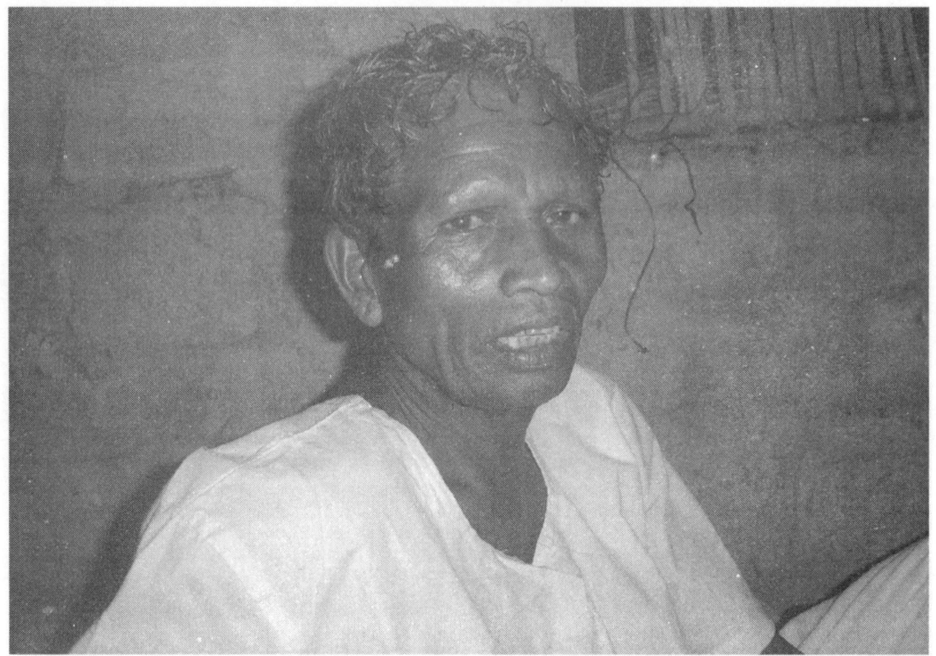

Kondla Gangaraju, Kathanapally

Where Nights are Nights, Dark and Deep

Repository of History, Illa Rami Reddi, Somalapadu

What a Dawn! Mist and Clouds over Kondamodalu

Godavari Journey 2009
Electoral Democracy, by People or Money, and Related Matters

Journey of Life

Sometimes life's jolts take time to recover from. While you do recover, several questions nag you – whether the 'within' and 'outside' (of Self), or 'individual suffering' and others' are really so unconnected? Do personal vulnerabilities impact on social 'responsibility', especially in the life of an independent/freelance journalist? What bearing does being a woman, being single, or source of income[1] have on your work and how you write? Does money figure in this picture? Does having it, or not having it, cripple you, or set you free? I let those questions be, for another time.

In the meantime, in January'09, my final month of the short stint in England, a friend[2] invited me to tell his neighbours and fellow environmentalists about Polavaram at a small gathering in Kendal, in the beautiful Lake District which poet Wordsworth made famous (or vice versa). You can't even begin to think what lyrical romanticism would do in a context like this dam! Little did I think of what was to come of it, but something did. For, the people present that day wrote and endorsed (as signatories) a letter that they faxed/emailed to all the major political parties in AP (and to the PMO, too, I think). The letter said, among other things, "We are concerned about the adverse ecological implications of the Polavaram dam on the region, the river, its natural flows and the forests and wildlife; in the light of the current global economic crisis…We are ourselves questioning the route of development we have followed (in the West) and we strongly reiterate the need to rethink an economic model more environmentally, socially and ethically…We support the tribal communities and other marginal communities of East Godavari, West Godavari and Khammam districts of Andhra Pradesh in their struggle against the Polavaram dam project…"

The temporary set-back[3], till mid-2009, forced a full-stop to my Godavari visits in that year, barring one. This time, there were just a few people I managed to meet in just one single visit which I still call a journey, because it was a sort of turning point in AP politics and Polavaram. 2009 also put me in the call of 'duty' (as journalist) as the Telangana agitation picked up momentum in Hyderabad. The year ended with a visit to a colony of the displaced in Hyderabad (on account of the GMR airport at Shamshabad) resolutely fighting their own, continually adjourned, cases in the Upa-Lokayukta. Many of these seemingly unrelated events helped join to dots, about the larger canvas, for the Godavari journey in my head and heart. Future would tell that what was to end with my Journey 2008 continued well through 2010 and 2011, thanks to that one eventful year.[4]

***

The Electoral Battle
April 2009

The elections happen. A couple of phone calls from familiar voices, about what was happening in the 'submergence zone' literally pulls me out. This is a crucial moment: the elections to the AP Legislative Assembly and the Parliament of India. Will these elections also decide the fate of the Polavaram project? Perhaps. People of Andhra Pradesh have generally shown 'freaky' results at these times (they showed the door to the former Chief Minister Chandrababu Naidu in year 2004). In April I make that short, but significant, trip to villages in all the three districts. And again I meet few familiar people, because this time round, I find some of them contesting in elections as candidates! But it fills me with hope, just the idea that a Kondareddi is contesting a Parliamentary seat from this region for the first time, a candidate put up by his own community importantly, to highlight Polavaram-project displacement as an electoral issue; and yet another from that village Mamidigondi of Boragam Rama Rao. I witness a massive public meeting addressed by Y.S. Rajasekhara Reddy in Kovvur, West Godavari District, giving his famous trend-reversing direction to the elections that the Congress had almost about lost. And I see him address people atop his campaign van at Rajahmundry town, which reveals a crucial aspect regarding Godavari waters. Incidentally, this was to be his last ever election process as he died in an air-crash in September 2009 within months of becoming the Chief Minister a second time. Even before 'paid news' became critical news in national media discourse, the Polavaram issue has seen much more of media sell-out in the local areas; I heard people in Polavaram and elsewhere mention the term *'packagi'* this time in a different context: that each political party or leader had given a package for coverage to some media houses (TV or print) depending on the column size – full page or half; front page or supplement, colour or monochrome, with or without picture. I did not investigate this particular exposé; though it could well be a deep issue, requiring all kinds of skills, to analyse media in the Polavaram context, right from the time the project's foundation stone was laid.

April 12th - Rampachodavaram

I have come to meet, well, Illa Rami Reddi and Kondla Gangaraju, outside their homes and in the electoral battleground, where they are both contesting as Independent candidates (supported by the CPI-ML Party). Rami Reddi is the first Kondareddi contesting the Parliamentary Constituency of Araku and has the torchlight as his election symbol and Gangaraju is contesting for the AP Legislative Assembly seat Rampachodavaram and his symbol is the drum. Appropriate, I think, signifying light and a wake-up call, respectively. They have for their vehicles, a few autorickshaws. A single cab has been hired, as well. Resources are in short supply, but naturally. And the contesting is more of a symbolic nature, since they are aware of the chance they stand against all the biggest names eyeing these parts (Congress is defending against majors like Praja Rajyam which is in this electoral scene for the first time, Telugu Desam Party, as the major contenders, though CPM has some chances in Araku and Khammam). I was lucky to get a ride that morning from Rajahmundry to Rampachodavaram along with some other reporters from local Telugu media. Bojja Tharakam of the Republican Party of India says, "Whether it is Congress or Mahakutami (the grand alliance) or TDP, there is no difference in their policies on SEZ or land. Whoever comes to power will continue with the same set-up. We are using the electoral process as a forum to expose their vested interests and lopsided polices. We have aligned with CPI (ML) on common agendas especially regarding SEZs, unemployment, education and health."

Araku parliamentary constituency consists of Devipatnam, Maredumilli, Y. Ramavaram, Gangavaram, Addateegala and Rajavommangi. Of these, Devipatnam and Gangavaram have direct and indirect links to the Polavaram project in terms of submergence and the R&R colonies in these mandals. Whether the Polavaram project impacts on the votes cast in these areas remains to be seen. And whether Illa Rami Reddi is able to bring a sizeable number of votes in these areas is another question. But for him, fighting with the spirit of resistance is the main battle. He says, "Instead of free rice, if they gave land for tilling it will help lakhs of farmers. In Rampachodavaram, the Kondareddis have fallen behind in education. It is the same case in East Godavari and in Khammam…We are saying 'no' to the Polavaram project. Many agitations have happened. There has been no prior example of such a vast volume of displacement anywhere else… No gram sabhas were held. What can ordinary people know about cases filed in the Courts on these issues?"

A Press Conference has been organised by the candidates and I notice very few media (mainstream) are interested in this one; so the candidates wait a while for some more to turn up. The conference is addressed by the AP Rythu Coolie Sangham members, including Nagaraju, the advocate,

State Committee Member of the CPI (ML), K. Kotaiah, Suresh and Bojja Tharakam (of RPI, which is fielding 15 Assembly and 5 MP candidates), and of course Rami Reddi and Gangaraju. The 'top' leaders (non-tribal) speak first and then it is time for the Kondareddi candidates to talk. Nagaraju immediately mentions Polavaram in the context of displacement, but I find the discourse slightly changed this time round. For he says, they are demanding that every tribal household be given land for land and forest for loss of forests within six months after which they may start the project. I think to myself, they were against the project not so long ago, why bring in the compensation terminology, like all others are doing? Well, things change, and especially if I have not been here since last several months, much water has flown under the bridge that I need to figure out. Illa Rami Reddi speaks of the Bhupatipalem Reservoir which was built without people's consultation and tribal people were arrested instead (for protesting) and sent to Rajahmundry Jail. Similarly, the girijans displaced by the Surampalem project did not get the promised lands, so was the case with the Yeleru reservoir (30 years ago) where people haven't yet received any compensation. "Where will they give compensation to so many lakhs of people displaced by the Polavaram project, if they have not compensated these smaller numbers?" he said. "They slap cases against us saying we are supporters of the Naxals, if we question them. Shouldn't YSR and Chandrababu Naidu also be charged with *desadroham* (Sedition) cases since they are fighting within the Assembly which is regarded as the pillar of our democracy? Their intention is to give water to the foreign companies...(They) are not protecting people suffering from the dynamite blasts by the Aluminum company in Araku but are protecting the elephants. We want to tell them we can protect the elephants ourselves! The government speaks of protecting elephants and tigers but not tribal people in whose habitat these animals survived for centuries...Our demands are 'stop mining, protect adivasis, Polavaram displaced people must get land for each land taken, and complete compensation..."

They serve soft drinks for the media persons (since it is hot and the reporters may need them!) Incidentally, this is the first time I am attending an electoral situation in this region, among the people here and closely observing the kind of hold media persons seem to be having. I was to see more of it, soon, especially in Polavaram. Rami Reddi, et. al., are hardly the 'important' stories in the newspaper columns. Incidentally, there was no Telugu TV Channel present at the PC (Press Conference). Whether these two candidates will change the situation, how could the mainstream media *not* consider it important that they are Polavaram-affected tribal communities, discussing Polavaram within the electoral debate, while no other candidate has even mentioned it (barring the celebratory passages by the Congress on their Jalayagam)? I cannot understand the media, I think. My idea was to cover some of the villages along with Rami Reddi and Gangaraju on their campaign to see the response of their own communities and others affected by the project whom I had met earlier in my journeys (especially the Indukur people), but they are not going to those villages now. I guess they have already covered those areas in the course of their campaign in the last few days. But they are moving on to Devipatnam, where a meeting is scheduled for tonight in the middle of the village. That is good enough and I decide to move along. Oh, and who should be the cab driver, driving us to Devipatnam, but that same angry old man who was cursing us (Saikrishna and me) all the way to Sivagiri! How things here keep moving around; sometimes it is like a musical chair that you see the same faces in different contexts and conditions, each time. I tell him, 'I hope today you will not crib as much?' He smiles, sheepishly. It is his livelihood as well, and he cannot avoid these roads, much as he may protest. I wanted to take the autorickshaw ride, but then it is moving along with campaign material and I am a journalist here, and I should keep my distance. So I move on ahead.

Afternoon, at Devipatnam (East Godavari District)

It takes a long time for the auto rickshaws to reach with their loudspeakers and campaign material; there are many women too, in the procession which has been following the two Kondareddi candidates and some of them have come from their villages far away to cheer their leaders. I watch the spectacle and wait, it seems like endlessly, for the evening meeting to start (at 6 pm, they had said; I would attend for a while and leave). I am also waiting for the arrival, as announced, of some candidate from the Praja Rajyam Party and people are very excited that he is to be accompanied by none other than the star (Telugu actor, otherwise called "mega star") Chiranjeevi. Devipatnam has become very busy indeed and many shops are making brisk business. I will be going to Polavaram this time, as well, because that is where much of the heavyweight contest in these parts is going to be – the Koya 'sitting' Congress MLA, Tellam Balaraju is contesting from this constituency. Then there is the former MLA, from TDP, Punnem Singanna Dora, another Koya. And I am curious about the CPI-M's decision to align with TDP. This one jars, though. One Party (CPI-M) has been against the Polavaram project, and the other (TDP) in fact inaugurated the project before YSR went way ahead with it! But in electoral politics, they say, you have to be prepared for the strangest of alliances and calculations.

Just walking by, I am invited to escape the hot Sun at a small eatery. And here I meet fellow Tamil-speaking people! This eatery belongs to Pir Mayadin and as I stepped inside, into the home which is an extension of the shop, I was struck by the sight – the mud-brick cooking space, with a door opening out to Godavari down below, and against the Sun outside, the play of lights and shade was a marvel. It is almost as if the kitchen extended into Godavari (which was, *ikkade*, 'just here', in literal sense)! I could sit there forever, oblivious to the din outside of elections – that moment's thought, of course. Hearing Tamil was yet another marvel. For the second time in all these journeys in these parts I heard my mother tongue spoken (the first was at Polavaram spillway site). I never imagined there could be Tamil people living here, in Devipatnam of all places (in spite of the fact that this region was part of the Madras Presidency of the colonial times). I asked the woman, Nagural Bibi, "*neenga Tamila*" ('are you Tamil?'). It was then, that she (Pir Mayadin's wife) told me they were originally from Tirunelveli, but had come down here more than thirty years ago. Incidentally, this was also the first time I was meeting a Muslim family in these villages ever since I started coming here in 2006[5]. Her husband came here through someone's reference apparently, and gradually established a small eatery, which is where his home is now. I am thinking, the R&R package does not mention Minorities, strangely. Why? There are a few Muslims in V.R. Puram mandal, too, I think, or Kunavaram (very few, but there). I then met their son and daughter, Syed Ali and Nysa, as well. Nagural Bibi said she was planning to go to Tirunelveli sometime soon after the elections. She will go to Rajahmundry and take the Sircar Express. They still have their extended family back in Tirunelveli. After a cup of tea, even in that heat, and something to munch on, I decide to walk around a bit. Just to take in the election 'air'.

I will go sit by Godavari, too, to set my mind at rest. The river is still at this time of the day, late afternoon. I reach her from a small bylane, through a couple of huts. Finding a woman on a boat, I reached the place and asked her if I could sit on the boat a while, which I did. And I was to forge another bond in a while, unawares. A glass of *majjiga* (buttermilk) at fisherman Gangadharam's hut (at Chapalapeta, Devipatnam) offered by his wife Manga (who was the woman on the boat) feels like nectar under the scorching Sun. I watch the placid waters of Godavari across their home, and observe Gangadharam's son and daughter sewing up their nets, a scene I capture in my old camera. This home has seen several Godavaris come and leave, but never left them homeless. Soon that is to change. This family does not figure in the list of those affected by the dam. But it is not the dam that they are immediately concerned with as we sit talking about who they think would win this time round – TDP, or the CPI (ML) or Congress. He is not sure about who he should vote for since he believes there isn't a candidate who represents him or the concerns (incidentally, this is a tribal constituency) of his community of Jalarlu, categorised as BC. With the Polavaram 'barragi', he will lose even the small thatched roofed hut he lives in, on the banks of Godavari.

Where will these elections take his community? Which end of the Godavari is really going to be freely accessible to them? Not a single candidate speaks of that aspect. No wonder he wasn't so enthusiastic about attending the late evening meeting of Rami Reddi, a Kondareddi, who he believes only represents the tribal interest and not his community's, which has been bearing the brunt of tourist launches on the Godavari since the last two years. The fishing nets, which cost them Rs. 1,500 (they usually keep two) get torn by the tourist launches which also affect the fish catch. For three to four months a year, 30 to 40 of them pass by. Petitions and requisitions have been useless. Sometimes his family manages fish catch of 120 kgs a day, sometimes none at all.

I thought, whatever else the dam will do, it will most certainly disrupt the diversity of the village. Chapalapeta here, for instance is the fishing community's hamlet; then there are other social groups in the village; in the R&R colonies the effort has been to house only the tribal groups in one place; in some cases there is disruption even among the tribal communities when panchayats are merged, or 'broken' up. It definitely means a disruption of social – cultural (and most times accompanied by occupational) complexity of these villages (not in the exotica sense of a typical Indian village) in terms of a lived history. At the moment, though, life may not be full of bounties for this fisherman in Devipatnam, but gives him enough to be peaceful in his stomach. Even as I take leave of them and thank them for the spontaneous offering of *majjiga,* I am thinking in my head, which I jot down in my notebook, "...To meet them again, with their pictures..."

Meanwhile, compared to the star Chiranjeevi, expected to arrive in their midst, the Kondareddis seem like a 'household affair' for the people. It takes a long time for crowds to gather, but they do, in quite a good number, by the local standards. But as long as I was there, neither the 'mega-star' nor his Party's candidate arrived on the scene. I was disappointed, like all the rest; for I was keen to hear what the new PRP was to say about Godavari and Polavaram. I am also looking at the nature of campaign: here, for the two candidates, it is a minimalist exercise and they do not and cannot of course pay money in return for votes. In that sense, they are absolutely 'free' to speak their minds, and stick to what they said, without the need to make promises they cannot keep. This 'symbolic contesting' is good for our democracy – you can speak the truth on a stage without fear and this is the one time you will not be arrested (unless, of course, you indulge in slander)! Basically, they had nothing to lose. So, with their *dappus* and a group of singing men and women, this made a visual treat to the eyes. By the time I reach the HPO, Rajahmundry (yes, again) it is pretty late in the night. Thankfully Nagaraju, a resident of Rajahmundry, and a few others from there who had come for the meeting agreed to drop me at the HPO. Had it not been that late at night, I would have refused that ride. But there was no place to stay in Devipatnam that night, without imposing myself, and no buses either at that time. Reluctantly, I agreed to take that offer.

***

13th April – Polavaram - The place has far more bustle than one I saw at Devipatnam. Am taking in the 'colours', as I see one right in front of my eyes, by the Polavaram MRO office, a jeep with a red flag and a yellow one, and I am amused at the politics of colours, thinking.

Cell-phone notes (back again) - "This shade of yellow and this shade of red just don't gel. Am not even thinking about mixing. The resultant shade then, in that case, would be saffron? Vermillion? No, there is a dash of pink here, as well. So, don't worry. Have I been sleeping too long? The world has changed meanwhile? A Tanjavur *bommai*[6] - you cannot put her together blindly; she will look a most awkward atrocity!.."

"Streetside economy, or the economy of the streets; a day with images. (Spotted) a chandelier (yes) on a pushcart and a keymaker's table, locked to (an) iron chain around a tree."

At Koyyelagudem (for a Public Meeting to be addressed by YSR, et.al.)

A large crowd has gathered at the *maidan* here in this town. I came here with Sitara master and a few others from different newspapers. It is hotter than it was yesterday. There is a separate space made for the "press" where we are all standing and waiting for the meeting to begin. I am keen to hear what will be said about Godavari in this meeting. This will be my first: hearing YSR at an electoral public meeting of this nature (have heard him at press conferences and events before). Apart from some scathing words for Chandrababu Naidu and Chiranjeevi being inexperienced in politics, the major part of his speech is about his ambitious jalayagnam and 48 irrigation projects, including Polavaram. He is campaigning here for the sitting MLA, Tellam Balaraju and MP candidate, Kavuri Sambasiva Rao. And he reminds people of loan waivers, Paavala Vaddi, NREG, Arogyasri and other 'achievements' of his government[7]. And he ends with a joke on Chandrababu Naidu, "A Sage cursed Naidu, that if you tell the truth, your head will break into pieces"!

Tellam Balaraju[8] stays in Jangareddygudem; as a sitting MLA, people tell me he never paid a visit during 2006 Godavari, nor has be bothered to address their grievances regarding Polavaram; he may be a Koya by identity, but his loyalties do not lie with the tribal people. I gather these tidbits from people gathered there and some of the cadres of rival Parties. But one piece of news that caught on, in this midst, was that people brought in as cheerleaders for YSR at the meeting were, in fact, daily wage construction workers of the Right canal of Polavaram project! They had been given a day-off and some money, along with Congress caps and flags, to be present there. A reporter found this out through a casual chat with one of them, who did not know Telugu and was speaking in Hindi. Some of these workers are from Madhya Pradesh. And two of the stringers present there had actually met them at the Right Canal in connection with the Polavaram project story! Many scribes (who had not already decided their loyalties) had found this news obviously juicy. I do not know if the TV channels picked it up. The workers had to wave the flags and clap at orchestrated intervals without understanding a word of what was being said. These workers, incidentally, were working for the Company called Progressive Constructions, one of the main contractors in the Polavaram dam right canal (at that point), belonging to Sambasiva Rao (the MP candidate).

Polavaram constituency has five mandals and a total of 1,66,290 voters. Among these, the people to be displaced come from 26 villages in Polavaram mandal alone – that is for the construction of the main spillway. These are not numbers of those displaced or losing lands to canal works. In Polavaram mandal, 30,565 voters will exercise their franchise. The number of project affected people is roughly estimated at (under conservative estimates) around 8,000.

Post-meeting, I met Tellam Balaraju, even as he was leaving the venue, to ask him about the Polavaram project. He said, "People are ready to leave; in fact they are waiting to go. 4,000 acres of land has been acquired. Polavaram is a dream project. Many people tried to build it, but could not. It is Y.S. Rajasekhara Reddy, who with his grit and determination made this possible. Since the last 50 years people from this region, in the dry-belt (*metta prantam*) have wanted this project. Tribal people are not opposing the project. In fact they are saying that 'we should get the best compensation and it would be good if the project is built soon and we are moved to some other place (*vere chotaki taraliste baaguntundi*). This is the best RR package in the country, but some people have yet to be included in the package; they want pattas for *konda podu* lands, and package for girls above 18 years. I have conveyed this to the CM. We will see that all this is implemented. Land acquisition is going on; it is in DD stage. There is sufficient land for the land-to-land package in my constituency. There are 29 villages and we have acquired approximately 4,000 acres land; we will give land to land once the processes are through."

Note: By 'DD stage' he meant, I presume, the Drafts have been prepared for disbursing the amounts. But I thought only cheques were being handed out? I could not clarify as he was being rushed into his car by his secretary.

People like Tellam Balaraju and Singanna Dora urge one to revisit stereotypes of Koyas and tribal communities (as a homogenous 'tribe') being oblivious to 'money' and power-seeking. Their rise within mainstream political parties is intriguing enough to ask, whether, in becoming part of a mainstream politics, identity and value of histories associated with that identity undergo such a massive shift. Their mannerisms, attitudes, and most intriguingly, the kind of Telugu they speak (with a certain educated flourish) are all at stark contrast to a common Koya person in the submergence villages. But in Tellam Balaraju's case, he was born and brought up in Jangareddygudem which is a town closer to the non-tribal belt and quite into the mainstream even reflected in the highway that runs through this town. He may thus be a Koya by birth, but possibly a non-tribal in his general attitude. This is not about the person, but an example of the nature of social conditioning that affects the ideas of the world around us, which need to be kept in mind. My initial reaction was the same as others, 'being a Koya, how is he oblivious to the nature of the project and its impact on his community?' This was my own stereotyping as well.

This was one long, long day without food and water (the bottle emptied in the first hour). I just had to eat. Have to return to Polavaram. Few party workers of the CPM and staff of Prajashakti newspaper whom I met on their way back, guided me to the bus-stand. Finding a small restaurant nearby, I ate some rice and *chapa pulusu* (fish in tamarind soup) and moved on.

***

Politics and its Many Faces and Polavaram Project in this Midst

At Polavaram, I met Punnem Singanna Dora, the Koya former (TDP) MLA of Polavaram, at last!

And I also meet (by sheer chance) supposedly the richest man in Pydipaka (as per earlier Pydipaka visits). He stays at Polavaram, too. He is the close aide of this former MLA and a TDP worker! I meet both of them in Singanna Dora's house in Polavaram, where, incidentally, I also bump into the founder of this NGO, supposed to be working against the Polavaram dam displacement, G. Anil Kumar[9]. I had always heard from people here that he is related to Singanna Dora, though the kinship is confusing: some people address him as *bammaridi* (brother-in-law) rather than by his name and some others say he is Singanna's wife's nephew and related by marital ties[10]. I finally meet him, too, in the field, for the first time, this time campaigning for his uncle. He is in a great rush, so I do not have the time to ask him the question if he finds it a problem (politically) at all that he is campaigning for a Party that was at the forefront of the dam building, which I thought he was against? But, there are many equations and local dynamics at play during elections. Just watch, observe, I tell myself. I will get some more surprises along the way during the conversation, I hope, and may be later. Punnem Singanna Dora has already served as MLA under the previous Naidu regime. In terms of local respect and credibility, he seems to have not fared too well, as all along the way I ask people about him and many a times in previous journeys they have blamed him for having been oblivious to their problems, especially with the dam. He does not fare too well in this electoral battle, it seems like. TDP laid the foundation for Polavaram dam, which the party has advertised several times. So there is no question of opposing the dam. I record this interview, with his permission. But since there are two people here, I am asked to speak to Kuncha Venkatratnam (the Pydipaka landlord) who is also a TDP local leader. I did not expect to find him here, nor did I know his Party affiliations. Now I do. I had met his partner Kuncha Varalakshmi in Pydipaka the first time before and after the 2006 Godavari. He is supposed to have sold around 125 acres of land (though in MRO records, he has only 26 acres; but the Returning officer informs me it is 22 acres). Dalits in Pydipaka say he owns other lands for which he was given compensation; they had worked on those lands as agricultural labourers. But when he talks, he seems like any other displaced person. Venkatratnam says, "None of us in the Telugu Desam Party, workers or leaders, are against the Polavaram project. We have sacrificed our own free and beautiful lives here and have been waiting to move out since the last four years. We sacrificed all this and were willing to assure all kinds of support to the barrage no matter which Party builds it. But the government turned it all into a curse and cheated the displaced people very badly, helping the contractors and earning from that (commission); they have done great injustice to the genuinely displaced people. Before the project started, they called people from the 29 villages, including the non-tribals, and promised that 'we will do justice to you, and give you a complete package, and resettle you all in villages exactly like the ones you are living in right now, with the same atmosphere and the government will bear all the costs, and help you to resettle without any problems or losses. Only after you have seen all this, after six months, we will start the works here'. They made all these promises…it is close to five years now and not even one village has been shifted to other lands and not one displaced person has been given the due compensation properly. Not a single displaced person is happy today. The sitting MLA (Balaraju) has not come even once here…He has not met a single displaced person in these five years…How will farmers till the lands in this condition when their lands are all being dug up in the name of the barrage? He has been making money giving lands on lease to contractors to dig up and dump soil on the lands. But since the otherwise helpless people today have a weapon called 'vote' in their hands, the displaced people and all those who have suffered at the hands of this government for Polavaram project, are saying that this government will bear our curse and will lose even their deposit…Even Chandrababu Naidu[11] has said that we will build this project, it is necessary. But we will recognise and acknowledge the people who have made sacrifices. We will give land even to the landless people, including girijans, to an extent of 5 acres. We will give pattas to those who have been cultivating the lands (which they do not have pattas for); Naidu *garu* (Chandrababu Naidu) has said he will give free house for one lakh rupees (unlike what is happening today, where girijans are building their own houses) and Rs. 2000 for every home, while Y. S. Rajasekhara Reddy is not giving a single rupee to the people here though he says that women will get money deposited in the banks in their name….If TDP comes to power everyone who is to be displaced, irrespective of caste, will get the package (compensation) – one hectare land for SC families through the SC Corporation and for STs through ITDAs. Though it is not mentioned in our manifesto, but there is a plan to amend the G.O (Government Order) 68 to that effect. Right now the government is only giving Rs. 1,30,000 for those who do not have land without any other kind of help…."

I asked about the cases slapped against people. "Yes, even our Anil Babu (Anil Kumar) was charged with Sedition. If our government comes, the cases against all the displaced people will be dismissed."

[I was calculating in my notebook about the compensation Dora Babu has received: if it is 22 acres, @ Rs. 80,000 per acre, from selling his lands for the project, he has made Rs. 17,60,000. But if he has sold 26 acres @ Rs. 1,50,000 an acre (as some government officials told me these lands were bought for Rs. 1,50,000 per acre), then he has made Rs. 39,00,000. But yes, if he has sold, 125 acres as the dalits and the Koyas allege, the figure would indeed be one crore (one million). But before I could ask him the question as to how much compensation he got, he rushed out to the SUV waiting at a distance, with Anil Kumar for company.]

Singanna Dora resumes now, "We (at TDP) want Polavaram dam to be recognised as a National project. In our R&R package we will give land to every tribal person, as well as employment protection, and we will also include minority communities in the R&R package. I was the one who set up R&R (package negotiation) committees in this area because we believed in the government's assurances and promises. All turned out to be false. We will cancel all the 124 A cases slapped on people here."

Singanna Dora says he and his extended family lost around 26 acres to the project. But the RDO Jangareddygudem (who I meet later that day) informs me that Kundrukota (which is Singanna Dora's village) is yet to be given compensation, and that Singanna Dora and his family held only 12 acres of land. The true figure is debatable. This village will be compensated in the second phase of land acquisition. Around 1,000 acres have been identified for acquisition in Kundrukota. In all, Kundrukota has 1,280 acres of land.

***

Through my conversations with people I can see that CPI(M)'s alliance with TDP has cost them their credibility in these parts. Until 2007 they were thick into the movement against the Polavaram project. TRS has not been making many statements on Polavaram, in the elections this time, I notice. Meanwhile, I manage to speak to the Praja Rajyam candidate, Boragam Srinivas over the phone as he is busy on a 'road show' (they call it!) and will not make it to Polavaram in time. And I find out that he is Boragam Rama Rao's brother's son! Was it this man I met in that ITDA school where I had seen children sitting under the canopy of a massive tamarind tree (in 2007)? Perhaps. He had then told me how the government was wasting all the resources as the school building was under construction at that point (and it would be submerged). He was worried about the fate of so many children studying in schools such as these, in environments such as these. I cannot say if it was him unless I see him, though Srinivas was also a teacher before he took to politics. He tells me (and I am surprised, considering the fate that his uncle suffered with his lands) "Polavaram project is essential. But we are demanding the best package, house for house, land for land, package for SCs. If PRP comes to power the package will be so good it will become a model for other States."

***

A Colonial Period School in Polavaram and Sitaramachandramurthy's Story

The school teacher in him urged Sitara master to pick up a unique angle in the midst of all the loud proclamations in the election battle. He wrote a story (dated 15th March'09) on the government high school in Polavaram and its students who later became important politicians of the State. He took me to the school which I had missed on earlier journeys. I took a photograph from the outside, since it was closed. I really did not want this story of his to be lost in the game of spaces in his newspaper. I do not know if it was published finally (in the Prajashakti district edition) but I got a copy of the original, written by hand in Telugu, with me. [12] I reproduce parts of it.

He writes, "The Polavaram Government High School was established in 1912 under the then ruling British Government. Since then, till date, many girijan and non-girijan children who got their education here have moved from local to State level politics. Is there any historic school such as this in the Agency Areas? It is indeed a matter of wonder. The Polavaram Indira Sagar project made me think - of Godavari river's vast expanse flowing between the East and West, its pleasant breeze (*challati gaali)*, thick forest area and Polavaram in the midst of the hills and this old school, built in British times, with a long history of 97 years....Students from West Godavari, from Polavaram, Koyyelagudem, Buttaigudem, Jeelugumilli, Tallapudi; and East Godavari, from Devipatnam, Sitannagaram, Rampachodavaram Mandals, some of these even 70 kilometers away, come here to study. This school has a great history...Students from the Polavaram Government High

School have entered into State politics, having imbibed the scientific training in this historic school. Following is a list of elected members of the Legislative Assembly who have studied in this school:-

1. Pusuluri Kodandaramayya - Polavaram constituency – 1955-62, MLA.
2. Syamala Sitaramayya - Khammam constituency – 1955-62, MLA.
3. Chode Mallikarjuna Rao* - East Godavari, Yellavaram Constituency – 1962-67, MLA.
4. Kondamodalu Rami Reddi* - West Godavari – Polavaram Constituency – 1967-72. Represented the Constituency (contested). His own village is East Godavari, Devipatnam mandal.
5. Karam Bapanna Dora* – East Godavari, Yellavaram Constituency – 1962-67, Represented this Constituency (contested).
6. Chinnam Joga Rao* – East Godavari, Yellavaram Constituency – 1983-85. Represented this constituency (contested). He studied in Polavaram Government High School from 1957 to 1962. He was born on 27.11.1947.
7. Midiyam Lakshmana Rao* – Polavaram Constituency – 1983-87. Represented this Constituency (contested). Studied in this school from 1957-58. He was born in Buttaigudem on 1.06.1945.
8. Punnem Singanna Dora* – Polavaram Constituency – 1994-99, MLA. Studied in this school form 1955-60. He was born in Kundrukota on 1.1.1942."

Sitaramachandramurthy's story was also accompanied by photographs which his usual friend of the local photo studio took which he sent to his newspaper, Prajashakti.

[* Tribal people who entered mainstream politics, and they all studied at this school. This part is not mentioned in his story, however. Was surprised to find that some people I met are also in the list!]

Life of a Stringer

Few stringers would pick up a story like this one. But publication of Sitara master's stories would usually depend on the Party's stance (which moved regularly between being vehemently opposed to the dam, soft opposition, to opposing it in 'its present form' and being silent when aligning with Telugu Desam Party in 2009 elections, a strong pro-dam party). In many cases, the money spent on the whole effort would be far more than the remuneration or honorarium received at the end of it. But he was better-off in the sense that his "word-rate" was standardised and fixed, and there the Prajashakti paper was prompt[13]; and his story would appear in the district editions, if not the State one. Of course, he was a family man who needed to get those 'word rates' and file a stipulated number of stories to run an establishment (and perhaps he was also doing other things to supplement his income, I never knew nor tried to find out). But when it came to the Polavaram dam, his character might find space as a part of an intriguing element in the dam discourse, but also for being a great local chronicler of other factors around Godavari and people of this region. Being a stringer for a Telugu paper is a task in itself, but being stringer for a Communist paper is quite something else – selling stories at the rate of Rs. 1.50 per word (it was 0.80 paise not so long ago) and not gaining more than a single column space or even less at times. This would include spending on phone calls to the typist at Eluru, or to the local photographer for his pictures. The Polavaram dam saga was an important point in his life; and he found lot of space when the Party was leading a popular movement against the dam. There was one part of him that I only learnt that day over lunch with his family; a man who reflects some fascinating aspects of the old world idealism in a town where things were fast changing with a tornado-like money-spinning activity commenced in the last three years where everybody could make their pie. Or, if they did not wish to, they remained quiet. At the end of it, they compromised in the form of silence, after initially raising their voices, or resigned to the inevitability of the situation. He named one of his daughters Priyadarshini because he told me he used to admire Indira Gandhi at one point. Both his daughters are getting the best of education, even though his income keeps dwindling with his word space shrinking in keeping with the local politics of the day. And his wife is not too happy with the situation, as she tells me when we are alone in the kitchen, and have a moment to ourselves. His story is somewhere the story of local media in the Polavaram issue. He used to work as a teacher in a small private school, but loved writing so began to send articles and later stuck to Prajashakti for many years as a stringer. His efforts to get a mainstream Telugu media job always boomeranged; he was either not 'qualified' enough to be called a journalist or that he came from Polavaram made matters difficult. Staffers usually based themselves in larger towns, such as Rajahmundry, Eluru, Bhadrachalam. 'Locals' could never attain the status of a 'staffer'/staff reporter! I did not know of these intricacies before. He had extensively written on the corruption and the scandals relating to the R&R compensation issue in each of the villages around Polavaram. He informed me of Gunjavaram R&R scam. There were also things he could not write, staying as he did in the place with people of whom he knew a lot; he would note them in a personal file. If someone could spend more time with people like him, it would make for an amazing historical document. It says a lot about how media works at these very local levels amidst more immediate fears facing a person up-close, especially one with a family and hopes for children's future. And there is the question of caste and community affinity which also plays a role. Life took many turns very quickly for me soon after I thought of working on this very 'theme' so his story remained incomplete, even if there was hope within that I will follow it up the 'next time' .

***

April 13th - Reflections in my Notebook

Characters I met in all these years seem like characters in a novel being written. It is a book in making; the characters have faced their trials and tribulations (just as I have) since 2006, battling circumstances...Each of these faces I see is in the process of writing its history. In every visit, every journey, I see a new chapter being written in the lives of these faces. Boragam Srinivas, Kuncha Varalakshmi, Kuncha Venkatratnam, Banerjee (who always meets me mid-way on his bike, in well-timed intervals, just as he is whizzing past – now he is a CPM activist, now he is a TDP worker, sometimes he is on parole; this is one man whose occupation is a mystery even to his neighbours in Polavaram), Illa Rami Reddi, Kondla Gangaraju (whose village is almost a haven, far removed from the political intrigues of other villages dragged into the Polavaram dam construction)...

Boragam Srinivas is now the PRP candidate. Kuncha Venkatratnam, an upper case (Kapu) landlord who became the earliest 'oustee/beneficiary/project-affected' taking a huge R&R compensation. I remember seeing him at the MRO office during 2006 Godavari. He now wears a thick gold chain around his neck and looks forward to brighter days if TDP comes to power. Anil Kumar is here too, campaigning for TDP, his NGO workers metamorphosing into TDP party workers. I met him at a meeting in Hyderabad two years ago where he opposed the dam. Now I get to know from Singanna Dora that he was part of the 'negotiation committee' - which in a way indirectly assisted the land acquisition process. At Devaragondi, Madakam Kondayya had told me in her village Tellam Balaraju was also part of the Committee. Well, Punem Singanna Dora, TDP candidate (with active footwork from CPM cadres who were once opposing him) is very hopeful of becoming the MLA again. All these characters, I just watch them, observe, and am amazed by them. Sometimes am angry about the intrigues, sometimes sad, but I cannot (and would not) judge them. Their stories are penned by me, and at these times I feel little better than a scribe merely, like I am meant to be one. They are all living their lives in the midst of a polity (politics) that is debauched and circumstances that overwhelm. Never do they give up! Not all of them are the innocent 'village-folk' beguiled by corrupt officials. They know on which side their bread is buttered (and why shouldn't they, when everyone knows it). And the history of this region has a lot to do with their life-stories. It has been a centuries-old history of many such political intrigues. And it is not a wonder that many tribal and peasant revolts have their roots here. There is just one name I seek so much to put a face on, however, but never succeed. Not his face, because his physical form is history (or a notorious legend in these parts), but that of his lineage. May be I will, in my next journey – Hota Sriramachandramurthy, a former Brahmin landlord of the Hukumpeta zamindari whose family (as per local legend) apparently owns (or owned) thousands of acres in the Agency Area in West Godavari district and sold hundreds of these to the government in the Polavaram land acquisition process, *apparently*. This news is local legend; you will hear it everywhere you go, including in the offices of the MRO, etc, unofficially. Most of their lands were acquired for construction of R&R colonies and as 'compensation land' for tribal people to be displaced by the dam. LTR (Land Transfer Regulation) cases had been filed against this family for owning land in Agency Areas (pre- and post-independence).

The colonial Gazetteer mentions Hukumpeta. "Gangolu: Eight miles west south-west of Polavaram. Population, 1,784. Its hamlet Hukumpeta is the head-quarters of a zamindari which was acquired from the Gutala estate by purchase about 40 years ago and is still held by the descendants of the purchasers. It comprises four villages and pays a peshkash of Rs. 1,240."[14]

By the way, Mallu Dora, one of the closest tribal associates of the revolutionary Alluri Sitaramaraju, was the first tribal MP from Andhra Pradesh (Vizag constituency) in the immediate post-Independence period, winning as an Independent candidate, supported by the Communist Party of India (united). Since then, there have been other tribal representatives in the Parliament yet the situation on the ground, though perhaps some notches higher, remains pathetic.

Voicing Dissent[15]

"When mainstream democracy fails to see or hear you, it is time to flash the torches and beat the *nagara* (drum), and march on… The 12 demands in their manifesto (of Rami Reddi and Gnagaraju) includes implementing the land ceiling act on the ayacuts newly formed by irrigation projects and distribute these lands to the poor; proper implementation of the PESA (Panchayati Raj Extension to Scheduled Areas Act); to regularise tenant farmers and protect their rights; to recognise agricultural labour as a skilled work, among others… For these two Kondareddis, the spirit of resistance is the main battle…"

Telangana – Some activists, intellectuals who were against the dam, mainly on the grounds that the waters would flow into the Andhra region depriving Telangana, are now campaigning for the TRS Party, either overtly or otherwise. But if Telangana State comes up and gets more water than has been promised today (If), the <u>basic idea</u> of the dam, one would wish, would be seen as problematic in the ecological, democratic and economic sense; it does not even concern itself with the idea of a river or forests.

***

Polavaram in the Politics in 'Displacement Zone'[16]

Neither the dam, nor displacement, figure in electoral debate here even if the Polavaram project, by itself plays the most crucial role as the biggest money-spinning exercise the constituency (or the state of Andhra Pradesh) has seen in the last five years. Polavaram constituency has far more intriguing politics than one that meets the eye in urban constituencies... There are 5 mandals in the Polavaram constituency with a total of 1,66,290 voters, comprising of Polavaram (30,565 voters), Buttaigudem (33,829, Jeelugumilli (17,965), T. Narasapuram (37,374) and Koyyelagudem (46,557). Voters directly affected by submergence to the Polavaram dam come from 26 villages in Polavaram mandal.

***

And, the Results Come In
May 2009

Godavari Waters and Return of the Congress in Andhra Pradesh[17]

Within a month of the Congress government taking over the reins in the State its Major Irrigation Minister (for second time in a row) has announced that the Government seeks "national" (project) status for five irrigation projects (related to Godavari waters). These include Polavaram (top on the list of priorities), Dummugudem tail pond, Pranahita-Chevella, Sujala Sravanthi and Sripada Sagar. He informed that Rs. 18,000 crores have been allocated for these, of which Rs. 4,000 crores would be spend for clearing pending bills. However, neither Polavaram (dam) nor Godavari waters were part of the electoral discourse this time round, barring the seemingly sudden 'shift' in the Congress' campaign strategy (in the last lap) in coastal Andhra districts – warning people about possible lack of access to the Godavari waters if Telangana State become a reality. Utilisation of the Godavari waters slogan was not invoked this time to the extent it was in the 2004 elections, at least by the TRS party or (to a limited extent) the Left (mainly, CPM). Water resources had a far greater share in electoral debates in 2004 than in 2009. Pulichintala, Polavaram, Dummugudem figured in most election meetings in year 2004. Both the Congress and CPM had raised the issue of Pulichintala in 2004. TDP, of course, was always all out for Polavaram and there was no charge in its stance this time round either...In a conversation at the time of the elections, the former engineer K. Vidayasagar Rao (from the Telangana think-tank) said, "Nobody is bothered about water issues. TRS was never against water being given to farmers, nor against Polavaram, *per se*, but only against it in its present design and form. We have been highlighting the need for several small structures – not a big dam – to minimise the extent of displacement. With the consent of Sonia Gandhi then, we got the YSR government to agree to initiate studies on alternatives but nothing came out of it. Ultimately, political parties will talk of what garners local votes. They are not interested in academic debate about these issues. TRS will (if it came to power) demand fort studies on Polavaram. Pulichintala has already started; but Polavaram can still be stalled. TRS should be asking the Government to set up a committee of outsiders with no stake in the issue. In 2004 even Congress was opposing the Pulichintala project but what they do right after coming to power?"

Now, in the face of absolutely no opposition coming to the Congress on these matters, there will likely be no assembly debates on the question of displacement, possible re-think on the level of displacement or even equitable regional allocation of Godavari waters. While the Congress government seems to have hit the bull's eye making a mention of the water issue in every meeting in regions where it mattered most (in terms of voting), other parties failed to even raise these questions. However, a former TRS MP from Warangal, B. Vinod Kumar said this was the fault of media, not presenting the right picture. He said (over the phone), "In all my meetings in Karimnagar (Parliamentary) constituency I spoke of Godavari waters, and raised the question of the Babli project." And he added, "We are not against Polavaram project but we are opposing the nature of displacement. We want Godavari waters for Telangana region first. If the Telangana State is formed, we shall not deprive people in other regions their entitlement to Godavari waters, but our priorities would be met first. A lot of water from Godavari is wasting into the sea. Canals and projects in Telangana must be built first; later delta farmers can get their share. Pranahita at Kanthanapally and Dummugudem in Khammam district must be completed…"

B.V. Raghavulu, CPI(M) (over phone) had said (during the campaign), "We want Polavaram project redesigned so that the submergence is negligible. Committee of experts such as Hanumantha Rao and Dharma Rao have submitted new designs where there is no need for a high reservoir of 150+. It can be reduced to 120… But where is the money for the project? They are estimating Rs. 2 lakh crores. Even Congress needs to rethink its priorities. The CWC has not yet approved the project design."

Rampant commercialisation and commodification of the Godavari river will be complete in the coming few years. And it will once again prove that industrial interests shall weigh over agricultural interests…Now that even the few remaining voices of opposition within the Assembly have been silenced; the implications on the spate of irrigation projects lined up in the State are worrisome. Several legal violations and socio-economic dimensions of these pending projects (not to mention the long-term environmental impact of restraining natural flows to the sea) will no longer be part of the Assembly debates, going by the interest (or lack of it) shown by parties such as Lok Satta and PRP in water issues, or Godavari waters. The second coming of the Congress government in the State means aggressive consolidation of a form of privatisation and total control of Godavari waters we have not seen before.

Where the Smaller Numbers too Matter

Among the candidates (from among those to be displaced) who stood for the first time, such as Illa Rami Reddi (for Parliamentary seat) and Boragam Srinivas (Assembly) the performance was not bad, considering that they had no money power to catapult them and were fighting veterans in their respective fields. Illa Rami Reddi polled 12,086 votes (most votes most likely from East Godavari region). Araku, incidentally, is a fairly large one, spread over four districts. The former CPM MP from this constituency, Midiyam Babu Rao (a Koya) polled 1,68,014 votes losing by a large margin to the Congress candidate Kishore Chandra Suryanarayana Deo Vyricherla who polled 3,60,458 votes. The contest was more interesting in Polavaram Assembly constituency where Boragam Srinivas the Praja Rajyam Party (Koya) candidate from Mamidigundi polled 36,483 votes while the TDP former MLA Punem Singanna Dora got 44,634 and lost to Tellam Balraju. Balraju got 50,298 votes. Total number of votes polled in Polavaram were 1, 66,290. Boragam Srinivas says he plans to continue working on the field and does not wish to return to teaching as of now. For a first timer who was given a ticket as a 'wild card' entry, his voting figures have a lot of significance, also since he has lost his land to the Polavaram project already. The dam-affected people may not have won a major victory but they have stood their ground in a battle of seasoned politicians and cast their votes against a government. These smaller numbers and their votes do matter.

***

June'09 - Some interesting news arrives

A journalist friend in Delhi sent me the following press release originally sent by the Deputy Director, AP Information Centre, New Delhi, P. Kiran Kumar on Friday, 19th June, 2009. I was interested in just some of that information. The press release pertained to the issues that the Government of Andhra Pradesh discussed with the Union Minister for Law and Justice and Union Minister for Water Resources. Some of it I paraphrase here below:-

"Issues for discussion with Hon'ble Minister for Water Resources: Request for declaration of Indirasagar Polavaram Project as a National Project – A proposal had been submitted to the Ministry of Water Resources through Central Water Commission for according the status of National Project to Indirasagar Polavaram Project under the guidelines prepared by the Ministry of Water Resources for implementation of the scheme of National Projects…The State Government has already spent Rs. 2,589 crores under this project till May, 2009. The status of National Project will help the

Government of Andhra Pradesh to get 90% of Rs. 5,974.52 crores eligible project cost that would expedite timely completion of this ambitious project."

"**Request to instruct the Government of Maharashtra to stop the construction of 11 barrages intended to utilize over and above their share of water** – The States of Maharashtra and Andhra Pradesh have concluded an agreement on 6.10.1975 over the utilisation of the waters of river Godavari in the basin area above Pochampad Dam. In terms of the said agreement, Maharashtra can use 60 TMC of water for its new Projects above Pochampad Dam. Further, Andhra Pradesh can go ahead with Pochampad Project with FRL+1091 feet and MWL+1093 feet. Pochampad Dam (Sri Rama Sagar Project) was constructed across river Godavari by Andhra Pradesh to cater to the irrigation and drinking water needs of 7 districts of Telangana and Godavari river is the only source of survival for the inhabitants of these districts. Government of Maharashtra is constructing 11 Barrages across river Godavari in between Jaikwadi and Sri Rama Sagar Project. The 11 barrages are (1) Amdura Bandhara (2) Digras Bandhara (3) Muli Bandhara (4) Mudgal Bandhara (5) Dhalegaon Bandhara (6) Loni Swangi Bandhara (7) Raja Takli Bandhara (8) Apegaon Bandhara (9) Jogladevi bandhara (10) Mangrul Bandhara (11) Hirvopuro Bandhara. With the construction of 11 barrages the entire stretch of the Godavari river for about 430 km would turn into a perennial source providing unlimited scope for drawl on both sides, much beyond the planned capacity of 7.20 TMC... Government of Maharashtra is going a head with the construction of 11 barrages at a rapid pace, without responding to the repeated requests from Government of India and Government of Andhra Pradesh. It is therefore requested to instruct the Government of Maharashtra to stop the construction of 11 barrages intended to utilise over and above their share of water."

The last point is important in so far as the construction of so many projects on the entire stretch of Godavari is most certainly going to impact the future of the river itself, as well as the ambitious, foolhardy design of the Polavaram project, as well.

***

Godavari in the Telangana Discourse – December'09

In late November 2009, the Telangana statehood agitation picks up tempo and one catches up with that agitation as a journalist; to look at the importance of Godavari to this whole discourse. It was a tension-filled atmosphere and the base of the statehood agitation was the Osmania University campus, which happens to be located closer home. What started off as an indefinite hunger strike by the TRS leader, K. Chandrasekhara Rao (KCR), turned into a violent situation after his forced removal and continuation of the fast at a city hospital. This was followed by a protracted agitation at the University campus joined by a large mass of people from

Hyderabad and other places across Telangana region, including working classes, students, teachers, lawyers and so forth. What sparked off the period of protest was, initially, a violent police repression on unarmed students inside the university campus. It was as a journalist, with violence erupting closer home, at a university campus, that I entered the campus and filed a couple of stories[18] mainly focussing on the need to understand the form and content of State, Region, Nation in Indian democracy and, as the days went by, locate Godavari in the entire debate. Incidentally, Godavari, nor Polavaram, figured in the conversations with students and others I had interviewed in the course of the protest.

River Waters and Irrigation in the Telangana Debate[19]

The larger the investments in the Godavari river (in Andhra Pradesh), the more tedious the path to a separate Telangana state. Politically, geo-physically, economically it is the nature of inter-regional flow of Godavari and the number of interventions on the river that to a large extent holds in abeyance the issue of the separate state of Telangana. Out of the total catchment area of the river (with its tributaries) which is 1,20,777 sq. miles (3,12,813 sq kms), the state of Andhra Pradesh has 72,346 sq. kms. Of this, nearly 79 per cent of area lies in Telangana region (across seven of its ten districts) while 21 per cent in the coastal Andhra region (3 of its 9 districts). Yet, there has been phenomenally minimum utilisation of the river for irrigation development in Telangana. When Sir Arthur Cotton constructed the anicut over the Godavari in the 19th century, it assured water for more than one crop in the coastal delta region which were then under the Madras Presidency. Post formation of AP, too, the ayacut irrigated under this anicut later named Arthur Cotton barrage at Dowlaishwaram was more than 12 lakh acres. In comparison, irrigated area under the Godavari for Telangana was hardly 5 lakh acres. According to 2001 official statistics, of the total gross area irrigated in Andhra Pradesh (5,916,147 hectares) 48.95 per cent is in coastal Andhra region, 37.89 per cent in Telangana and 13.16 per cent in Rayalaseema. This disparity has meaning. The ambitious 'Jalayagnam' launched by the former Chief Minister Y.S. Rajasekhara Reddy ostensibly intends to extend irrigated ayacut in different parts of the state but at the same time harness river waters – specifically Godavari – towards development of SEZs along the 'coastal corridor', power projects and industries along the way. Even while extending irrigation potential the Jalayagnam prefers the coastal Andhra region to the other two. Of course, since he belonged to Rayalaseema, late Dr. YSR took pains to please his constituency. The total irrigation potential creation under Jalayagnam (as per official website) is again a significant pointer. 22 projects proposed will develop an ayacut of 36,37,719 acres in Andhra region, 26 projects proposed in Telangana will develop an ayacut of 29,13,388 acres, while 11 projects in

Rayalaseema will develop an ayacut of 17,60,500 acres. In between, they will feed the SEZ and their water needs. According to reports, prior to the merger of Telangana (Hyderabad state) with Andhra state in 1956, irrigated area of Telangana was 17,55,000 acres, which increased to 18,40,000 acres fifty years after the merger. At the same time, within fifty years, the irrigated ayacut of coastal Andhra region was 34.5 lakh acres before the merger which increased to 99.9 lakh acres. At the time of AP formation, more than 11 lakh acres in Telangana region were tank-irrigated – in fact, historically, the region was covered by a network of tanks (rainwater harvesting structures) under the rule of various kings including the Kakatiyas. Subsequently it decreased to 6.5 lakh acres owing to neglect on account of reduced budgetary allocations to so-called minor irrigation projects. In case of Godavari, there has also been the issue of inter-state riparian rights. The Godavari Water Disputes Tribunal, set up in 1969, which published the award in 1980 permitted AP to utilise 1480 TMC ft (thousand million cubic feet) based on 75 per cent dependability of water. Since Telangana region is disadvantaged on account of the gravity flow of Godavari (the river flows downstream all the way to the Bay of Bengal) lift irrigation projects were considered more suitable to make use of the river waters. However, barring two projects, the Devadula Lift irrigation project commenced in 2004-05, and Inchampally (which has interstate implications) successive governments failed to initiate projects of this kind. On the other hand, especially in the YSR regime, Polavaram Indira Sagar multipurpose irrigation project was given utmost importance…to divert 80 TMC of Godavari waters to the Krishna barrage besides extending an additional ayacut of 7.5 lakh acres in the coastal districts. How does Telangana lose further from this project? More than 200 of the 300 tribal habitations that will be submerged due to this project lie in Khammam district in Telangana. And it will also submerge around 1 lakh acres of cropped area in the district. Of course the project also submerges villages in east and West Godavari districts but the ones who stand to lose there are the powerless tribals and dalits. Pulichintala, Dummugudem and Polavaram projects will divert Godavari waters to Krishna while the plans to divert Krishna waters to Rayalaseema via Srisailam assured the former CM of support from his own constituency. In July this year it was announced that the State government has added two more projects, Kanthalapalli and Dummagudem-Tailpond, to the list, seeking financial assistance from the Centre. So far, 12 projects are said to have been completed and water provided to 22 lakh acres. At least Rs. 46,000 crore has been spent till date on Jalayagnam. Projects in Telangana were consistently neglected or their original purposes modified. Singur, which was constructed under the Nizams to extend irrigation to Nizamabad, was diverted towards drinking water needs for Hyderabad.

If Telangana state does come into existence, what would be its blueprint for Godavari or its idea on Polavaram? K. Vidyasagara Rao, retired Engineer from the pro-Telangana group, says "There would be a survey of the entire area to identify potential for creation of several smaller structures rather than a big dam. Several smaller lift irrigation systems will benefit more and there will be no interstate issue there…; we need to improve tank irrigation either by creating new structures or tapping the ones that already exist…Polavaram in its present form cannot continue…"

Conscious underdevelopment of the region and inadequate harnessing of Godavari waters has been a prime concern for proponents of Telangana state. The way a region articulates itself politically, aspiring to form an administrative, political unit, in this case, also depends on the nature of political control (conjunct with economic exploitation) over natural resources, including river waters. It is important to look at the Telangana statehood issue from this angle. The current nature of resistance to Telangana perhaps has more to do with the investments in irrigation and real estate projects and fear as to future access to Godavari waters for the industrial corridor (and the promised 5 per cent water from irrigation projects for industry) than with ideals of 'brotherhood' and 'unity', which were the refrains of the Telangana statehood opponents – that creating a new state was like dividing two brothers, or breaking up the unity of Telugu people, as those who were opposed to the Telangana statehood demand kept referring to: that 'dividing the state' would kill the unity of Telugu people.

***

Land Acquisition Progress[20], Polavaram Project – 2009-10

June 2009 – Shows no progress, no figures in the charts, but April 2009 1211.74 surveyed. Nothing acquired. August 2009 = 124.89 acres acquired.

(May-August was election and post poll times in AP, hence these figures. September-October – post-YSR's death, no land acquisition is shown. But November 2009 = 65 acres. Telangana agitation too picks up by the end of 2009, which affects the land acquisition work)

April
Survey Done & Sub-division record prepared (acres)
1211.74
Handing over land to dept.
0.00

August
1.49
124.89

Cumulative Progress
25378.48
23183.26

***

Notes

1. After the last installment of NFI Fellowship had come in and been spent on the election visit, this would perhaps have been my first longest non-income and non-media fellowship phase – so far as the Godavari journeys were concerned (till early 2010) – save a chance to translate a Jaina treatise. 'Work' meant intermittent stories at long intervals in the media, on the elections and the Telangana agitation. 2009 (until May 2010) also saw rejections (fellowships, grant requests), each to do with Godavari-Polavaram. Help came from the best few friendships.
2. Environmental activist Peter Bryant's totally unexpected invitation also meant I understood the differences in perception in Europe about hydel power projects being relatively more 'eco-friendly' than coal, which he and his friends back then were fighting against. I spoke there only by way of sharing what I had seen on the ground. Though the letter was sent to some of the print media back home, it was not something they considered of too much relevance and it never appeared.
3. Adding to other woes, a crippling nerve condition virtually imprisoned me in immobility between July and October. Fortunately, found an Ayurvedic doctor who cured my condition without a fee and set me back on track, to accept it only when I had work and was cured. Adversities also open your eyes to humanity and lack of it.
4. Is it justified or important to mention the personal? Perhaps only so far as to accept the entire possibility of limitedness of one's being, weaknesses within, and to accept that reporting too (or responding) happens in some conditions. Those must be heroic who can and do break conditions of this kind of limitedness, as a routine.
5. Though in 2006 I remember speaking to one Sonde Veeraiah, a Koya, but converted to Islam – he was part of an Adivasi activist group waiting to meet the CEC team at Bhadrachalam. I learnt he was Muslim much later.
6. It is a craft peculiar to the Thanjavur district of Tamil Nadu. It has disjointed parts which need to be assembled just right for the dolls to move their individual body-parts imitating that of a dancer – it is an amazing piece of art. Even one piece out of synch place gives the doll an awkward appearance, and renders the doll immovable!
7. Various government schemes on loans, employment guarantee – National Rural Employment Guarantee, now called Mahatma Gandhi Rural Employment Guarantee, health, etc.
8. Balraju, I am informed, has now joined hands with YSR Congress (started by YSR's son, Jaganmohan Reddy, from 2012) and is now in opposition against the ruling Congress Party for which he campaigned in 2009.
9. Anil Kumar, with his own NGO, ITDS, in the area, who was until recently a

TDP (and briefly Praja Rajyam Party) supporter is today (as in 2012) with the YSR (Jagan) Congress Party. According to many people I met in Polavaram and Singanapally, he was an active participant in the negotiations with the government for compensation for the dam, along with Singanna and convinced people to accept the RR compensation. In NGO circles, meetings and conferences he was anti-Polavaram dam. When I met him, in one meeting, he mentioned he was arrested on charges of Sedition for opposing Polavaram. Since he was in a rush campaigning for TDP with Dora Babu, I could not get his interview. Would have been interesting.

10. Cross-cousin marriage being common in these parts. Whatever may be the case, he is part of the extended family and a very close kin at that.
11. Former Chief Minister of Andhra Pradesh; he has become Chief Minister again of the newly constituted Andhra Pradesh in June 2014.
12. I wanted to send it to the Sunday supplement of an English newspaper where I had visualised it. Of course, no saying it would appear. I was to get it read out to me, translate it and send it in his name. He seemed happy with the idea. But I failed him; too many things caught up after my return home. I still regret not having translated few other stories of his on Polavaram into English for a wider audience. Of course, he never sent me his other stories; life must have caught up with him, too. His daughter was to join college soon after I met her. The Telugu text was read out to me by Ms. Uma (Lakshmi) in Hyderabad.
13. For at least two (or was it three) articles of mine published, I remember being called on the phone over and over again by the Prajashakti staff in Hyderabad to send my address or come pick up my remuneration - Rs. 150; but my most precious Rs. 150 that felt warm and nice. But more importantly, it was for a Telugu newspaper. Andhra Jyothy would give Rs. 250 for an opinion editorial. Again, much cherished moments (sometimes those moments would also not happen often!), receiving a letter from them with the cheque. Nothing to beat the joy that it appeared in a language people I met read. Who then cares about money?
14. *Gazetteer: Godavari Districts*, 1915, p. 278.
15. An edited version of my article appeared in www.indiatogether.org as "Jumping into the Fray Themselves", 27 April, 2009. Sharing parts of that article
16. Some of these thoughts were published in different spaces in that time, including Andhra Jyothy, the Telugu daily, www.indiatogether.org.
17. Reference given in an earlier section, some of these thoughts, including the "smaller numbers" (written in May'09) appeared in the article "Empire Flows Again" in www.indiatogether.org (dated 2 June 2009).
18. This piece appeared in the New Indian Express, and in the Telugu paper Andhra Jyothy.
19. This is the original version of my article. The edited version appeared in the New Indian Express as "River water holds the key" on 28 Dec 2009.
20. Source:http://ppms.cgg.gov.in/DetYearProjLA (Centre for Good Governance). Only few months taken here, to see the general progress. No explanations given for Nil figures - September – October show nil perhaps because of the demise of the former CM, YSR (September 2009). Following this, from November 2009, the Telangana agitation (for separate State) had caught on.

Electoral Democracy - Related Matters (April 2009)

Optical Illusion? By the banks of Godavari, at Devipatnam

Poster of Kondareddi Candidates

The candidates Illa Rami Reddi and Kondla Gangaraju at the Press Conference - Rampachodavaram

An Autorickshaw campaign team at Devipatnam

Nagural Bibi's kitchen that extends onto Godavari!

Syed Ali with Nysa-Nagural Bibi-Pir Mayadin's children

Manga on the boat

Malladi Gangadharam on his nets by his home

Gangadharam's children at the other end of the nets, repairing

Women attending the night election meeting of Rami Reddi and Gangaraju at Devipatnam

Congress Party's Tellam Balraju addressing the crowds at the campaign meeting at Koyyelagudem

Congress Party's Dr. Y.S. Rajasekhara Reddy addressing the crowds - same place

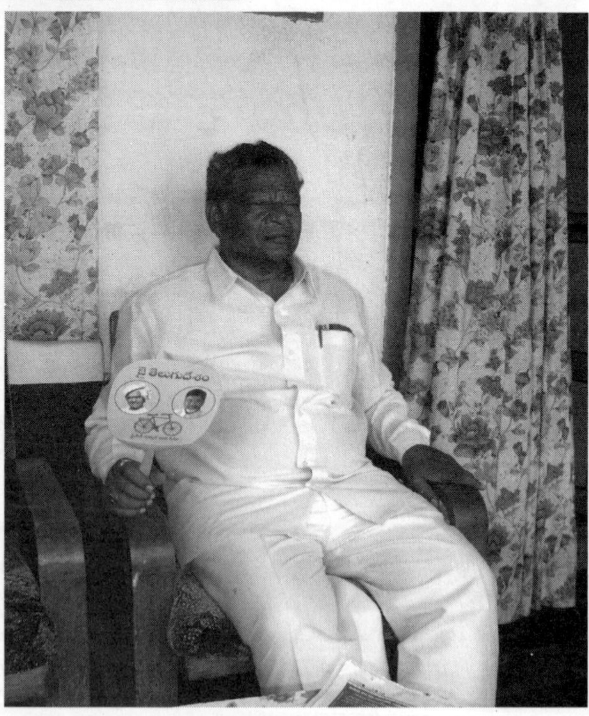

TDP's Punnem Singanna Dora at his house

TDP supporter Kuncha Venkatratnam (Dora Babu)

Singanna Dora's campaign poster on a jeep.

A Colonial Period School at Polavaram

Godavari Journey 2010
Meeting the Fish-hunters, Seeking out the Metaphor-giver, and Other Accounts

Back to the Source

It was like coming full circle. After all, didn't the idea of Godavari come from one of them? I touched base with them again – the fisher communities or fishworkers or fishermen and their families[1] – whom I saw from a different angle, as people dealing with their own kinds of dislocations. They do not figure in the R&R statistics of the Polavaram dam as 'project-affected', either in government lists or in the list or purview of NGOs working on the subject. And this time round, the Forest Rights Act is at the implementation stages; with its own paradoxes. And in this meanwhile, the Polavaram league (as I refer to the state and its agents – whosoever they may be) has brought in another order by Gazette in place making Papikonda Wildlife Sanctuary a National Park and hence out of access to tribal communities. This will be the first time perhaps that most parts of a National Park will be submerged by a dam with the whole-hearted support of the Forest Department. This time round, too, the visits were pretty intense and I managed to cover much ground. A lot went into making this year's journey, which also makes it special.[2] This year's journey was entirely based on some incidental small savings[3] and the FES-Infochange Media fellowship[4] on Common Property Resources. I am thinking, increasingly, of perspectives on nature, life cycles, inter-connectedness. Why is any talk of nature itself exclusivist or exclusionist? Why isn't wildlife, or nature itself, relevant to any understanding of the communities that our democracy (or caste, or class, or any other historical structure) has rendered marginal? Is there need to reclaim and re-invent the idea of nature from the perspective of the marginalised, who have been closest to *both* the redeeming *and* damaging aspects of nature? In the Indian context, Buddhism and Jainism as systems of thought and praxis looked closely at the human-nature interface and ideas of life-cycle where every organism had its place and in this thought was to permeate the everyday lives of the followers. Whether it did, is a question, and this is not the place for that discourse. But we may remember that these ideas came essentially from non-Vedic systems, and as protest against excessive consumption of nature and exploitation-driven Vedic religion. Nature in these systems was not an elitist enterprise. Today, 'nature' has become increasingly the preserve of the rich: at one end in terms of 'enjoying' what is considered 'pristine', 'pure' natural habitat as an eco-tourist with a certain idea of nature being necessarily 'unsullied' by the human habitation (especially tribal communities hitherto displaced in order to build an 'eco-resort') and the other as preserves of resources to be exploited. Perhaps in many senses, nature has been re-envisaged with a brahminical purpose. I have a problem also with the negation of communities having their own ideas of 'enjoying' their natural environs; as if being overwhelmed by the Sun setting over Godavari, or being quiet and reflective over mist over the Vali and Sugriva hill ranges at a distance, or pondering over life in the moment between laying the net and catching fish at dawn, are necessarily the 'stuff' of the urban elite, or an unconcerned romantic poet. Why is the language of protest against displacement itself from these environs so couched in the language of state power and productivity-driven, life-systems-negating economy? How many times have I seen people in their quiet and reflective moments? Is beauty itself the preserve of the rich? Does it not have any value for the poor? Is money or the very term 'adequate compensation' (again money and cemented structures) all there is to the people we have pushed to a corner even in our urbane discourse on politics and 'people's movement'? Is there no value for animals, birds, or a childhood rich in the knowledge of these or an old-age in relative peace of community living? Why should every tribal or other marginalised community be measured in terms of what works for an urban idea of living?

I retrace my steps to the fish-hunting communities'[5] settlements (huts) along the banks of Godavari whom I observed in my very first journey on the river. It was time to go into their homes and understand their perspectives on life and the river. And there was yet another context that literally presented itself to me in a newspaper advertisement, supplementing my idea about Polavaram and its long-term vision. Now I could put some numbers to the game, which worked so contrastingly, with the lives of the fish-hunting communities of Godavari: both were 'activities' on the river, but catering to radically opposing interests. There was subsistence on the one end and greedy consumption on the other. Several stakes had developed on Godavari all over again, within the last two years. It was also time to trace one's steps towards one of the important headlocks of the Cotton barrage at Kapileswarapuram. It was virtually a journey that the fish-hunting or fishing communities took, upstream and downstream, and I was merely following their tracks. At Kapileswarapuram, again, through their eyes I was to see, in the rusted headlock of the colonial system, a symbol of a river's stagnant, dead, end, and I heard of a fish species that I had encountered in the Gazetteer of 1915. It was time to go back to those pages of history. The flow of the river and its stagnation meant life or death of a traditional community and its livelihood. River fishermen can never fish in dead water without losing a bit of their selves and identities in the process, besides losing their benign hold over their Godavari.

***

Seeing Godavari from Tungabhadra? 'Flood Gates' and Power January 2010

I happened to take a trip down a different region, river-zone in this time, which gave me a significant perspective – reinstating some questions about the entire spectrum called "water management" and what a distance it creates between people and a river system. I reverted to Godavari some months after this trip with those questions.

24th January - Bhimavaram and Kakanur, Nandyal Mandal, Kurnul District[6]

The Kurnul-Cuddapah Canal was built between the years 1863 to 1870 by a Dutch company, originally for navigation, but later converted into an irrigation system. In 1997, the Japan Bank for Irrigation Corporation (JBIC) gave financial assistance for what is called the Kurnul-Cuddapah Modernisation Project to the government of AP. The entire project area received irrigation and drinking water from 2005. The "objective and scope of the project" (officially) was the usual – 'increase agricultural production, livelihoods of farmers, assured water supply for the farmers' fields' in the command area', etc. They also constructed a new anicut at Sunkesula and the Alaganur Balancing Reservoir at an expenditure of approximately Rs. 1100 crores. But after all this expenditure and 'modernisation', while nature took its course, the politics of 'cusecs control' and 'release' between two States inundated Kurnul town and other areas downstream in October 2009. The concept of modernisation (in spite of adjuncts such as "participatory water management") with foreign aid essentially increased the administrative apparatus around ecological flows. In simple terms, the reason for the floods in 2009 in Karnataka and Andhra Pradesh was caused by lack of coordination among the States in opening and closing of gates in their respective reservoirs. Added to this were breaches and broken structures all along the irrigation projects making the situation worse, inundating lands which had never seen that volume of water ever. A media report said, "While Karnataka built Almatti and Narayanapur dams in upstream Krishna, Andhra Pradesh constructed three major dams on the Krishna – at Jurala, Srisailam and Nagarjuna Sagar – and a smaller one at Sunkesula on the Tungabhadra...Whenever heavy rains cause floods in upstream Krishna, Karnataka opens the Almatti dam gates to prevent a submergence of places in Bijapur and Bagalkot. The water's release is periodically monitored by the Central Water Commission (CWC)...It sends bulletins every six to 12 hours to the respective governments, alerting them about floods, so that they monitor the water release at their dams...It takes about 72 hours for the water released from the former to reach the latter. Similarly, it takes 60-65 hours for water released from a dam on the Tungabhadra at Hospet (Karnataka) to reach Srisailam (Andhra Pradesh)...But on the night of September 30, there were heavy rains in the catchment areas. The Krishna basin in Mahbubnagar received 30 cm rainfall and the Tungabhadra basin received 28 cm, leading to flash floods. Against an inflow of 1.16 lakh cusecs forecast...the water flow multiplied to 5 lakh cusecs. The Srisailam dam authorities first opened four gates and later 11, to let out water on October 1. On the same day, the CWC's next bulletin predicted 5.95 lakh cusecs. But the actual flow turned out to be 9.72 lakh cusecs. In the meantime, the Tungabhadra went in spate because of a breach at the Sunkesula dam...Several small and medium tanks along the catchment areas of the Tungabhadra were also breached. Added to this, the swelling backwaters around Srisailam flooded Kurnool on October 2. The situation worsened in the next 24 hours as the inflows into the Krishna touched 25 lakh cusecs at Srisailam. This was almost double what the dam is designed to hold. All the gates of the dam were opened soon after. At Nagarjuna Sagar, too, the authorities lifted all 26 gates on October 1 to release 13 lakh cusecs of water...At Prakasam barrage, further down the Krishna at Vijayawada, the authorities discharged 10 lakh cusecs for the first time in 50 years...On October 3, Karnataka also started releasing 3 lakh cusecs from Almatti dam..."[7]

If this is not a devastating construction of river control, what is? A farmer named Y. Balasavuri, at Kakanur[8] (Nandyal mandal, Kurnul district, AP) gave a simpler analysis of the same event - "Karnataka released the Alamatti waters and informed the officials here; for a week they opened the gates there, we lost our fields and crops. Officials at the Srisailam dam kept waiting, while the water levels kept rising; they suddenly released all the water (over the level needed for power generation) into the Pothireddypadu reservoir. The badly-maintained gates of the reservoir could not hold all that water and the backwaters flowed through into the Kundu river which rushed in, causing several breaches and destroying all our lands and so much in its wake...This time it is the officials who have caused this kind of havoc." But the Assistant Engineer at Kurnul, Mr. Madhusudan said, "This was nature's fury; entirely unexpected. Instead of 5,25,000 cusecs, the level of maximum flood discharge, 9.8 lakh cusecs were realised at the barrage, causing destruction, erosion of earthern bunds, etc. This was not official neglect."

The debate then is a perennial one. I understood that natural *upstreams* and *downstreams* perhaps produced fewer disputes than artificially made upstream and downstream of dams and irrigation projects. Meeting the farmers and Water Users Association members in the Command Area raised several questions of the crops it make a farmer grow, and creating canal systems as means of surveillance by the state which has also has its methods to create needs where there aren't any, or deny needs where there are substantial through the game of 'opening' and 'closing gates'. I would go later to Godavari (and the anicut) armed with fresh insights. But enroute a trip to the Nagarjuna Sagar dam command area, Godavari and one of the tales of resistance encountered me unexpectedly in the strangest of manners.[9]

***

Fishing in Contested Space
May - July'10

25th May - Revisiting Manturu

I am back in Manturu, the very beautiful village, with a bluish-green pond by the Godavari, and in that same beautiful old house overlooking the pond and the river. I am staying with Veerappa Reddi's family – wife Ammaji, children Surekha, Bharat, Kiran and Udaya. The youngest, who loves to dress up and pose for the camera, is a new entrant, with wavy hair and is an absolute Kondareddi beauty. She is Ammaji's niece, whom they have adopted since the child lost both her parents (she does not know). She must be around three or four years old. And yes, there is another new entrant this time, who sits all day within a pot hanging by the main entrance: a two-hooded Krait that Veerappa Reddi has caught in the forests and plans to sell. No amount of conservationist jargon will work with him on this front. The children use the pot to scare the daylights out of their friends as they did, in my presence, once. Today I met the fishermen here. There seems to be some celebration in the village and I can hear drums. Veerappa Reddi's daughter is cooking; a typical teenager with her temper-tantrums and demands. Ammaji is away probably attending the celebration, a wedding, I think. I wonder if I should ask for some tea…Finally did. Surekha too wanted some tea, so I got lucky. Ammaji is from Talluru, a village some distance from here. I notice now that more people in this village have a TV than before (when I came in 2006) when only one house had a TV.

An Evening on the banks of Godavari with the fishermen (25th May)

I am introduced to Malladi Posi. Others joined in: Kottabattula Kondaiah, Gopi, Pandu, Patabattula Nageswara Rao, Malladi Adinarayana, P. Sreenu, and Dokkada Bapiraju. I note with interest that they call it *'veta'* – hunting, fish-hunting, not fish-catching. They do not refer to themselves as *matsya karmikulam/karmikulu* ('we are fish workers' – *memu matsya karmikulam*), in these villages. The *karmikulu* term is more prevalent in and around Kakinada, Rajahmundry, perhaps from political socialisation. Posi informs me that he is from Yanam (in Pondicherry administratively, but AP geographically[10]). He came here 20 years ago. "With Polavaram project we will lose our livelihood because water levels will increase here. We will not be able to fish anymore. We found it convenient here so we settled down here, while many of our people are still in Yanam. We do not fish in the months of July to August or September (sometimes). For those three-four months we have to live on whatever we earned in these months. We sell fish in the nearby villages…No official has come here so far (to meet us)… We have seven families (from this community) in this village. Others are from Rajahmundry, some from Pattiseema."

[Without my even asking, almost all of them spoke first about Polavaram and the tourist launches, as they saw both threatening their spaces on the river; I just told them I was there to find out about their *vrutti* (livelihood)]

He continued, after a while, "Service launches which were meant for people have now been converted into tourist launches. There are as many as 60-70. We lay our nets in the night time to avoid these tourist boats but nowadays they also return down the river this way even at night and our nets get torn. We lodged complaints at the Devipatnam police station several times, but nobody cares. They do not give us a receipt of our complaints lodged, so there's no way to follow up. Twenty years ago, there were plenty of fish in Godavari at this place. Now the numbers are very less. We moved to these parts because in Yanam huge sea-bound ships usurped our livelihoods. Here, we earn Rs. 1,000 to 2,000 per month, but it is uncertain; sometimes it is much more. The income varies..."

They call Godavari Gangamma, as well, interchangeably, at times. Locally, the fishing community belongs to the Palli sub-caste. Their caste title is Agnikula Ksatriya, under the category BC (Backward Caste). I found it interesting that in this village, living in close proximity with the fishermen, the tribal communities have also learnt the art. At times it is difficult to tell that they are Kondareddis or Konda Kammaras doing the job of the Pallis. And some fishermen have learnt the art of agriculture, for they have to supplement their incomes as agricultural labourers when fishing is not so fetching, sometimes during the lean periods. This is particularly so with the fisher communities who settled down and stopped being itinerant as in the past, the change in this socio-economic setting is interesting to study. For instance, Kachala Veerappa Reddi is a Kondareddi but started fishing, after buying a boat and a net (which is provided on subsidy by the ITDA). So the traditional fishermen consider him their own. They were unanimous in denouncing the tourism launches as result of which, they believe, the fish are not laying eggs. The tourists dump garbage into Godavari – plates, glass, plastics, etc. The water is polluted and they have to drink that water.

The boats sometimes disturb the fish which go hiding by the stones and do not enter the fishermen's nets. Posi says, "We kept raising questions, gave petitions to the Collector (East Godavari). They stopped for a while but resumed again...When water is at higher levels, in the peak season, there are lots more. There is no resting time for them; even fishermen and the fish have times of rest; not the tourists nor the tourist boats."

The Sun sets on Godavari and darkness is about to envelop the village by its banks. After spending some time in relative silence, watching some of the fishermen readying for a night of fish-hunt, I head back home, to Ammaji's.

It's Evening Time: TV

What might be the connection between Polavaram project and TV? Not the radio so much; for some reason, All India Radio has not drastically transformed the sense of community, or other contexts, as much as TV today has. Veerappa Reddi has a new gadget. And so do others in his neighbourhood. That explains why I do not see the regular stream of visitors who used to come by when I last came here, to check out the 'stranger', or outsider, to talk. Ammaji cooks the food in a hurry and leaves to her neighbour's house to watch a Telugu TV serial. With TV and numerous private channels the quiet and darkness of the night of/in this place has disappeared. Associated goods have come in and the bamboo platform under the tamarind tree is relatively emptier by late evening where men in groups, and women, used to sit down to gossip or discuss. TV has individuated the place in other kinds of ways. Indirectly, the compensation discourse and need to procure gadgets may have stemmed the rebellion that was loud and visible the last time I stepped in here, where people would only talk about Polavaram dam and express anger (which the Pallis still do). Now the anger is muted or has disappeared. Technology can, I sensed that night, alter the discourse. And sometimes Telugu mainstream films can be great 'levellers' – only, it depends of what form and nature, and to what purpose that 'levelling' happens, for levelling itself is not unproblematic, unless it is known if it is to a mainstream, unequal universal or an alternative, redeeming one. Things do not stop at SEZ or Polavaram. It begins a whole process, goods that cash can buy; private TV channels through cable connections. In this once-quiet village, Manturu, when I came last, no cell network worked, barring BSNL, at just one far corner at a higher altitude, by a government school building, facing the river. Today a private telecom company (Tata) and TV channels have taken over. But then, why should we overrule the agency and choice in these changes? This is what people have brought into their lives based on their choices. The problem lies in the extent of freedom in that choice and the manner in which it is given to them as a universally accepted norm.

Morning of 26th May

Veerappa Reddi's daughter has just had a fight with her mother. She wants the latest pair of jeans. She fought and got her way this time, and left for Devipatnam, the nearest market town to buy her jeans and also get the dish-set box repaired. Meanwhile, Reddi has arranged for me to be taken by a small boat to cover Kachuluru and Tuthigunta. But today evening, if there is time, I shall cross over, across, to Vadapally. All these villages have fish-hunting communities.

By the river banks, I find two men and a woman busy cutting vegetables readying them for some kind of a meal. Bags of onions are being pealed. This is a new sight for me. One of the men peeling the loads of onions is Venkata Babu, from Rajahmundry. They are readying the meals which will be picked up by the tourist boat on its way to Perantapally (the Papikonda circuit). Since the river is at her low lying season they are cooking meals at Manturu these days. It is easier for the launch to anchor at this point. In fact, across the river on the other end Vadapally is where three tourist boats are anchored. Back at Manturu, the cook Dora Babu is busy slicing onions, other vegetables and giving instructions to his two assistants for the day's recipe – a full plate 'Andhra'[11] meal – for the boat which will arrive that afternoon. None of them are locals; most of them are from Amalapuram and Rajahmundry. They start cooking by 8 am and by around 2 pm that the tourist launch comes to pick up the food, stuffed in huge steel boxes for passengers who will eat it in the boat (and throw the leftovers in the river as they eat and move though dustbins are provided in some of the boats). Since water levels reduce substantially in these months (April through June) the arrangement has been made at Manturu. At other times, the food-halt happens at Devipatnam where a local non-tribal café owner prepares the food and keeps it ready to be picked up.

Kachuluru (East Godavari) –"If the dam happens, we have to go looking for a river elsewhere" says Eswara Rao. Kachuluru is in the Tunnur Panchayat of East Godavari district. There are around 150 households, owning 85 acres of land, and forest for *konda podu* land. Sangani Eswara Rao's is the single fisher family here.

Or one of the two, since his son also has a family of his own now. He belongs to the Palli sub-caste. He again speaks of the changes in recent times to his Kapileswarapuram some 25 years ago, built a home for himself and stayed back. "I came from Kapileswarapuram because there was no water there (in the river, enough for fishing). So we came here." There are four fishermen here, two from the Konda Kammara community (who learnt to fish from Easwara Rao). "The best time for fishing is from Sankranti (January) till June. But they are building a barrage in Kovvur and have stopped water there; so no water is flowing in, this year, and no fish have come by" he says. "*Touristlu* (tourists) have added to our problems. Fish don't lay eggs anymore. Godavari has become so dirty; all the dirt (*chatta*) is being thrown into the river by tourists everyday. Water is polluted now. The fish are moving away and not coming here. We spoke to the Collector (East Godavari); for a month tourist boats stopped but have returned. There are four tourist boats plying here every day these days...On Sundays, there are nine or ten boats a day (in this season). Our nets get torn, our boats are hit. My boat was hit once by one of their boats. We went to the Devipatnam police station and registered a complaint; the tourist boat owner gave us Rs. 1,000 as damages. But let me tell you all this is happening mainly because of the Polavaram *barragi*. It was not this bad earlier. If our nets are damaged by the tourist boats, we need to invest Rs. 70,000 to 80,000 for new nets; or Rs. 30,000 to repair the old. We need Rs. 60,000 to make a new boat."

If fishing is good in the four months they earn between Rs. 15,000 to 20,000 per year. Other times, they work as agricultural labour on others' fields, including those of the Konda Kammaras. He adds, "If *pedda Godavari* comes, we get good fish, not otherwise. See, we can't fish in the sea; we can't manage. We start throwing up if we go to sea..." Tourism affects change in the culture of fishing in these parts. Easwara Rao says, "Nowadays we have to fish by night." Earlier they went fishing by dawn and only rarely at night. "Now we do not go daytime-fishing because of tourist boats. We hunt at night from 6 to 8 pm."

He informs me that there are around 370 families of fishermen (Pallis) settled in Kapileswarapuram. He suggests I visit Kapileswarapuram and meet the President of the Fishermen Cooperative Society, Ponnamada Ranga Rao at Satyanarayanapuram or the Secretary, Lanke Sattibabu. He gives me the route – from Rajahmundry to Jonnada Bridge by bus and from there, take a service (shared) auto to Kapileswarapuram.

It is time for us to take a boat to Tuthigunta. Durga Rao, Durga Prasad and two of their friends will 'captain' the boat (*nava*) to the other bank. From there one has to walk to Tuthigunta... We walked, and walked and walked! Up and down the hill, on hot sands and through stony paths and by rocky outbursts, under a blazing Sun for nearly two hours. Even my water bottle was now empty. Midway, we crossed a mango grove, which was the only shaded place in that entire stretch. I asked the 'guides' to stop a while, so I could sit for a moment. The shade and the grove were enchanting, with a small shrine, *paata sivalayam* (a Siva temple, supposed to be ancient), but next to the Siva linga was placed a photograph of Shirdi Sai Baba, the Sufi saint of Shirdi in Maharashtra. The Siva linga was but a small rock without the characteristics of the usual linga. After fifteen minutes (or was it half hour) of rest in this quiet place (with the hope I would go another time, alone, and spend an entire day), it's time to move on. Soon, not too far away from here we could spot Tuthigunta far down the hill, a tiny speck, by the river, but a downhill trek is never half as tough as the uphill we had been doing all this while. Once there, we found only two fishermen in their makeshift thatched hut on the banks of Godavari. At this time of the day nobody ventures into the river. Or anywhere, for that matter.

Tuthigunta (Tuthigunta panchayat, West Godavari district)

Here I meet Karri Bhagyam and Votala Durga Rao. Bhagyam gave me some 'refrigerated' Godavari water from his bottle – the bottle is covered with a thick cloth, to protect it from smouldering heat but what makes it cooler is the river where they dip the bottle for a while. Many boatmen do that. And in that heat it feels heavenly to drink cold water. I suddenly remembered, this is the first time I am drinking Godavari water by the banks of the river, at a fisherman's hut, in these five journeys (from 2006). A sense of déjà vu (of 2004 at Warangal). At Manturu, too, Reddi's household consumes Godavari river water, which is not boiled. But what adds to the taste of this cool water is the ingenuity of the fishermen in devising ways to beat the heat.

Life, in a Thatched Shelter by the River

Votala Durga Rao's father came here from Chintapally Lanka (Mummudivaram, near Amalapuram) many years ago. Durga Rao was born here. He says, "There are no fish here; there are too many weeds. You would get prawns too, often, but the weeds have affected those, too. There was no *pedda Godavari* last year, which is why the river is full of weeds. And they release the effluents from the Sileru factory (a paper mill) into Godavari, which is also affecting the fish population. There are very few fish now. We fish from Sankranti (January 13th/14th) till June 15th. We then go to Kundrukota nearby, where my home is, or Madipalli to do agricultural labour. Sometimes there is no work so from July to September – in those months we stay home…When I was younger, there were few others fishing here and fishing was very good. Now we have to contend with the tourist boats which affect our vocation (*vrutti*). We gave complaints at the Fisheries office at Kovvuru, about our nets being destroyed by tourist boats. They did not give us any compensation. We also went to the MRO office at Polavaram, but nobody listens to us. Now if the dam comes, we have to go either to sea or do labour (*cooli pani*). We will be left with nothing…"

Karri Bhagyam's grandfather had come here from Chintapally, all the way by river route, on his boat, many years ago. Bhagyam today has a home in Kundrukota and a voter card as well as a ration card. Durga Rao's and Bhagyam's families were in Kundrukota. They get their food in 'tiffin' boxes from home. For three or four months they live this way, away from their families, fishing in these troubled waters, for whatever it may be worth.

Bhagyam says, "Sometimes we get one or two kilograms of fish; when there aren't any fish, we have rice with vegetables which we buy from the shop. Sometimes we earn as less as Rs. 5 or 10 a day or Rs. 50 a day and at other times, nothing at all. If we go for agricultural work we do get Rs. 100 a day. Nowadays, we are habituated to do that work, even though we were not born to do it. Since we are in the Agency Area since childhood, we have learnt that work, as the tribal people have learnt fishing. But we prefer to do our own *vrutti*. There are nine families of Pallis in Kundrakota today. Previously there were few launches (bigger boats) and even fewer tourists. Those launches have reduced in number with the coming of these tourist boats. The tourist operators make at least Rs. 1 lakh per day; since their tickets are priced at Rs. 450 - Rs. 500 per head; and they carry nearly 300 to 400 passengers, each day."

Bhagyam and Durga Rao go fishing from 10 pm to 2 am everyday; and again from 7 am to 8 am the next day. At around 3 pm they lay their nets again, and wait. After fishing for about three to four hours, they stay back in their thatched roof shelter by the river watching the dusk settle in; and also the tourism boats return with the passengers, who are oblivious to all this. Before we leave, I see women coming down to catch the *seer meen* in their saree. *Déjà vu*!

The men called out to a boat. Happened to be Posi's again, but the young boys took long to respond; there are no boats coming down this way at this time of the year. If we do not get a boat we will have to walk all the way back again until Kachuluru and wait for a local boat to ferry us to Manturu. Finally, the boys relented and drew closer. Back to Manturu again!

Back at Manturu, In my Diary[12]

Ammaji has made some lovely *seer meen* curry and *pulusu* and rice. As it always happens, they put lots of it in your plate. But today, was amazed as to how much I ate, and finished the plate! There is some respite now, not from the heat, but another long walk uphill and downhill in the Sun; for we will start for Vadapally, which is right across the river on the other bank, within the next two hours or so. It is perhaps little later than 1 pm now. Am here when most school children have come back home on a holiday (many of them study at the ITDA residential schools far away from this village). It must be lovely for them to come home to the village, I think. And lead a carefree life for a while before they are packed off to school again. You can do a lot of things here during holidays; run around, take a swim in the river, go to the forests, play hide and seek, and a special local kind of 'cricket' with a thin wooden stick with not one but two 'wicket keepers' behind the batsman. And a stick for a bat. And sleep under the tamarind tree of the coolest breeze on the hottest day….Am back to my old camera, as a singular record of these moments. But couldn't help recording a bit on the other (digital) camera and voice recorder, which, by strange coincidence (as if understanding my sentiment and that resolve) have stopped recording. Am just recording some moving images as though one has to 'freeze' all these now …Incidentally, the *seer meen* was the same that I had first seen in my 2006 journey and a rare sight – a mother and a daughter holding on to the ends of their sarees and going knee-deep in water, where they stayed for a while, and slowly, like the moves of sychronised swimming, emerged from the river with some fish in their sarees…"

***

At Vadapally (Kundrakota Panchayat) - Again, Pallis here - 7 households, 7 boats.

I meet Malladi Sreenu (from Tallapudi). He used to stay in Gutala, Tallapudi (West Godavari, near Kovvur) earlier. He says, "Every year we come here, if the fishing is good – around January or November-December. We stay for nine months here and return to Tallapudi. Most of us – barring two – have votes there. If we do not get a good catch here, we go sometimes up to Bhadrachalam. For eight months it is a struggle – repairing nets, etc…Even if we get Rs. 200 worth of fish, we have to spend as much or more to buy rice. The nets cost us Rs. 20,000 to 30,000. We take loans from contractors who sell fish at Devipatnam, Tallapudi and Polavaram…We have to give our fish to them sometimes as loan repayment…When we were in Tallapudi, we used to supplement our income doing agricultural labour. Here we don't. We have to forget about our traditional occupation if the barrage comes up. Whatever else happens to others, fish-hunting communities will definitely suffer."

Sreenu says that there have been times when some of them have gone all the way till the borders of Madhya Pradesh (Chhattisgarh). "Today there are not enough flows in the river, so we hire a lorry, keep our boats in them; on the return journey we come via Godavari. We can't say where we will be next year…"

How wonderful would it be to join them on one of these long trips on the road and back on the river! Just a thought.

Malladi Peddakondaiah says, "Polavaram project will mean that…we cannot fish anymore. Waters will rise up to the hills here…Even though we pay *pannu* (tax) and we have receipts, nobody cares…"

Cell-phone Notes - 26th May - Full moon night by Godavari; immense depth of feeling. That moment. What could be better?[13] That same night bird (the first time Manturu one), kabbilam, calling in the background. Met old fisherman Arjayya, who poured out poetry of life, but I couldn't record his voice. Just wrote. Missing the old cassette recorder, my reliable old personal stereo.

The Old Man and the River

In his younger days Malladi Arjayya had taken on quite a few adventures down the Godavari (or Gangamma as he and his clan call her), as he fished in her waters not caring for boundaries of State or districts or regions. He went where Godavari led him. That explains his perspective – expansive – of fishing where Godavari takes him. In other words, the river and the fish in it decide his direction, until someday he decides it is time to stay back and settle down, as Arjayya did, at Vadapally on the banks of Godavari. Never mind that the fish in Godavari waters on this side of the physio-geographic zone have consistently shrunk in numbers. But Arjayya has a home there, and today his son and grandson follow the same profession. Talking to him as the full moon rises on the banks of Godavari, with one or two fishermen on their *nava* (fishing boats) laying nets late night, was like getting enlightened with a whole philosophy of the river, the fish and life, and the larger changes perceptible only to someone so closely tied to it all. There was a simple philosophy of life in what he said - "Those days were good. Now it feels like this occupation has been hit by drought. No matter what we get, it isn't enough, as we have to spend that much more. Those days we got more fish in our nets, too. Now it's not enough for the family...There has been a sharp decline in the birth of fish in the Godavari water. There aren't enough eggs now. They keep the Dowlaiswaram *barragi* (the colonial one) gates closed and open them only during *pedda Godaari*. If they would open the gates in normal times there would have been ample fish eggs and prawns as well. If Godavari comes, we get lots of fish, too. When the fish is pregnant, it needs air, and it needs to move. When Godavari comes, it helps the fish turn and move around and lay eggs and new fish are born. Fish eggs are laid near pebbles, stones etc. they do not lie on it. With floods these eggs get washed and new fish are born which move along with the flood waters. After *enabhai aaru* Godavari (the 1986 coming) the river was full of fish. Where do we even have that kind of Godavari? The water has to go to sea. Are they allowing it? Farmers and government only think of their interests, building canals and dams. Where is the water going to sea as it used to? How will new fish be born?"

After a second of silence, Arjayya said, philosophically, "Actually, if you see, we are all living on a *lanka* (river bank/island); the sea is all around us. And rivers go to the sea."

The system cannot fathom the language that people like Koppanati John Babu of Vadapally speak – "we go where Godavari is."

A Report on the health of river systems supplements Arjayya's point on fish birth, from a different angle - "In the river systems of Godavari, Krishna and other rivers, several large pools called *madugus* are formed during summer. When there is no flow of water, large fish settle in madugus and rest until the next monsoon, when the rivers get flooded again. They then migrate for spawning. However, during the period of rest, the fish are caught indiscriminately, sometimes even using explosives and other destructive fishing methods and the brooders never get a chance to migrate back. The Zoological Survey of India has issued a list of madugus and recommended that at least some of them be protected as fish sanctuaries."[14]

Official Figures of Fish, Pollution, Decline in Fish Catch in Godavari – "In the AP rivers, the freshwater shark 'goonch' (*Bagarius bagarius*) has long disappeared. The migratory hilsa has been affected due to the barrage on the Godavari river. Such indigenous species as *Labeo fimbriatus, Labeo calbasu, Tor khudree*, which were abundant in the earlier years of impoundment in Nagarajunasagar, declined over years due to habitat loss and breeding failure. These were replaced by minnows, which are of little commercial significance."[15] And, according to the APPCB, the Godavari river water quantity is within safe limits upto Mancherial but polluted further downstream. Along the Godavari, the major polluting points identified are Ramagundam, Mancherial (AP Rayons), Bhadrachalam (Paper Mills) and Rajahmundry.

***

27th May - I left Manturu in the morning, again on Posi's boat, crossing over to Vadapally, walking uphill on the sand beach for some distance to get to the road. I will be getting a bike ride till Singanapally from here.

***

At Singanapally, Chegondapally Panchayat, West Godavari District - 50 households of Pallis here. These families came in 1949 from Yanam.

K. Ballaparaj[16] says, "People from my community came here, moving up along with the river from Gautami and settled down. Fishing is our sole livelihood source. There are no loans for us in the BC quota… Fisheries department officials come to us only to collect *pannu* (Rs. 120 annually per person), once a year...We sell fish in Singanapally and Polavaram. We also go for MGNREG works for one to two weeks. Even though we are Pallis and fishing is our traditional occupation, we are forced to do other kinds of work."

At one level, there is a certain de-skilling of traditional occupations, where every person in a village is expected to give manual/physical labour for work such as "bush clearing", "digging", etc. under the NREG schemes, which renders everyone a daily wage earner living on physical labour alone, and physical labour is restricted to the people in the rural areas (as of now) and mostly the lower castes; somewhere it has not broken away from the connection between the caste and occupation.

Ballaparaj continued, "Forty years ago, fishing in Godavari used to help the entire family survive. Today, there are no fish in the river…There are schemes for fishworkers but they emphasise more on sea-fishing since those communities are part of the BC "A" category. We have been given BC 'D'... Our life moves in circles... We are itinerant, but now settled. Isn't that a strange predicament? If they consider us itinerant, why collect taxes as people settled down here?"

I wonder, too, perhaps the fisher people returned their caste status as Agnikula Ksatriyas in the past? But in the traditional caste hierarchy they do not figure at the level of the so-called 'ksatriyas' of mainstream caste system, both in terms of ritual interactions and the homes that are traditionally at the extreme corner of the hamlet, by the river, removed from the others. The only homes that are usually in close proximity to theirs belong to the SCs, as far as I have observed in the Godavari villages. But how, when and why did these communities claim the ksatriya status?

Ballaparaju continued, "The tourist boats begin operation during fishing season and dirty the river. Many fish lay eggs at that time and water pollution is affecting even smaller fish....The sound of boats drives many fish away. Even prawns have reduced in number. In the night we lay nets which are torn by the tourist boats. They give us just Rs. 200 in compensation if we complain...Nothing came of our petitions to stop the tourism boats... From Annavaram the river is moving all the way to the sea. If it is stored in dams, fish will decline. Dowlaiswaram barrage also holds Godavari. Those fish do not travel down here..."

***

27th May - Singanapally to Devipatnam first time from this route, on a passenger boat

At Devipatnam – Fishermenpeta/Chapalapeta (again) - 30 households of Pallis.

The last time I met Malladi Gangadharam was during elections 2009. Meeting him a second time, giving him a print of photographs of his family I clicked last year, in fact returning back to his home, seemed a single sequence of a long time span in which not much changed in his marginal existence but somethings changed in the surrounding din. For one, the older daughter has been married off. The boat with a plastic sheet for a wind sail isn't in a good condition. There is a general bustle and some joy as we are all pouring into the photographs I had last taken in 2009. I am so happy to give them back their 'property', which is being passed around. We are all sitting outside someone's locked house, with a traditional sit-out, in a small bylane in this hamlet. Quite a few have gathered together here. It is always so nice to retrace to familiar people and places. Rambabu, Ankadi Ramu, Malladi Posi (not Manturu Posi), join us. Gangadharam says, "Living by the Godavari is our dharma... Wherever else they take us, we have to survive on Godavari; we know no other craft...There have been no surveys of our houses so far...The dam will come (*memu kaadanna barrage ostundi*) even if we say no, so if it does come up, we ask for our houses to remain by Godavari."

The general discussion continues over the tourist boats which ply here between 8 am and 10 pm. 'They move about when we lay our nets. During Karttika (October-November) month, there are around fifty boats a day leaving no space for fishing...Fish are moving upwards towards Papikondalu. We complained at the Devipatnam police station and they said these boats have government permission and they cannot be stopped.'

Gangadharam says, "We gave a petition at Eluru – the launch owners said we have invested lakhs in tourism boats, how can we stop? They only stopped once when a mishap took place and someone from the city drowned, near Sivagiri. All this is the result of TV advertisements which tell people that all these sights will disappear forever once the dam is built, so this is the only opportunity to see it all one last time."

Everyone pitches in with their bit as there are so many problems. During 2006 Godavari they were paid Rs. 450 for ferrying marooned people across. It is the same story since a decade. There is apparently an insurance scheme for inland and sea fishworkers, to be given to a family if the person dies in a mishap. One of them says, "So, only death gives us something!"

Incidentally, the months of June, July and August are meant to give the fish, and the river, some rest here. Even as it is time for the monsoons to set in.

Official Records[17], etc - Mr. M. Anantharaj, Assistant Director, Fisheries, Rajahmundry division (East Godavari) says, "We issued 619 licences but actually there are at least a 1,000 fishermen families depending on the Godavari."

These are numbers for his division alone. He says it is a "continuous process"– issuing licences, because the numbers keep varying, as new migrants keep coming. About Polavaram, he says, fishworkers are not affected; only "girijan fishermen" not the others. He was not aware that some of the villages where the fishworkers families are settled fall under the submergence zone of the dam. "We are not enumerating them."

It's time to leave, but Gangadharam and others are having some discussion amongst themselves. And suddenly Gangadharam says, "*annam tindaam randi*" (come let us have food). In spite of all my protests, they all lead me to Gangadharam's home. I am seeing Manga again. His daughter who has been married off is also visiting, with her husband. I realise the importance of food and water. I have had enough food and water on these journeys to fill my soul for a lifetime. Here, it is part of a tradition followed ever so religiously, to feed the hungry visitor. And if one is a woman, and travelling alone, all the more so. What lovely food, though they call it 'humble' the big fish that was caught today morning by Gangadharam (by sheer chance, for they did not know I was arriving, he said), made into a *pulusu*, a pile of rice on my plate (as usual) and some love. Last time it was *majjiga*. For me this is an emotional moment, to have retraced my steps to the same house, in two different emotional contexts within a span of one year. The home is decorated with large posters of film stars – there is Chiranjeevi's, too. But did they vote for him? I forgot to ask. Manga keeps feeding me until I can eat no more. They cannot believe how 'little' a person can eat!

Post-lunch, I return to Singanapally and from there, towards Polavaram.

***

Cell-phone notes - 27th May 2010 - Annavaram Sattibabu, Durga Rao, Raju, my first 'wind-sail boat' drivers Devipatnam to Singanapally. Now to get the service auto to Polavaram.

[They navigated the boat on a particularly low-lying river, against a heavy gust of wind; three young boys and me on the boat, wind on our faces and the blazing Sun, and sounds of the wind-sail against water.Thankfully, I could record the moment on the digital camera I so disliked. Moments like these come by rarely.]

A Rough Boat Ride and a Question - What Would Life at School Be Like for Them?

Sattibabu, Durga, Raju…three boys sitting on their *naava* with a wind-sail made of plastic sheet, which refuses to open up until we are midway on the Godavari, with a heavy wind threatening to topple the boat, or take it in a different direction altogether, and three boys taking turns… One wondered about life, and their lives. They were here on a vacation. Going back to school, for these boys, must mean recounting all these adventures of the river and the fish and the boats and the nets, surely far more varied and filled with life than the holiday homeworks of kids in public schools in cities where you just know your immediate life and nothing more. Theirs were smiling faces and managed the boat with a typical early teenage attitude to life.

***

A Break in the Journey (28th-30th May) - I had to take a detour from Rajahmundry before resuming my second phase of the trip to Kakinada, Nedupuru, Rekapally downstream Godavari.[18] Upon returning, on 30th May, a freak fall,[19] forcing me to return home to Hyderabad by road from Rajahmundry.

***

Godavari: a Common 'Property' Resource? Placing the Idea in Context Inland Fisher Communities of Godavari and Polavaram Dam[20]

Whose "commons" are we talking about? A river resource that is "common" to the politico-administrative divisions of States in independent India or a resource "common" to agriculture, industry and fishing activities? The 'common' must be inter-read as 'contested'. Whose stake becomes paramount when it comes to ultimately building irrigation projects on rivers? Whose Godavari is it ultimately? People whose lives are affected on an everyday basis are not consulted in the matter of resources that should, ideally, be common in terms of being 'communally' perceived and treated – the way the tribal communities perceived their land and forests until they became marked administrative divisions in the colonial state; or the fishermen/fishworkers perceived rivers (or seas, as the case may be) until the time they were issued licenses as 'permits' to define their access to these. In a way, one tries to look at these questions through the perspective of unnatural control and claim-hegemony. Posi and Eswara Rao's ideas lead to deeper questions – for whom does the Godavari flow? Just as the tribal communities seek land for land and forest for forest; can the fishermen seek a river for river in compensation? Is Godavari a river meant only for industry or agriculture and not for the fishermen/fishworkers? If the Polavaram dam is completed, these communities might forever lose their identity tied to their vocation to become the hundreds and thousands of wage labourers on construction sites or agricultural fields. The fisher communities are not counted as those of the Agency Areas – a strange paradox and irony of the system. While they have fished in these waters for centuries the subtle changes in their pattern of settling never seemed important enough for statistical officials of the census. But they do seem to count when it comes to voting and they do have ration cards; though not when they lose their livelihoods. If a river were to be compensated for, perhaps the calculation should begin with consultation of the fish-hunters/ fishworkers/fishermen of Godavari. The government plan to "link the Godavari and Krishna, thus reducing pressure on the Krishna waters; recreation facilities and pisciculture, etc" (as stated in the DPR) is bound to affect their traditional rights. Malladi Posi's is the only boat that connects Manturu by river (which falls in East Godavari) to Vadapally (in the West) from where you take the road to Singanapally or Polavaram. If the monsoons arrive in time, within a week or two you would not find Posi and others on the banks of Godavari. They will stop their fishing activity for the next three - four months, to resume again by early September. In that time they only survive on the income they raised in the fishing months between September and May. All that, of course, depends on the rains and the volume of water the Godavari would 'churn' up. In the entire debate for and against Polavaram dam they are the most visibly invisible community. It is almost as if their very presence in physical terms is marginal in the submergence zone. They are unlikely to come into the line of the debate as they are also small in numbers and under-represented or not represented at all in political terms. There is also the issue of their being categorised as BC (D). The term 'ksatriya', perhaps a status they claimed for upward mobility at some point in the past obviously worked against their favour it shows after all these years with their category being BC on the face of it, yet being ineligible for some of the government concessions and support schemes available to the SCs and STs. Being in the Agency Area also works against their favour. But the fact remains that even this caste status keeps them from benefits of the Polavaram R&R package, besides being seen as communities who are capable of moving along where the river takes them, hence not in need of any special package. The license given to fishermen is a 'permission' (not a Right) to fish in the river and collecting annual taxes from them for the same falls within this commercial paradigm of ownership of common natural resources. Just as they issue permits to tourist boats on Godavari to throw any amount of garbage in the river and the accompanying diesel and oil and excreta because they have 'paid' to do so. Ecological well being – an all-encompassing idea – is a far cry. By the time I go there next, more names and faces will be added to the list of people who spoke and perhaps the list of woes will remain. I remain a mere chronicler of their woes.

***

Khammam District
July 2010

The monsoons are on; many fishermen in these parts left for their 'original' dwellings – either at Yanam or Kapileswarapuram. Realised that in Khammam district part of the fishing area of Godavari, not many have settled down in the villages as was the case in East and West Godavari district where a few have made permanent the villages they earlier used to come to fish and have voting and ration cards there as well.

21st July - Reached Bhadrachalam. Stayed a night. Looked up the temple of Rama and Sita by the banks of Sabari-Godavari which was relatively quiet with very few visitors today.

22nd July - Reached Rekapally by bus from Bhadrachalam. It costs Rs. 23 now. Found out that there is a tourist launch between Kunavaram and Papikondalu, daily at 10 am. They charge Rs. 250. May be it would be good to take this boat back towards Rajahmundry/Kondamodalu someday. Took an auto to Jeediguppa. The only boat available at this time is this tourist boat from Jeediguppa to Kolluru and it is on its way to Papikondalu. They charge Rs. 250 per head from here.

In my Cell-phone – "22nd July - Life is Good! Imagine to suddenly hop on to a boat owned by the man you had wanted to meet so long! And the place you wanted too see since so long! Kolluru. Unprepared. Unexpected!"

To Kolluru – The boat employees are Satish and Narasimha Rao from Kunavaram and Bhimavaram (near Kunavaram). The day I took that boat (which belongs to Kaigala Satyaranarayana – had to, no other means left that day) there were a group of sarpanchs and MPTCs from Medak district visiting these parts - Anji Reddy, Anjaiah, (a stringer from Vaartha newspaper) and others.The conversations meandered to Telangana statehood on the boat. Suddenly, (who would have thought?), Perantapally! At 1.35 pm.

Met Rambabu here and 'Vennela' Sumitra. So her name was not Vennela; it was her *inti peru* (house name), which was Vennela, indeed! Nice, both ways! This time I went up and into the home of one of the Kondareddis, who is member of the cooperative here. Rambabu says, "The tribals are getting no income from the Svami's lands in Dummugudem, V.R. Puram (Rekapally), as the lands have been leased out, we hear." I tell myself, I must find out more; stay more days in Perantapally next time. The silence still envelopes one, but there is a strange, dense, undergrowth of some kind of intrigue here that the Kondareddis do not know of. It is just a sense that one gets in this place and I can't fathom why.

In my cell-phone, I note – "22nd July – Met Vennela again! (Of the faraway look) Her name is Sumitra. Vennela, my imagined name for her, is actually her *inti peru*! Whoever thought would come to Perantapally!"

The boat moves on and I get dropped at Kolluru. The tourist resort which I have heard so much about (and the man, a quintessential landlord I have heard so much about from people – nothing very positive). By the way, his name is Kaigala Satyanarayana and not Karikala Satyanarayana as I had wrongly heard in the 2007 journey. He has a huge house here and is an upper caste landlord. For someone looking at the small bamboo huts by the Godavari, surrounded by hills and forests, this is indeed an idyll that hides a not-so-pleasant past and present. Again, full of intrigue and the sad paradox that eco-tourism itself rests on in our country today. Most of the eco-tourist spots have a history of conflict between the owners and the local people, mostly tribal communities and many have come up in violation of several rules. So this place is set up in a village that is in the submergence zone, built on tribal land (in Schedule V area) by a non-tribal, utilising tribal labour. I reach his large wooden house with tiled roof and a ferocious looking, but otherwise meek, black Doberman (seemed cross-bred) dog sitting by the entrance. The boat employees Satish and Narasimha Rao took me to Satyanarayana's house. He offered me some tea. I noticed he had a TV and other gadgets though the bamboo huts down there are supposed to be run on solar energy. And everything here is supposed to be eco-friendly.

Kaigala Satyanarayana says, "I started this venture five years ago." [His visiting card reads, "Bamboo huts in Papi Hills, Kaigala Satyanarayana, Kolluru, Khammam district."] "When the Polavaram project was proposed, AP Tourism people encouraged us to start this. Instead of people just coming, seeing this place and going back, they can have a place to stay. I started with six huts in the first year and many tourists came. Gradually, we have increased to forty huts. Tourists should spend a night in peace. No cell signals reach here; there is no electricity here."

He continues, "There is no sound pollution here as we don't allow people to play music systems. From here they can trek up to the Kondavagu stream and Pamuleru (from Sileru Hills) where they can bathe in the hot springs. We provide snacks, tea and dinner. We get fish from the river here, and chicken is cooked. The Kondareddis are employed to serve the tourists, not to cook though – the tourists do not prefer their cooking, they do not cook well; so we have cooks from outside (Rajahmundry, etc, Andhra cuisine is served mostly). In winter they light campfires. They enjoy it. In the morning they have breakfast and leave. We charge Rs. 600 per head per night including breakfast, and lunch. We are originally from Tadepallygudem. But my father had land here; he was trading in auctions with the forest department. He was a great believer in the Perantapally Svamiji. His name was Kaigala Rama Rao. We are here since 65 years. AP Tourism collects fees from the tourists; accommodation and food is our responsibility. The Tourism department pays us. Local people are supportive of us. There is no problem from their side. The present house was re-built after the previous one was drowned in the 1986 Godavari. October to February is the peak tourist season. There are few who come later months also, but they are far less in number"

About Polavaram, he says, "The project is important; it will increase irrigation and power (electricity) though my cashew orchard will drown in it. But many will benefit; a few will lose out. So we don't mind. Godavari has changed; there is less water these days. In future it is possible there may be even lesser water. Therefore, storing water in a dam will always be beneficial and address that problem. May be it will save some water for future use."

The tourist boat leaves Jeedigupa each morning, these days, moving towards Perantapally and Papikondalu and back. Narasimha Rao tells me there is a "syndicate" of 20 boats, belonging to different owners that ply on Godavari; each day a different boat is put into service so that each owner may get a share of the business. The employees get Rs. 3,000 as salary. Satish gets Rs. 150 per day. I am on my way back by the same boat. It is late evening by the time I reach Rekapally where I plan to stay at the Forest Rest House of V.R. Puram.

And here I encounter a strange fear in the midst of the drama of a rainy night, an empty house, power cut and a knock at the door[21].

23rd July'10 – Fisher Communities at T. Pochavaram, V.R. Puram Mandal, Tumileru Panchayat (Khammam)

Karri Danamma's is the only blurred silhouette visible as one walks down to Godavari at Pochavaram in search of the fishermen in these parts of Khammam district. The season has changed since I last met the fisher community in East and West Godavari barely a month earlier. It was peak summer then. This time I met them on their journey downstream – right after the heavy showers of June and July as the river rises. The next four months will be their most vulnerable-to-circumstances period. Danamma, says, "We are ten families here. We came from Kapileswarapuram. Some of our people live by Kondamodalu, Kolluru and Pochavaram. We come here around Dipavali (October-November) and leave when Godavari comes (July). We stay three months at Kapileswarapuram, at home...Since childhood I have been coming up here. There are not as many fish now as there were earlier. We used to get lots of fish in the past...Good fishing happens especially in May."

Karri Surichandar, Karri Ramakrishna, Karri Prasad, Karri Tirumurti are the only ones left here and will leave in a day or two when the river rises a bit. I was lucky to meet them. All of them belong to the Palli sub-caste. Tirumurti says, "Earlier we lived right here in Pochavaram. But there are problems in terms of health care, etc. So we leave for Kapileswarapuram for three months of the year. In the past we had built our homes here. We had ration cards, too. Now we have them our cards at Kapileswarapuram. We have to go there every month (when we are staying here) to get our rations. The President (ex-Sarpanch Ranga Reddi, a Kondareddi, whose wife, Chandravati, is current Sarpanch, but he is still referred to as "president") told us to go back to Kapileswarapuram and make our ration cards there. We had votes here earlier and even voted for him..."

Karri Subbi says, "We will lose our livelihoods when the barrage comes. We will have to leave. We came from Kapileswarapuram to settle down here because there were too many of us there. There are around 200 boats and decreasing water levels of Godavari makes it difficult for us. But these days fishing has changed here, too; tourist boats and sound of the boats are keeping the fish away. Even birth of fish has declined...We do not have fishing license since the last twenty years. The government took away our licenses. They said we are in Scheduled area of Khammam district and hence ineligible for fishing licence...Fishing is not a trade or commerce; tourism is commerce. Perhaps that is why they give the tourist boats permits and licence. For us, it is our traditional occupation we earn from."

Here they fish through the day, unlike upstream (Manturu, etc) in East and West Godavari districts, because 'water is flowing here, not stationary' (they tell me). There are no problems with boats and nets either.

At Pochavaram – This Time I Stopped By (and Found a New Forest Issue)

Ranga Reddi (ex-Sarpanch, but Sarpanch anyway) says, "They surveyed fields here. There are 108 families here, in this village, 8 are BCs and rest STs. Fishermen had cards here; they cancelled them on their own because they want to be beneficiaries of the Indiramma scheme in Kapileswarapuram. They come here seasonally. So we said to them it is their choice to either stay here or in Kapileswarapuram.They stay in thatched huts (*paaka*) here." He then said, "We opposed Polavaram, initially. We are just fine here with our forests, and stuff to build our homes with; outside, everything must be bought. How long will the money stay in our hands? We know we are losing out. People in Valairpadu and other places are accepting cheques. Everyone is taking them and if we do not, we will be left behind. Where will they dump us then?"

Suddenly, he turned to another issue. "We have other problems now facing us here. Earlier we had *konda podu* lands in the forest. Now they have made it a National Park and are not giving us pattas for *konda podu* saying this is a National park, in spite of the *adavi chattam* (Forest Rights Act). They are not even allowing us to take beedi leaves this year…The Forest guys (Department – *forest-vaallu*) say they are going to fence the forest. *Polavaram to modati debba, ippudu abhyaaranyamto rendo debba* (the first slap - *debba* - was Polavaram; the national park is a second one across our faces). We set up committees, conducted surveys and gave the list (of *podu* lands) when the Forest (Rights) Act came up. And now they made this a National Park!"

[It is almost as if they surveyed their forests to make an estimate of the area of forest cover that was already in the pipeline for National Park status.]

"Kolluru, Gonduru, Kondepudi are all part of Pochavaram panchayat. Tribal people have between 5 to 10 acres land, had some kond podu, and cashew trees. If they lose all this here, what will their status be?"

About Kolluru, he says, "Some tribal people feel Satyanarayana's being there is good for them as they are getting some odd jobs. But on the other hand, we have no sanctions for roads or buildings because Polavaram is coming up. National Park has further hurt tribal people."

***

Meeting the Naikpods at Mulakapally (Jeediguppa panchayat, V.R. Puram Mandal, Khammam District)

Mulakapally, on the way down from Pochavaram to V.R.Puram, is one of the few hamlets of Naikpod settlements. People here face many problems – lack of drinking water, water for their fields, dysfunctional pumps, etc. They trek at least three kilometers uphill and downhill to bring Godavari water on *kavadis* (two pots strung to a pole they carry on their shoulders).

These pots are used for drinking and washing, for people and cattle and livestock. They bring 20-25 pots a day sometimes, usually going to the river in the morning by 6 am and again in the evening by 4 or 5 pm. Everyone in the family pitches in. Requisitions to ITDA to repair pumps in their village have been useless. Although 60 to 65 acres of land are identified here as tribal lands, most Naikpods and Kondareddis are landless, I am told. Others own small pieces – 2-3 acres. They find work for only six months in a year (agricultural labour, mostly on others' fields). The Forest department gives them work but pays irregular wages. Since the last two months they have been waiting for wages for work last done on some plantations. Almost the last nail on the coffin has been the warning issued to them not to enter the forest to collect beedi leaves (*tunukaakulu*) or any forest produce, since the forest is now *abhyaranyam* (National Park). Muttuboyina Jaya says, "We had sown cashew saplings when that forest *chattam* (FRA) came. They made us plant saplings saying we will get pattas for these under that Act. Now they are not allowing us to enter into the forests."

The National Park and Polavaram Dam Connection

23rd July – V.R. Puram Range Office, Forest (Bhadrachalam South Division)

I met Mr. P. Babu, Forest Range Officer of this Forest Division, who informed me that "Compensation has been given and area (of forest) allotted in Anantapur and other areas[22]. Management of the National Park has not yet been decided – whether it will be under Territorial department or National Park – Wildlife department. If villages come under National Park, as a rule they have to be rehabilitated but that is the responsibility of the Revenue department. The people cannot collect anything from forest; we have stopped beedi leaf collections this year."

He also shared the following information regarding the extent of the National Park and how much will be submerged by the Polavaram dam:-

Extent of National Park

Bhadrachalam Division: 207.4049 square kilometers comes under VR Puram (Vararamachandrapuram) range.
Chintur – 69.1808 sq.kms.
Valairpadu – 163.8240 sq.kms.
East Godavari: Rampachodavaram range – 230. 5445 sq.kms.
West Godavari (Polavaram range) – 261.8280 sq.kms.

Total – 1012.8588 sq.kms under Papikondalu National Park (declared vide Gazette no. 678 dated 26th November 2008)

Area of National Park coming under submergence

V.R. Puram range – submergence 16.8957 sq.kms.

Chintur – 3.2570 sq.kms.
Valairpadu – 10.2773 sq.kms.
Maredumilli – 7.8700 sq.kms.
Rampachodavaram (RC Varam) – 20.9118 sq.kms.
Polavaram – 55.4341 sq.kms.

The list of villages that come under the National Park— Gonduru, Kolluru, Kondepudi, Thummileru, Pochavaram, Ippuru, Venkatanarasimhapuram, Kotaragummu, Maredupudi, Mulakapally, Mutyalammagandi, Bhimavaram, Darapally and Buruguwada.

I realise that all of these are Polavaram-submergence villages. While tribal communities not being allowed to pick beedi leaves since last year on account of National Park, the tourist resort at Kolluru continues (hence, I underlined it). Unfortunately, not many people know if their village and forests come under the Park. Many depend on hearsay.

***

Cell-phone notes – 23rd July - "The name is Puli, stick-thin, auto driver. From Mulakapally to Kunavaram. I find pictures with calendars given to auto drivers. If they refuse to display these, fine Rs. 100. Police department has given M pics."

[Puli is Telugu for tiger. The police has found a new way to keep a tab on suspected Maoists' activities here; they have made small calendars with photographs of suspected Maoists which are given to the auto drivers. Each auto driver has to keep it with him, in case some passenger may recognise the faces and give information. If the auto driver refuses (for his own reasons of safety) he is fined Rs. 100. So says Puli.]

23rd July – At Kunavaram – The fishing community here is from the Jalar sub-caste. They set up huts at Polipaka, Kapavaram and Nellipaka to fish for some months in a year. C.H. Umamaheshwara Rao here says, "In Chapaladibba and Ginnaladibba there are a hundred fishworker families. In Kunavaram there are 1,500 fisher families. We went to participate in a dharna against the Polavaram dam two years ago. Today we feel powerless…We still hope the dam will not be built…"

Golla Venkatesu, the President of the Matsyakarmikula Sangham here (Fishworkers Union) says, "They should allow us to be by the Godavari. They have not shown us anything so far…Though I don't know if there are any more fish left in Godavari at all….Fish are killed using medicines. Farmers using chemicals in agricultural fields in the *lankas* are killing fish. They are taking away the remaining fish for cultivation in *cheruvus* (tanks)...They kill newborn fish using medicines so that they do not have to work hard to get fish; these are not traditional fishermen. If the dam water comes people like us will have to leave with empty hands…Whenever

Godavari comes our officials (Fisheries) don't bother to come and enquire after us. They do not give us compensation for boats or nets lost. How many times we have faced this. Do you think they will listen to us now (if we talk about package for Polavaram)? Boats cost us Rs. 60,000 (new one); teak ones cost anywhere between Rs. 80-90,000. Nets cost a minimum of Rs. 3,000. And the ones that are used by seven fishermen (bigger nets) cost about Rs. 1,50,000 (which we pool in money to buy). Every year boats demand minor repairs. Every five years we overhaul the boat through a carpenter (a skilled one) who comes over. All this requires money. The Cooperative bank officer tells us we will not get loans, because they are afraid they will not be able to recover if Polavaram dam comes."

***

23rd July – At Kunavaram MRO office I get some updated information. I reproduce it here verbatim since for me, names, as histories, matter.

List of Villages Being Affected due to Indira Sagar Project (Polavaram) Kunavaram Mandal[23]

Fully Submerge

1. Pedapolipaka
2. Potlavai
3. Gommuayyavarigudem
4. Chuchirevula
5. Bheemavaram
6. Kunavaram
7. Lellavai
8. Laxmipuram
9. Kondrajupeta
10. Sabari Kothagudem
11. Walfordpeta
12. Komuru
13. Tekubaka
14. Sri Rampuram
15. Chinnarkuru (Z)
16. Chinnarkuru (G)
17. Mulluru
18. Bojjaraigudem
19. Tallagudem
20. Koderu

Partly Submerge

1. Pochavaram
2. Gunduvarigudem
3. Chinnapolipaka

4. Kachavaram
5. Kudalipadu
6. Dugutta
7. Karakagudem
8. Marrigudem
9. Potlavaigudem
10. Gommugudem
11. Ayyavarigudem
12. Paidigudem
13. Chuchirevulagudem
14. Palluru
15. Jaggavaram
16. Pandrajupalli
17. Vankataipalem (G) Venkanagudem
18. Bhagavan puram
19. Repaka
20. Pedarkuru (notification of land acquisition published in Sakshi newspaper of 23rd July '10)
21. Kuturu
22. Abhicharla
23. Regulapadu
24. Lingapuram
25. Narasingpeta
26. Venkataipalem
27. Kumaraswamigudem
28. Kondaigudem

Fully Submerge = 20 Villages
Partly Submerge = 28 villages
Not affected = 8 villages (Seetharamapuram, Tekuloddi, Mettaramavaram, Kuturugattu, Gandikothagudem, Bodunuru, Bairavapatnam, Haravaigudem)

How Lists change!

I noted that, strangely, when I went in 2006 all these villages – Pochavaram, Gunduvarigudem, Chinnapolipaka, Karakagudem, Potlavaigudem, Marrigudem, Gommugudem, Paidigudem, Chucirevulagudem (and there may be others as well) – which are now mentioned in the Partly submerged list were supposed to get fully submerged; perhaps they have changed it to avoid paying compensation.

Statement showing the Tribal Patta land being effected under IS (Polavaram) Project and alternate land to be provided in Kunavaram mandal

Sl. No.	*Name of Village*	*No of Girijan Pattadars*	*Tribal Patta lands Submerged (acres)*	*Alternate land shown to the STs (acres)**
1	Chinnarkuru (G)	234	1005.33	775.56
2	Chinnarkuru (Z)	66	121.62	118.81
3	Pandrajupalli	138	818.63	533.32
4	Vankannagudem	65	52.03	45.91
5	Abhicharla	131	888.48	525.80
6	Tekubaka	-	-	- +
7	Dugutta	108	390.66	312.50
8	Narasingapeta	109	549.97	390.28
9	Kudalipad	110	414.95	324.70
10	Venkatayapalem	82	263.22	235.27
11	Marrigudem	7	6028	6028 **
12	Gommugudem	2	8.83	7.95
13	Chuchirevula	-	-	-
14	Kumaraswamigudem	-	-	-
15	Potlavai	22	52.72	51.35
16	Karakagudem	49	132.19	118.32
17	Potlavaigudem	27	115.04	88.82
18	Gommu Ayyavarigudem	-	-	-
19	Pedarkuru	244	728.55	728.55
20	Repaka	151	592.22	4592.22 [?]
21	Bhaganpuram	81	333.59	333.69 [sic]
22	Thallagudem	20	130.56	67067 [sic]
23	Lingapuram	124	745.19	453.91
24	Tekulaboru HOKunavaram	78	206.01	178.59
25	Lellavai	1	1.70	1.70
26	Kachavaram	-	-	-
27	Jaggavaram	87	337.75	291.08
28	Kondaigudem	34	138.84	112.37
29	Kondrajupeta	87	215.95	186.64
30	Kothuru	41	107.06	98.47
31	Valfordpeta	72	156.79	144.49
32	Sabarikothagudem	200	570.05	424.35
33	Regulapadu	170	532.94	491.58
34	Bojjaraigudem	90	372.46	255.04
35	Mulluru	101	339.64	256.68

* There is no mention of exact place where the land is.
+ Why??
** Amazing! (pattadars are less, land more, all of it compensated for?)

24th July – (At Rekapally - ASDS office) – Gandhibabu says ASDS is trying to get the government to allow tribal families to cultivate lands bought from non-tribal upper caste landlords. He has the list of the extent of land

and people affected. He also informs me that the Bhadrachalam paper company (ITC) has already been given a lot of land – taken from tribal people for the project – for eucalyptus (*jaam* oil) tree plantation. He then says, "I am a member in the R&R committee. We plan to raise the issue of fishermen; we have been talking of it... We will lobby with the ITDA and Revenue department about this...By organising the community we can raise this issue with the Pollution Control Board...."

***

A Unique National Park in Submergence Zone: Turning Pages of History

Life here is about negotiating for an inch of space of mere survival. Policy still talks of conservation of green cover in exclusion and mainly for serving the timber needs of private companies. The Forest Department is not interested in rehabilitating tribal communities (as per normal rules) in case of the Papikonda National Park-affected people and is passing the buck to the Revenue Department, since people are being displaced by the Polavaram dam, even before the National Park was declared! And anyway they are getting R&R compensation for the dam. So the Forest Department owes them nothing for loss of livelihoods from collection of forest produce such as beedi leaves, or cashew or soapnuts. *Konda podu* of course is a far cry, and 'sacrilege' under the National Park rules. I am turning pages of history to look at the paradox of the forest department and find a reference that closely resembles the departmental problematic that I just observed here. "The general effect of the First World War on Indian forestry was to accelerate the transition to an extractive and commercial orientation...The war also had an effect on the mindset of foresters, engineers, and other civil servants in British India" says Benjamin Weil.[24] Further, "By the early twentieth century, forestry had become a state enterprise. From 1890 to 1920...the recorded amount of timber removed from the forest increased 300 per cent...The annual revenue from grass and fodder accounted for about 27 per cent of the total revenue surplus of the Indian Forest Service in 1916-1917...With the expansion of irrigation in the Punjab, the general management of forests in all their diversity began to give way to the business management of irrigated tree plantations. Meanwhile, two million acres of forests in the southwestern plains of Punjab disappeared under the colonisation scheme. Moreover, with plantations, the forest department could charge more and better regulate tree growth. One could reasonably argue that the Forest Service forsook its role as environmental protector to become a client of the irrigation department."[25]

The idea of forests, for the forest department, did not change post-Independence. Thus, a 1979 government document says of the East Godavari District (and might be similar for other districts too), "The forests of this district are managed with the objects (1) to obtain a sustained yield of timber, fuel, bamboos and minor forest produce to the fullest extent possible..., (2) to regulate the supply of fuel and timber with particular reference to the requirements of both the urban and rural areas, (3) to meet the agricultural and pastoral needs of the people living in the vicinity of the forests, (4) to improve the composition of the crop by supplementing natural regeneration and by converting suitable areas bearing comparatively less valuable growth into teak plantations, (5) to protect and conserve, as effectively as possible, large areas of reserved forests in the inaccessible Rampa Agency that form the watershed for the tributaries of the Godavari, (6) to obtain the maximum financial return from each hectare of the forest consistent with the established principles of conservancy...Several thousands of hectares of plantations of economic and fast-growing species such as teak, bamboo and eucalyptus have also been raised to meet the ever increasing demand for timber and pulpwood for paper..."[26]

In 1901, the extent of forests reserved in Bhadrachalam was 460 square miles and in Polavaram, 111 sq.miles. But, "They do not include Rampa, which though containing large areas of jungle, has for political reasons been excluded from the operations, and yet it will be noticed that 737 square miles of the total 942 square miles is situated in the agency divisions... In Rampa, the muttadars at one time claimed the right to lease out the forests, and large quantities of timber were removed by the lessees they appointed. But it was eventually ruled that...neither the muttadars nor the mokhasadars[27] had any right to lease out the jungle or fell timber for sale, and that the Rampa forests were the property of the State. As however these subordinate proprietors had hitherto been enjoying a considerable forest revenue of which it seemed harsh to deprive them absolutely, it was resolved in December 1892 to pay them an annual allowance amounting to half the net average of this revenue during the previous three years, on the understanding that they would assist Government in the future administration of the forest...In the Bhadrachalam division the forests and waste lands in zamindari estates...after a liberal deduction from them (called the *dupati* land) had been made round each village to allow for the possible extension of cultivation, were declared to be State property...The Koyas and hill Reddis lived in villages situated on the borders of, and even within, the proposed reserves, and for political reasons great care was considered necessary in dealing with them. Dissatisfaction with the new forest rules in Rekapalle was apparently the reason which had led the Koyas of that taluk to join in the Rampa rebellion of 1879. Both the Koyas and the Reddis lived by the shifting (podu) cultivation...Reservation, to be thorough, necessitated the exclusion of this class of cultivation from the reserved blocks and meant a considerable curtailment of the old privileges of the hill men, who had been accustomed to wander and burn wherever they liked."[28]

Let me quote further to highlight the attitudes towards the Koyas and Kondareddis, adding that not much may have changed in the past so many years even after independence in how they are perceived.

"The Koyas proper are chiefly engaged in agriculture. Their character is a curious medley. They excite admiration by their truthfulness; pity by their love of strong drink, listelessness and want of thrift; surprise by their simplicity and their combination of timidity and self-importance; and aversion by their uncanny supersitutions...Their intemperate ways are largely due to the commonness of their *ippa* (Bassia latifolia) tree, from the flowers of which strong spirit is easily distilled, and are most noticeable when this is in blossom..."[29]

About the "hill Reddis", "They call themselves by various high-sounding titles, such as Pandava Reddis, Raja Reddis and Reddis of the solar race (surya vamsa), and do not like the simple name Konda Reddi...In character they resemble the Koyas, but are less simple and stupid and in former years were much given to crime. They live by shifting (podu) cultivation..."[30]

***

25th July – I received information from the VR Puram Mandal MRO office[31] about the List of Villages to be Submerged from Polavaram dam. This is a 'fresh' list again, and hence important.

Chinamatapally Panchayat: Chintarevipally, Chinnamatapally, Gunduru, Kanaigudem, Pattipaka, Tushtivarigudem

Ramavaram panchayat: Ramavaram, Choppalli, Somalagudem, Koppalli, Sabarivarigudem, Advaivenkannagudem

Mulakapally panchayat: Jannuvarigudem (partly)

Peddamatapally panchayat: Gurrampeta, Nutigudem

Thummileru Panchayat: Thummileru, Pochavaram, Kondepudi, Kolluru, Gonduru

Jeediguppa panchayat: Jediguppa, Isnuru, Bhimavaram, Mulakalapally, Kotaragummu, Venkatanarasimhapuram, Maredupudi, Ippuru, Chilakapally, Somanamallu

Sriramagiri panchayat: Sriramagiri, Chokkanapally, Kothuru, Kalthanuru

Rajupeta panchayat: Rajupeta, Thotapally, Sitamapeta, Moddulagudem

Rekapally panchayat: Rekapally (partly, will become an island), Ummadivarem, Annavaram, Narayanapuram, Siddagudem

Vaddigudem panchayat: Vaddigudem, VR Puram, Dharmatallagudem

No. of panchayats in V.R. Puram mandal = 11
Submergence = 10
Except Chintarevipally and Chinnamatapally all the rest are tribal villages.

Vaddibalijas (BC 'A') reside in Chintarevipally, Sitampeta, Rajupeta, Vaddigudem and VR Puram villages.

Meeting with a Former Karnam[32] of VR Puram, and Revisiting Some Pasts

I met M. Ramachalapati and his wife, Radhakrishna, an elderly couple almost living on the cusp between the past and present. They live here only a few months a year, they said, in this house of the old times. Ramachalapati was born in 1932 but his family has lived here from the time his grandfather arrived here, for a reason. "In 1856 the anicut was built in Dowlaishwaram by Cotton dora" says he. "In 1864, on 1.11.1864 a cyclone struck and around 30,000 people died. Our grandfather was brought here. Machilipatnam and other villages were drowned. They resettled the affected families in Rajahmundry, all the way up to Dummugudem by giving them lands and work. Many non-tribals came to these parts (V.R. Puram, etc) in that manner. In our times they only had two distinctions, 'poor' and 'rich', not 'non-tribal' or 'tribal'...In 1959, Bhadrachalam and Venkatapuram taluks were merged with Khammam district; before that this area was part of East Godavari."

He then told me another story which I could not pursue later, yet I recorded it since the Kondareddis of Perantapally seem to exist in a different zone amidst all that is happening, and not many have mentioned their status with the dam. The generic 'Svami' of Perantapally about whom I have read and heard so long, was apparently called Balananda Svami. Ramachalapati said, "He died in the 1970s. He had started the Poverty Relief Service Society, they say. In 1945 the Svami bought some 25 acres of land...The Settlement records in 1985 show an extent of 4.60 acres for which pattas were given. Chella Sundar Rao in Jeediguppa used to manage that. He appealed to the Directorate, Court of Settlements in Hyderabad who removed the names of people because there were no records. There were no documents those days. 4.60 acres were in possession. Some were occupied illegally. In Dummugudem the Society had 104 acres which gave the Kondareddis of Perantapally 10 bags of paddy. Today they get nothing...The Perantapally Svami's land is in the submergence zone, so the Kondareddis should get compensation for it. The land in V.R. Puram which the Svami bought, where the BC Hostel is coming up, is going to be submerged under Polavaram dam too. Rightfully, the Kondareddis of Perantapally should be getting that land."

***

This time I was to take a bus all the way to Rajahmundry from V.R. Puram which passes through a lovely valley – which I may have touched. Is it through Maredumilli? I waited for long but no signs of the bus. People in the small eatery at the bus stand tell me there is some problem with the roads there; the Rajahmundry bus will not arrive. Others ask me to wait. In the meantime, a Bhadrachalam bus arrives. The conductor advises me to take this instead. Says the other bus is not 'safe' for me; it would take hours and will reach late in the night. Going by his wisdom, am off to Bhadrachalam, from where I will take another bus to Rajahmundry. A long detour! Strangely, the bus from Bhadrachalam to Rajahmundry is filled with women, barring the driver, conductor and two men. A large group of chattering women from Rajahmundry is returning from a pilgrimage to the Rama temple here. Through the journey they sing bhajans and one of them plays Telugu film songs on her i-pod. They feed me, too, some sweets and snacks. The road to Rajahmundry is beautiful at one particular point, forested on both sides. Five and a half hours of bus ride. Suddenly, mid-way, the women want to pee but the driver wouldn't stop the bus, saying it would get dark by the time we reach. For the first time I see a group of women literally threatening the driver to stop the bus by some fields, cursing him for his insensitivity. He obliges. I am glad.

26th July'10 - At the HPO Rajahmundry

It is pretty late in the night, and the attendant has been given instructions to open the doors when I arrive. I dumped my bags and rushed to the nearest hotel to eat. Some developments here – Mr. Radhakrishna is taking voluntary retirement; so next time (if) I come here, whenever, I will miss this extremely helpful and kind man I have seen since 2006. He has daughters to marry off, he says, so the 'VRS' (Voluntary Retirement Scheme) money will help. He gives me, for the first time, his mobile phone number, asking me to get in touch whenever I need any help about town. The HPO will always be open for me, he says. He is always about 'help', this man. Only I have not given these people even a single token of thanks for rendering help on their own accord each time I have come here. And it has been a long while since my friend's husband left the department and relocated to Tamilnadu. Can't help the sadness that wells within; will I come again? As of now, this seems to be the last lap. Mr. Radhakrishna has introduced me to a cab driver who will charge me very less to take me to Kapileswarapuram tomorrow. It is an old Ambassador, 'non-AC' (that is why). On the way back I plan to stop over at Dowlaishwaram – I have a thought in my head.

Reflecting on the Issues at stake

Pollution; several stakes on Godavari – commercial, non-commercial and subsistence-based; equity; access and denial of access are the major issues that came to the fore in the two-month long engagement with the question of Godavari's fishermen/fishworkers being displaced by the Polavaram dam. Tourism seems to be a major source of contestation in the fishing zones of the Godavari upstream in East and West Godavari districts. In the few fishing settlements one visited on the Khammam district side there were issues of identity as fishermen/fishworkers in citizenship terms. Speaking of Common Pool Resources and Common Property Resource or Regime, when it is not land or forests that we are talking about, but a river, especially one that flows through six Indian States, the issue becomes more complex. There is a fluid sense of 'ownership', and more so, non-ownership. In that sense, there is just a usufruct proposition to it. Especially so long as the state/Government has all the power over that resource. The fishermen's families make their temporary shelters along the river banks to fish for five to six months a year. The tribal communities do not question or lay ownership rights over the sand and the banks which may physically be part of that village. Sharing a natural resource, such as a river, a mountain, a forest, have been for years almost an extension of their lives, an aspect currently being questioned in the making of Polavaram dam and the entire R&R exercise. For, perhaps, in the government's narrow vision the fishermen of Godavari can fish anywhere at all where the river exists (provided it is as 'free' as it seemingly is, at the moment, not a State resource for power generation and leased out yet to private corporations, at least in the geographic environs of these three districts of East, West Godavari and Khammam, the submergence zone of the Polavaram dam project. That they can 'move' (as they have moved indeed with the river for centuries) also works against these communities in terms of fixing the responsibility on the government to compensate them nevertheless for their now-settled existence, and thereby, livelihood, threatened by displacement at the least, not to speak of the larger economic changes that will impact nature and even the fish lifespan in the Godavari, with dams at every State, old or new, in the present and in future. Meanwhile, I find a discarded newspaper supplement in Suite I. Do these things happen, really? How am I led to the most significant questions when I least expect them, setting in motion a new stream of thoughts? The advertisement supplement is dated 30th June'10 (came with the New Indian Express). It is a four-page pullout (glossy) issued by the Andhra Pradesh Industrial Infrastructure Corporation highlighting the "10 Things Good with Godavari" (including – "Timeless river, rice bowl of south India, rich agriculture, natural resources, social infrastructure, connectivity, access to sea ports, big players already there, a willing administration and peaceful politics) and points out the "big Guys Already There" – including "ONGC, Reliance and Cairn Energy…"

And it helps place into a wider economic context the whole issue of displacement on account of Polavaram.

The Industrial Context in East and West Godavari district

There is a larger economic context to the challenges of identifying different communities' share of Godavari and establishing its validity and legality.

Let me focus on some economic developments that are going to have a long-term impact on the socio-economic profile of Andhra Pradesh as a state. The investments are big; so is the concept and design of the Polavaram dam. Who will it feed, once it comes up, if it does (taking if as a possibility of organised protest, even if it seems weak and inconsequential today) but these "big guys already there" and ones on the way. Other points are mentioned in APIIC's promotional message in the four page pull-out which bear testimony to the fact that the Polavaram dam will cater to all these categories and stakes on Godavari which are far higher and of a more powerful nature than what people (even those opposing) do not realise.

Some of these are reproduced below:

"APIIC has secured four blocks of oil fields – two offshore and two onshore – covering 4,587 sq.km. The proposed Pertoleum and Petrochemcial Investment Region (PCPIR) covering a sprawl of 603.58 sq.km is coming up with an investment of Rs. 3.43 lakh crore between Kakinada and Vishakhapatnam. An astounding 12 lakh people will be included in the employment footprint of this grand project. The state's chief facilitator is the shaping hand in 300 industrial parks, covering a cumulative extent of 1.30 lakh acres. In just the last five years, APIIC made available 30,000 acres of land to entrepreneurs, besides accumulating a land bank of 82,000 acres."*

"Land bank" is meant for some future purpose, but acquired. Incidentally, just about 4 years ago the small airport (it was an air base) at Rajahmundry then had just one private economy airline operator, the erstwhile Air Deccan operating from it, with just two flights (one arrival and one departure) in a day. Today other operators, including Jet have started operations. Which means air traffic has increased in just the last four years, especially after` commencement of the Polavaram dam project.

"There are 62 industries with an investment of at least Rs. 10 crore each in East Godavari, with a cumulative total investment of Rs. 9,940.91 crore and job potential of 18,502. There are 18 industries with an investment of Rs. 5-10 crore, with a total investment of Rs. 133.44 crore and employment capacity of 1,520 persons. There are 62 small industries with an investment of Rs. 192.16 crore, providing jobs to 4,900 persons. Single window clearance has been given to 2,500 applications involving an investment of Rs. 33,933 crore and job creation of 22,759....East Godavari has 15 industrial estates with 1,600 plots and 381 sheds. Of them, as many as 1,548 plots and 381 sheds have been allotted to various industries. As many as 108 industries are scheduled to be set up at an estimated cost of Rs. 60,290 crore with a

* Also see the Section, Reflections, later in this book for a related discussion in the context of the Right to Fair Compensation and Transparency in Land Acquisition, Rehabilitation and Resettlement Act, 2013.

capacity of 21,805 jobs…The Kakinada Special Economic Zone is coming up over 10,000 acres in East Godavari straddling Tondangi and Uppadakothpalli mandals. In addition, many other projects are in the pipeline, prominent among them being a 138 km Petro Corridor with an investment of Rs. 2 lakh crore between Kakinada and Vizag."

The Godavari river-grab has the capacity to turn what is called the "rice bowl" today, into folklore for all the agricultural land.

***

26th July - At the Department of Fisheries, Katheru Fish Farm, Rajahmundry

The office in-charge here informed me that between October 2009 and April 2010, 619 licences were issued to the fishermen of Godavari and Rs. 51, 340 realised (in the form of *pannu*/tax) from them. In East Godavari there are three divisions –Rajahmundry, Kakinada and Amalapuram.

Number of fishermen in Rajahmundry division

Fishermen families population - 30,652 ("approximately", according to Government records).

"Active fishermen" - 7,583.

In Rajahmundry division fishing communities are settled in 23 mandals. I observed a general ignorance among the officials about the caste groups. There is no information about number of fisher families to be affected by Polavaram dam though the government has asked the department to prepare a list for R&R. Devipatnam, Kondamodalu, Polavaram, Pinikelapadu, Lingavaram, Manturu, Singanapally, Mulapadu, Midipally, Agraharam are among the villages with fisher families to be affected by the Polavaram dam (submergence). The Fisheries department has prepared a proposal, a list, meant for R&R for fishermen affected by the dam. I was handed over the same "unofficially" (still in preliminary stages). Was told the government does not want the fisheries department to mention anything about rehabilitation yet. They have just asked for names of affected areas where Fisheries Co-operative Societies exist. Mr. Ramakrishna, Senior Assistant, Fisheries, said, "We plan to ask for craft (that is boat and nets) for fishermen who will be affected…We give loans only to ST and SCs and STs…Two years back, there was a centrally sponsored scheme for river fishemen. We gave nets and 50 per cent subsidy – 5,000 nets were supplied. This year too, we proposed the same for 240 units in east Godavari. We have taken the survey records from Revenue office and submitted to the government. The Deputy Director, Fisheries, in Kakinada is in charge of the same. ITDA looks after tribal fishermen in the tribal areas."

***

Kapileswarapuram Panchayat and Mandal: the Colonial anicut, *Pulasa* fish and Sand mining[33] - 27th July

I am sitting in a temple (a meeting place) in Satyanarayanapuram Hamlet. There are three hamlets of fishermen here – Satyanarayanapuram, Durgampeta and Salipeta. These fisher families spend nine months a year at Pochavaram, Thumileru, Koluru, Gonduru, Kondepudi (Khammam district), Perantapally, Tadivada, Kondamodalu, Sivagiri, Nadupuru, Sivaduru, Tekuru, Kachluru and Manturu (East Godavari district). The remaining three months they stay here. It has been this way since last few generations, they say. All of them here belong to the Palli sub-caste. Sangani Peddamaidi (who goes to Kondamodalu) says, "I have been going there since 1971. There is not enough to eat or for fishing here."

Kamada Suribabu says, "Here livelihoods was always a problem; no fishing happens here. Only when Godavari comes we get the *pulasa* fish, not otherwise. We just have to sit idle if she does not come and live on whatever we earned there (at Pochavaram, where he fishes)... Once the Polavaram barrage comes, we cannot survive there; if the girijans go away, where will we stay?"

There is a symbiotic relationship between the two, even though there is an element of conflict only in one case, in Pochavaram. Incidentally, the Gazetteer mentions the fish that Suribabu is talking about – "The large sable fish (*clupea palasah* or hilsa) are netted in very large quantities near Dowlaishwaram anicut, when they come up the river to spawn...."[34]

I note that this hamlet is located just by the headlocks of Cotton anicut. I take a picture of the rusted colonial headlocks with a by-now worn-out cell-phone. Lanke Sattibabu (secretary, Satyanarayana Fishermen's Cooperative Society), says, "Water will be stocked if the barrage (at Polavaram) is built. Fish will not come into our nets. It will all become a plain area. There will be no flowing water in Godavari...At least 200 members from our community go upstream every year to the Agency (Scheduled Area) from Kapileswarapuram. There are 400 familes here (members of this Society)...Nobody cares about us (*mammalni pattinchukune vaallu leru*). Siltation is a problem here; the *pulasas* (one pulasa can earn them Rs. 2,000 or more) have stopped coming as they used to earlier. Sand mining companies and contractors have started putting leaves and pebbles into the river for sand mining. There is no space, nor possibility, for nets to be laid in the river anymore. We spend Rs. 10,000 for nets for *pulasa* fishing alone. Now we are suffering a loss of over Rs. 20,000 because there are no *pulasa*. We manage with loans from money lenders on high interest which we repay with the income from fishing in the Agency Area for nine months. Some of our folks went and settled down in villages where the local people had no problems. They have their voting and ration cards there too. Others come here to cast votes or get rations. There are three ramps

(Kapileswarapuram, Kedaralanka and Korumilli – here too you find fishermen settlements) of sand miners. They are mining 24 hours. They place ramps between November and June months. They stockpile sand before Godavari comes. They do this business for nine months. The government is now laying roads for them to bring in bigger machines. Earlier they used to get sand in boats, with help from labourers. Now they are taking the machine right into the middle of Godavari. They bring huge cranes and are being helped by the officials. Nets get stuck and tear. We plan to oppose this in a big way. Twenty years ago there was much more depth in the water, now the depth has reduced... If the dam is built we will be reduced to *sunyam* (zero). Because of siltation here, there is no use for fish seeds either."

Over the years the anicut has produced problems of siltation and swamps in canal areas. I ask them about the anicut. He tells me, "They are not opening the Dowlaishwaram barrage headlocks since four years now; it has rusted. So we cannot bring our boats through when we come here from upstream. We have to leave our boats there by the lock and walk with all our belongings to our homes here. The headlock superintendent is not responding to our pleas. He should be opening the lock."

This time I see the logs being collected in the river, on the rafts that the dalits in Pydipaka mentioned. Some of them are tying them up on bicycles and taking them to town to sell. Unfortunately my camera is not with me at this moment (left it back in the room; no film roll left; no money to buy one). I feel empty-handed without it. The cell phone camera just about manages – who would have thought I would actually see people collecting logs of wood on rafts in these circumstances, by Kapileswarapuram?

***

The Kapileswarapuram-Pochavaram Connection: Citizenship and Stakes on Godavari

It's been pouring incessantly in V.R. Puram mandal over the last week, causing the river to rise (by the night). Life moves with the seasons for fisher communities and as a result, affects many things, in its wake, including bigger ones such as ideas of citizenship and access to basic rights. The river of course is one single entity flowing downstream to the sea, and each of the administrative distinctions (the Districts) make no particular sense for the fishermen. The Kolluru fisher community had left for their homes in Kapileswarapuram in the first fortnight of July. The four in Pochavaram stayed back, fortunately for me, since "*ninna koncham teesindi* (yesterday she took herself back), almost like a woman playing hide-and-seek with them. One of their homes had the impression of Godavari's 'visitation', and a woman starts to move her things up into the village where she will wait for others to leave for Kapileswarapuram – some are taking their boats while others (especially the older ones) will take buses and shared (service) auto rickshaws. Shared auto rickshaws take you to Rekapally – an hour's journey – from where you get the Rajahmundry bus through two different routes. From Rajahmundry it is a bus ride to Jonnada and then by a shared auto to Kapileswarapuram. It makes better sense for them to take the river route, though it will take them two to three days to get home. Once they are back in Kapileswarapuram, even if it is for three months in a year, at the most four, they will deal with several other problems until they are back in Pochavaram around Dipavali. Some of them are invisible in the records and live a life between a place they call home and another they used to call home not so long ago. Of course, their votes were counted last year. Thankfully, that makes them citizens enough. How are rights to be defined and guaranteed to fishworkers who follow the river and have fixed territories for nine months in a year? On rivers that flow, what about seasonal fishermen and what is behind their movement, why is it not recognised? Their movement may be through regions and perhaps states with the movement of the river. In case of Godavari fishermen, especially in the context of Scheduled Areas, how are their rights defined? How are the licenses issued? The officials themselves say 'unofficially' that it is difficult to keep a count of seasonal fishermen who are itinerant for some months as they move. But anyone who has observed them closely, even for last four years, can see the pattern of their movement and their places of settling down. These are almost certain either because of the history of visiting the same place and bonds with peple, or based on availability of fish. Even decline in fish population has not yet affected their movement patterns beyond these few villages. So, between Kapileswarapuram and villages upstream in Scheduled Areas it is a question of movement that needs to be recognised, along with access to the river as a Right. For them it is a question living between two aspects, or sides, of the same river. There is one side that still provides enough fish to make ends meet and the other which is a highly contested zone and leaves no scope for fishing. But the problem in their case would be, like Malladi Posi of Manturu had said, "Godavari cannot be compensated for Godavari." While portions of Godavari have of course been leased out to Reliance (Ambanis) in the previous Naidu regime, which seemed such a simple transfer of papers, fishermen have to make do with tank and canal fishing in restricted and reserved regimes if the dam is built at some point. How can citizenship questions be defined on a river? Seasonal fishing on Godavari in these three districts is unique. One, this is a living tradition with a long history. But over the years many changes to the way the river is being used, the stakes on the river, policies, change in fish production, etc, may have affected the nature of movement and new settlements and fishermen constantly in search of new spaces to put up their fishing base nine months of the year. They need to be identified as traditional craftspersons for whom movement is intrinsic. A river is not land or forest; rivers flow and cannot be physically demarcated in exact terms of visible boundaries. There is just a loose understanding of which part of the river is for which state (in terms of inter-state dynamics) and its riparian allocations – a category too technical. Godavari's fisher communities, too, have moved along the river; they cannot file petitions in courts over 'river rights'. Or can they, perhaps? There seems to be no scope for fisher communities to assert their rights, which do not seem to be inalienable (by what one has seen so far) over a river-based ecology, which is as much their traditional right and occupation as the forests have come to be acknowledged (at the very least) as the traditional right (of access, usufruct, etc) for the tribal communities (even though in practice we know how much of that is truly the case). In all such cases, what are the minimum protective legislations in place for fishermen to protect their fishing turfs from being alienated or converted? Which part of nature is theirs?

Ramaswamy Iyer has written that, "a change in the state's position from one of total control and ownership to trusteeship in relation to natural resources is very desirable for ensuring resource – conservation, inter-generational equity and ecological sustainability, and for building a constructive and harmonious relationship between state and civil society…"[35]

Godavari's fish-hunting communities operate on territories that they have fixed for themselves over generations. Thus, over sustained interactions with the community one gets to know that fishing community from Manturu are from Yanam; or those from Pochavaram are from Kapileswarapuram. Or that they are *Pallis*; the fishing community in Kapileswarapuram call themselves *agnikula ksatriyas* and fishermen settled in Kunavaram are Jalarlu. There is no universal homogenous group here. The interesting pattern of their movement or fishing history is not recognised as a living historical tradition. What one observed was that nobody decides where they 'ought to' fish, in the process of movement for better hunt, settle down, and gradually establish an unexpressed sense of belonging, not ownership, to that fishing ground. There is an unexpressed (unwritten) synchrony about where nets are laid; no two fishermen lay nets in opposition to each other, in competition to each other. They have separate timings of laying nets and bringing their catch.

***

Retracing the Metaphor Giver, Muttu Rama Rao

At the Fishermen Colony at Dowlaishwaram — Almost reaching his house, with great expectation, I have already visualised what I will say, and decided that if now his son Satti Babu asks for a 'walkman' (the personal stereo which he had requested that I buy him) I shall give mine to him, since I have not managed to buy a new one for him these many years. That was

2005; it is 2010 now. But he will remember. I shall also tell him I received his inland letter sent to me in 2005, and asked my friend to read it out to me (it was in Telugu, with a drawing of a fish being caught with a bait, on the space reserved for the sender's address!). I would tell them, '*nenu occhesaanu choodandi*' (see, I have come by, finally). We (the driver of the cab I had taken from Rajahmundry and I) knocked on the door we were directed to, but who do we find in there? Not Muttu Rama Rao, not Satti Babu, but someone with the same '*inti peru*' (family/house name). We went to quite a few homes in that colony seeking Rama Rao. Finally, we met a woman who told me something which increased my sense of fascination for Rama Rao; I understood why he was the lone fisherman. One of his kind. Fishermen of Godavari usually have their culturally passed on (through generations) traditional fishing grounds that they usually never cross. A Dowlaishwaram fisherman will only fish in the area upto Rajahmundry on the river, for instance. The fishing zones and community bonds seem to coalesce. But this woman told me, "*memu Warangal atu pakka vellamandi; wallu okkare velli poyaaru*" (we do not ever go towards Warangal side, but their family left alone to those parts to fish). Their family, she said, was 'different' and they do not keep in touch with the community here. I could not stop building in my head an entire story of a fisherman who decided one fine day to go far away, upstream rather than downstream, with his family, making his own boat and his own roads on Godavari, trusting her flow, though it must have been rather tough to row upstream against the wind at some point. It must have been something to find a corner for his family so far away from people of his own community, away from his coastal language, to go into the Telangana tribal region and settling down there by the river, to become part and parcel of the place. For him and his family, the boundaries of 'Andhra' and 'Telangana' meant nothing. For them, the boundaries, if at all, of a certain measure, are made of a certain measure made on the river notionally in their minds – never too far away from their settlements – temporary – these are not boundaries of State or a region, but made in accordance with the river's flow. They could, if fish were to come by, go really far on their boats, but rarely. Muttu Rama Rao must have decided or planned for months, or on a whim, or a hunch, decided to take the boat upstream to find his perfect fishing ground. On the AP-Chhattisgarh border within a district called Warangal. At Guttala Gangaram. The relationship of acceptance between the tribal community of Koyas there and this itinerant fisherman of Godavari played its part as he settled down and sold fish in the local market at Eturnagaram. But his sense of identity and belonging still remained in that fishemen colony where he asked me to meet him; the next time after Godavari comes. He was always saying he would go to Rajahmundry 'when Godavari comes' (in the period where they do not fish). I realised that day - Rama Rao never came back to this colony! My respect and regard for him grew immensely for having been a man who charted a new course and went upstream leaving behind his clan and community and its protective system. For, the fishermen at Rajahmundry will always tell you, as do the others from Kapileswarapuram, that their community gets many protective benefits from being here, including schooling for their children, and other schemes that the Fisheries department floats for them from time to time. They seem to be relatively better off if not really in the best of conditions. Hemingway's Old Man and the Sea kept recurring in my head as a theme on the way back from that colony even if I could not find the man I went looking for. I do not know now where I shall see him or when. But the visit to Dowlaishwaram, more than spotting the structure that defined deltaic agrarian regime for centuries after it was built (the Arthur Cotton barrage/anicut), was important for adding an element of poetry and drama to the life of Muttu Rama Rao and others like him (if there were) whom I may not have known. You cannot always pin down people to their community and lived historical contexts. There are always those that break away and make their own paths; those who never 'belong'. Who are not seen in statistics and perceptions built around statistics.

***

August 2010

News Arrives on 9th August – (over phone from Kunavaram) 'Godavari came two nights ago - 20-25 families on the road now; nets lost; boats they could save. Water entered homes in the entire colony of fishermen – Chapaladibba. We told you we have stopped approaching our officials, they do not listen to us. Vadapally, Devipatnam, Manturu, Kondamodalu, Tuthigunta – all these inundated today.'

Surely, the fishermen's boats must be ferrying supplies, again, while they themselves remain unaddressed when it comes to compensation for any nets or boats lost.

Another News in mid-August'10 – The media reported that the AP Government issued an order clearing about 3,725 hectares of forest land for the Polavaram project (3,731 hectares of notified forest area and 258 hectares of deemed forest land). The State government would implement a 'wildlife management plan' prepared by the State Forest department in consultation with the Wild Life Institute of India.

***

17th October'10 – Thinking - In between journeys, it is the story of development. The distance between two journeys being symbolic of the distance between fact and fiction, reality and make-believe, government and people, construction and destruction. Of these, the distance between fact and fiction can yet be dissolved. Others? Not so sure.

***

Land Acquisition Progress[36], Polavaram Project - 2009-10

January 2010 – 56.25 acres are acquired.

Cumulative Progress in this volatile period is 23,183.26 acres acquired; Rs. 31,093.35 lakhs award disbursed.

April 2010 – 260.37 acres acquired.
July - 125 acres acquired.
December - 41.76 acres acquired.

Cumulative Progress – 2010-11 – 23,916.42 acres land acquired; Rs. 31,093.35 lakhs award disbursed (Note this is the same figure as in 2009-10)

What about R&R?

As on September 2011 – R&R (as seen from the figures) shows NIL against all columns, which include, community infrastructure, number of families rehabilitated (land and housing). Though they claim that Rs. 7,500 lakhs was spent in year 2006-06.

***

Notes

1. I am yet to find an appropriate term; workers has its own connotation which can also be problematic; folk has its problems, fishermen did not represent the women , but in this particular context women are not involved in either catching or selling fish (unlike those who live by the sea). I use each of these terms interchangeably at times.
2. For instance, I remembered a long forgotten, non-operational PPF (Public Provident Fund, of the Government of India) account of mine and an old LIC policy a father may have foreseen use for in future (it did give me a loan). They helped me make the visits. One cannot thank enough these Public Sector entities in India which, one hopes, will resist the onslaught of profiteers. They do help in times of need. I was allowed to withdraw half the amount. Yet, from zero to 15k is a giant leap. But in familiar villages, retracing steps, money is never a handicap. Though the money and freedom that comes with it did not last long.
3. An anomaly, really for a person who scoffed at the term. But it humbled me accept, it helps.
4. It helped me make two visits, in end of May and then in July'10.
5. This is term they use to refer to their occupation – *chapa veta* (fish-hunting). They would refer to a person as having gone to hunt fish, not 'catch fish'. They did not use the term *matsya karmikulu* (fish-workers) in any of the villages I went to. The term appears only in petitions, or requisitions made on behalf of

the community by the more educated or in a political context. Fish-hunter, as a term can also take care of the gender. Though here it is only the men who go fishing; though Kondareddi women have their own way of getting fish from the river which has been mentioned earlier.

6. An assignment came in after months - an organisation called Centre for World Solidarity in Hyderabad asked me to write a study paper (which was to be part of a series planned) on the KC Canal Modernisation project and its lessons for the ongoing Nagarjuna Sagar modernisation. I do not refer to my full paper here, which was not published ultimately (as far as I know). But I share useful aspects of that journey to the command area KC canal (in Kurnul district) and Nalgonda district. I thank Mr. Sastry and the CWS team for this work which has given me valuable insights.
7. Max Martin and A. Srinivasa Rao, "Were the southern flash floods avoidable?" www.mailtoday.in, October 6, 2009.
8. Meeting the flood-affected people was not part of my study paper but having gone to Kurnul one felt duty-bound to meet the affected people. With the help of local members of the farmers' organisation, AP Rythu Sangham, I visited some of the affected places in the short time I had with me.
9. See in the Section *Reflections...* – Idea of Forests.
10. Apart from the fact that people in Yanam speak Telugu. Since it was under the French rule, as with Pondicherry or Puducherry Yanam continued with the latter administration post-Independence. Geographically, it is close to Gadimoga where the Reliance gas pipeline is located.
11. Andhra meals comprise the cuisine typical of the coastal region – mainly the settled agriculturists, landlords of the upper castes. It would ideally include *vepudu* (fried vegetables), *pacchadi* (pickles), *pulusu* (spiced tamarind water), side dish (brinjals, usually, or any seasonal vegetable), and *perugu* (curds or buttermilk) – this is the usual meal served in restaurants and cafes. The true full-course would of course contain other items. Incidentally, while *Chapala pulusu* (fish cooked in tamarind water) is part of some Andhra meals, the food in the tourist launches do not serve 'non-veg'.
12. After a long time I picked up a 'diary' again, apart from the reporting notebook. Strangely enough, this was the last entry in my diary. All the rest just became entries in the notebook recording people's interviews and conversations. But this diary turned into my Kolkata diary in year 2011!
13. Perhaps, Malli's (my dog's) presence. Thought of her for the first time in all this while.
14. AP Water Vision Draft II, p. 3.12.
15. Ibid, Vol. I, p. 22.
16. These interviews appeared in an article in New Indian Express, as "For whom the Godavari Flows", dated, 24th July'10, as a Feature story, and the pictures of Ballaparaj and Malladi Posi and others in Manturu were also featured; later when I had the chance to show the same to Ballaparaj (I think Posi too), he was very excited and got me to give him a copy of the same. Raju (Ballaparaj) met me much later in Hyderabad, too, where he brought his uncle for a medical problem. "There are several problems with the RR package" and asked as to when will I come there again. I sincerely do not know, but ended up saying, "next month". It has not happened.

17. Over a phone communication, on July 14, 2010.
18. An obligatory attendance at a wedding on 28th night-29th morning (in an acquaintance's family) in the villages Adavur and Gomada on the north Andhra-Orissa border. Another problematic landscape.
19. Described in *Personal Memorabilia...*
20. Some of these ideas were published in infochangeindia.org as part of a series under the FES-Infochange fellowship.
21. See *Personal Memorabilia....*
22. Already suffers as a drougth-prone district. Do not know where the "other areas" are.
23. Source for all the data from Kunavaram: MRO office, Kunavaram mandal, Kunavaram. Tahsildar (MRO) Venkireddy; Deputy Tahsildar – M.A. Raju; RI – T.V. Ramanarasaiah, Assistant – Ramesh (as on that date).
24. Benjamin Weil, Conservation, Exploitation, and Cultural Change in the Indian Forest Service, 1875-1927, *Environmental History*, Vol. 11, No. 2 (Apr., 2006), pp. 319-343, Forest History Society and American Society for Environmental History, accessed through JSTOR, (http://www.jstor.org/stable/3986234), p. 334.
25. Ibid, p. 333.
26. AP District Gazetteer: East Godavari, 1979, p. 12.
27. These titular heads from the local communities acted as mediators and negotiators on behalf of the colonial masters.
28. Hemingway, cf. cit, pp. 93-4, Emphasis added.
29. Ibid, p. 62.
30. Ibid, p. 66.
31. Source: Kalti Bakkaiah, attender at MRO office VR Puram mandal at Rekapally, Revenue Inspector. Others – attendant G. Ramu; Dy Tahsildar - G.L.N. Swamy; Tahsildar - K. Uppalaiah.
32. Karnams used to be repositories of all the information regarding land dealings in the area. Their services were utilised by the Revenue department.
33. By this time I was through with my film roll and no money to buy one; the feeling is hard to describe for here I found a live example of what the anicut ultimately led to; found ramps placed in the river for sand mining, saw the dalits collecting logs of wood (as the Pydipaka people had described), but no camera. By this time my mobile phone camera too is not its old glorious self, yet, I clicked with it, all I could. Felt handicapped in a sense. If I had not taken the cab, perhaps the film roll was possible.
34. Hemingway, op.cit, p. 15.
35. Ramaswamy, R. Iyer, *Towards Water Wisdom: Limits, Justice, Harmony*, Sage, Delhi, 2007, pp. 157-8.
36. Source: http://ppms.cgg.gov.in/DetYearProjLA (Centre for Good Governance). Only few months taken here, to see the general progress. They give no explanations given for Nil figures – September – October show nil perhaps because of the demise of the former CM, YSR (September 2009)? Following this, from November 2009, the Telangana agitation (for separate State) had caught on.

Revisiting Manturu (2010)

Udaya

The Krait (apparently) in the pot at Veerappa Reddi's

An evening on the banks of Godavari with Sreenu, Pandu, Kondaiah, et.al

Malladi Posi (seated) with Gopi

Dusk sets in, nets laid to rest

It's evening - TV time; few left on the bamboo platform

A picture from the previous journeys on the bamboo platform

Morning of 26th May - Cooking a meal for the tourists

26th May - At Kachuluru and Tuthigunta (Life in a Thatched Roof)

Sangani Eswara Rao with a friend from the Konda Kammara community

Eswara Rao with family outside his home by Godavari's banks

Karri Bhagyam

A child by a hut on the banks

Votala Durga Rao

Back at Manturu, in my diary - a different kind of cricket

In my Cell-phone Notes - Full moon night by Godavari (26th May)

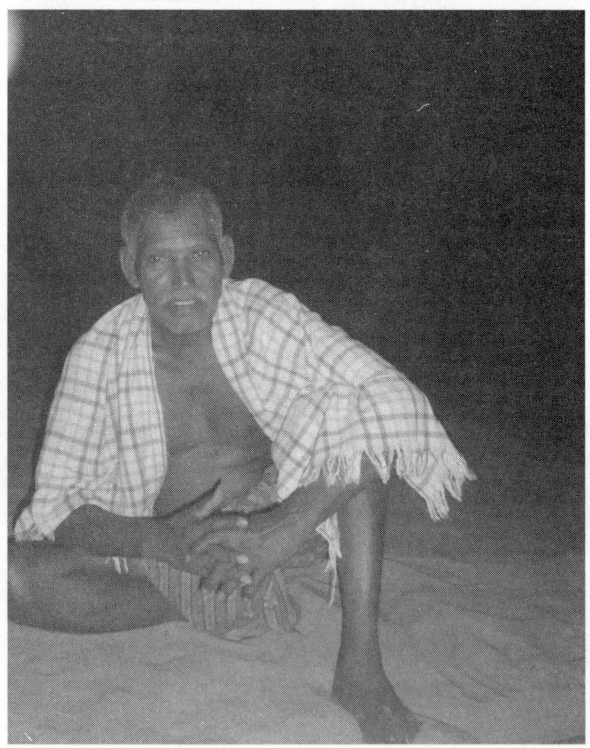

Old Man and the River - Malladi Arjayya at Vadapally

At Singanapally and Devipatnam (27th May)

Ballaparaj and others at Singanapally

On a passenger boat (Singanapally to Devipatnam)

Manga and her daughter

At Devipatnam, Gangadharam, Ramu and others

Sattibabu, Durga Rao, Raju and a friend (cell-phone notes - 27th May)

A rough boat ride and a question

Fishing in a Contested River

Perantapally Ashram

Children by the Tamarind tree (where I had first seen Vennela)

A Tourist Launch passes by Perantapally

Kolluru Tourist 'eco-resort'

Kaigala Satyanarayana, owner of the resort, at his home

Danamma's boat, Pochavaram

Karri Danamma, up close

Karri Surichandar, Ramakrishna, Prasad and Tirumurti at Pochavaram

The Headlocks of the Colonial Anicut, Kapileswarapuram

Stockpiles of Sand, Kapileswarapuram

A truck load of Sand by Godavari, Kapileswarapuram

Personal Memorabilia for Life: Adventures and Accidents

A journey is incomplete without them. I share some of mine as I think of each of these as somewhere connected to certain basic ground realities and at some level, having a deeper meaning in the larger journey of life in the very nature of these accidents. The contexts, timing and form of the mishaps, misadventures, miscalculations, etc. leave questions even as they remind you of your limits as a human. I share from some of these from my 'diary' moments.

Journey 2006

7th June - On Nocturnal Life in a Colonial Edifice, and Fear (Kunavaram Forest Rest House)

Within minutes of my lying on the bed, I saw several eyes watching me curiously from the ceiling above my head – a 'ceiling', weather-worn 'thermocol'-plastered, infested with the biggest bandicoots ever. I froze. A few would peer at me, disappear, making scurrying sounds and another group would arrive to look at me with the same eyes. Had I read about bandicoots that chewed on people's feet? No way would I spend another moment in that room! But how was I to leave that room? For, within a moment, there were two huge, fat, wall lizards[1], right on the wall by that impossibly heavy teak wooden door! One of them would definitely fall on me the moment I touched the door – or so I believed. I called two of my friends in panic, one after the other (each in a separate city). I can still say with certainty, they may have never forgiven me for destroying their respective 2 am deep slumbers and keeping one of them awake till 5 am, constantly seeking an answer to, "what should I do?" What a paradox – here I come all the way, having walked so far through every circumstance but am stalled by two lizards on the door and several bandicoots on the ceiling! With great effort (with my very pregnant friend on the phone and her words of assurance) I virtually dashed out of the room and after a lot of coaxing by her, knocked on the door of the Forest official in the room next

1. Perhaps these were the Golden Gecko, in the list of Endangered Species, found in these forests.

to mine. I started the sheepish, intermittent, knocking 'process' at around 4 am waiting in a dark corridor. Finally, at 6 am, precisely (as I checked my wrist-watch), the door opened. The officer kindly exchanged his room (with an air cooler, and a definite wooden ceiling) for mine. Ah! For that elusive sleep! But wait, just around the corner of the room, perched on the door knob, what should one see but a green frog (was it a tree frog?), watching me, almost waiting to leap, any moment! With a feeble plea to the frog in my head, not an ounce of energy left in the body, I didn't realise when sleep overcame me. Only to be woken up promptly at 8.30 am with a loud knock on my door! Another long day ahead. Sleep is just a moment's illusion, on these paths.

I realised in the morning what a historical monument I was put up at – at the Kunavaram Forest Rest House, built in 1911 (with the Godavari Flood level of 1915 marked as a testimonial)! The waters came right up to the 'Suite I' that was offered to me. At that moment, an immense pain welled within – if the dam were ever to be built, would the waters devour all those creatures (I was afraid of last night) and along with them, innumerable other smaller ones, the frog, the crickets, other organisms that have been part of this habitat since ages? What on earth was my fear all about? Of creatures that intend no harm, unless in self-defence? I never visited that rest house again in my following visits. But I do see it now and then, when I pass by the Kunavaram bridge over Sabari-Godavari – a quaint colonial edifice, a historical legacy of the exploitative forest department of the colonial times. But home/haven to species of reptiles and amphibians in the true old conservationist paradigm!

***

11th - 12th August – The Road, and Life!

As for me, hitting the road from Rajahmundry at 8 am on 11th August 2006, until 5 am the next morning when I reached Rampachodavaram, was one of the most intense moments in my Godavari journeys. All those boulders to manoeuvre through, an aged driver who got drunk on *ippa* toddy (while I was wading through the backwaters with people and witnessing the Panchayat office fight), a rickety old jeep, slush and rain, and a pristine forest – who can be so fortunate to have it all at one go? I felt more intimately closer to the people than ever. At a more mundane level was a realisation that one must either walk barefoot in these circumstances or wear the most sturdy shoes (though people manage with rubber slippers most times in these places). The walk with four young Kondareddi boys through streams and stony paths in the depth of dark (none of us had a torch) was something! You felt closest to either humanity or an 'expansive universe' – there are no midways. The jeep broke down midway right in the midst of the pebble-stoned stream. It was pushed up with great effort across the stream back on to the road uphill. It was then that we realised the driver was drunk. Once on the steep uphill, as he revved up the accelerator, the jeep moved in the reverse direction, all of us seated within! This process was repeated a couple of times. None of us knew driving. The ultimate drama happened as with one last effort, the jeep moved ahead and then quickly started sliding backwards, downhill, and just as it was to hit a tree by the corner of the road-bend, I shouted like I never did before, asking everybody to get off while the driver shouted at all of us to remain seated. The tree protected us all. We did get off. The men somehow pushed the jeep back safely to a corner and leaving the driver to his (at least safe) destiny, we walked on. I shared with others some dry dates and apricots (which I never ever carry otherwise) – the only time I ever ate that day. Any other time, that walk might have been beautiful, quiet for the entire stretch, none of us talking, sound of a stream down below, night birds and insects and unfathomable darkness. Yet, the silence, the sounds, kept me outside of a self within limited bodily pain. Suddenly (or was it gradually?) a biting pain moved up from my ankles up to my calves. It was certainly an insect bite (perhaps from the stream we waded through) that left me with swollen feet the size of an elephant's; the pain almost numbed me out of my senses. I thought I couldn't walk no more. Thankfully, in the midst of that forest, or partly out of it, we managed a short night halt – those four young men (my guardian angels) and me – in a hut around a kerosene lamp at Lankapaka - a visual memory frozen in my heart and phone camera.[2]

By the way, I was to find out later that the jeep driver was found the next morning sleeping under a tree. Nobody knows when he returned home and how! If there was just one regret, it was not having encountered a single wild animal, something that amazed even the forest official who gave me a cup of tea the next morning (when I reached Rampachodavaram), as he enlightened me with a list of wild animals of these forests (in that particular stretch), panther, for one! Back home, the exhaustion of the journey (physical) caught me unawares, a virtual black-out and sickness followed which committed my pen to recount the pain of the road to Kondamodalu – having truly understood, and felt, what it must mean for the people in those villages.[3]

****

2. Till date, whenever I managed to visit villages around Kondamodalu, it has been that road journey with the young men that is recounted. They never forget! Not a trip would be complete without joking about it. And in 2008 I actually met one of those young men 'of that road' (as he introduced himself to me!) who guided my way briefly.
3. Even as it was time to bring to a logical (?) conclusion my own PhD thesis, on a different historical, cultural, political realm.

Journey 2007

5th April 2007 - On the Way Back to 'Charaka' at Lalacheruvu

For the first time on these journeys, I encountered a situation of this nature, making me cautious as a woman. It was dark by the time I reached a bus-station from where I walked back to the guest house. Suddenly an autorickshaw driver began to follow me. I saved myself with some quick thinking, running into the first lane I could spot, and into another, waiting till he left, after waiting for a long while. This is the highway that leads to the Rajahmundry airport, which is becoming busier by the day and I cannot help linking every new development I see in this old town/city (or midway) to Polavaram. Anyway, safely back in the room, to realise the nearest eating place is again through the same old road and truly far off. Guess have to sleep hungry.

***

Journey 2008

22nd July '08 - a Night with Bats at the Panchayat Office Guest Room, Polavaram

I stayed the night here. I wished later though that I had left for Kathanapally by the last bus to Sivagiri. What a night! Between a host of bats disliking the human presence in their midst after a long gap (this room adjoining the open terrace did not get too many visitors, and certainly not a woman, alone), hovering over my head and a wall lizard that fell on my bed just as I was to sleep my tiredness away. The bats were better, at least I could run. It is pitch dark outside, though I can see the silhouette of that by-now-familiar statue of Alluri Sitaramaraju holding a bow and arrow and Dr. Ambedkar on the street below. Not a soul moves out in these parts after 8 pm and it is nearly 10 now. What do I do? I left the room open to bats and other forces of nature, and down in the street tried to call out to someone in that thatched roof hut where I usually have tea. But none responded. No doors opened. But finally, someone came out from her dwellings and responded to my call from a distance. She seemed like an elderly woman. I thanked my stars. She then asked her son to fetch some help. He went to the attendant (at the Panchayat office) Jyothi's house close-by. Soon, the young woman (the same person who brought in some tea that evening) accompanied me to my room upstairs and shooed the lizard off the bed and then asked me a bit reluctantly - "Would you sleep in my house?" (She was probably reluctant because it was a tiny tenement, even by Polavaram standards.) Need I ask for more! For me, it was the home of a woman with a big heart. Jyothi is a contractual safai karamchari (a cleanliness worker) at the Panchayat office. She lost her husband not so long ago and was offered the job after his demise. She has a young school-going child. We spoke until we both could hold our sleep no longer. I slept like a child, after all that chaos. The next morning I was to realise another aspect to her tenement – she paid 'rent' for using a toilet and a bathroom, some houses away from here. Would they make her pay if I used it? Wish I had not put her in trouble. I could not hold myself but didn't wish to look like a typical outsider either. I told her I would go back to the Panchayat office and use that place – however dirty. But she wouldn't listen. But I had to move on soon lest she plead with me to have breakfast as well. We held each other's hands tight that morning, before I left. Since I was an outsider and it was an emergency situation, my 'rent' for the toilet was waived off by the owner of the place. I could not thank her enough. Jyothi said her home would be open to me anytime I came back. I should not need to stay in that Panchayat office all alone. Sometimes, you make friends in strange circumstances!

***

Journey (or non-Journey) 2009

July-October'09 - Caught in a Wrap (or Splints) – Soon after the April electoral journey, a bizarre nerve impediment induces immobility and forces a temporary halt to the Godavari route. A misadventure in its own right!

***

Journey 2010

22nd July'10 - Some Drama in another Forest Rest House, Rekapally

I am staying the night at V.R. Puram Forest rest house at Rekapally – with a long corridor, dining room and two tiny guest rooms, a plantation nursery behind and trees in the front yard. The rest house is located by the road. In the evening not a soul walks on this road leading to Vali-Sugriva hills on one end, all the way to Chintur and Chhattisgarh border (Kotaragummu, Mulakapally, Pochavaram villages too on the way) and Kunavaram-Bhadrachalam on the other. It is a long road built by the British colonialists to transport timber and check tribal revolts. It has been pouring persistently and (so) there is a power cut. There is a thoughtfully placed lantern, candle and match in the room. Mosquitoes are aplenty today, and even a repellent doesn't stop them from whining in your ears. But the real drama is the sound of rain against closed window panes, a house away from the others in the village, not a soul around. It isn't too nice to sleep in a tiny room with all the pores closed. The last time I came I stayed here (in 2007) at the ASDS office guest room. Strangely, it had rained then, too, and I was listening to rain drops and recording them. Today is different. Suddenly, I heard a hurried knock on the bathroom door (which is attached to this room in a strange way), not once but thrice. Nothing can beat that feeling that welled up at that time of the night (10 pm or after)! Animals, birds and even lizards are not half as scary as men are, in these situations. And it is the latter one has to be most apprehensive about. Anyway, after much thinking, I called up Gandhi Babu, who sent some of the women from the training camp (who had also refused to sleep in the rest house as they found it eerie, as they told me) to step out in the rain, umbrellas, torchlights and all. They took me to their guest room, sharing it with four other women (they said too many people?) for that night. Felt silly, but, well...Until the next visit (if), the question will always remain - who could have knocked that late in the night? In daylight the place looked as serene, normal and friendly as ever.

***

29th-30th May'10 - A Break in the Journey...

I plan to take a two-day break at Rajahmundry and then resume the second phase of this trip to Kakinada and the mangroves again, and Nedupuru, Rekapally, downstream. Before that, I must attend a wedding at Gajapati District (in a village called Adavur), on the borders of Odisha[4]. The brother of an acquaintance in Hyderabad gets married post-midnight today (29th). The invite and my attending is an emotional issue for them. I cannot avoid though ideally I would continue downstream. Rajahmundry burns at 40°plus Celsius.

And 1st June - Home with a Memento

When you stretch your body beyond a point, it gives in. And how! In the wee hours of the morning of 30th I returned from Palasa, to check into the hotel by Kotipally bus stand. Thought I would rest my limbs a bit before hitting the road continuing my journey. Was feeling queasy since morning and thought two cups of tea and rest would help. Everything thereafter happened in quick succession: non-stop vomitting, calling the reception for medicines, and falling on my face right as I stepped out of the bathroom. I could feel I hit my teeth, thought had broken all of them; what a fall! I dragged myself on to the bed and lay near-dead. By a quirk of good fortune the boy who brought in medicines was a trained paramedic (another guardian angel) and the room was open! I learnt his name was Raju. He quickly prepared some sugared water and offered to call anyone I might know in Rajahmundry. The advocate, Nagaraju was intimated – from the

4. Incidentally, this village sends most of its families to Hyderabad to work as watchmen in apartment complexes, I find out; so a wedding or ritual in a village shrine is the only time you actually see the homes opened up; it is a time nobody misses. That story, of the village, and these friends, is for another moment.

random numbers on my mobile. Nagaraju took me in an autorickshaw to a friend's clinic. The doctor took a while to arrive. He apparently told Nagaraju (I learnt later) it was a wonder how I had survived on those levels of BP and haemoglobin (6, did he say?! I did not believe it) It was a freak 'all-in-one' case of food poisoning, low BP, heat stroke, etc. I was put on a bottle of saline for the next two hours. As for the teeth, I hadn't broken any, just a chipped front tooth for which the doctor gave me a medicated ice pack. He didn't take any fee, strangely. I was advised to return home. As for Raju, who came by to enquire after me, he worked on night shifts in the hotel and during the day as a nurse. Not for a moment had I thought I would face sickness on these paths. Was it the water; all those stories of how polluted Godavari is becoming by the day? I do not regret drinking water there. But my mind goes back to the nature of access to a hospital or a doctor for a person in those villages, were something similar to happen to them. As for getting home, it was another experience requesting two reporters of a media channel (finding out their numbers) who were to attend a wedding in Hyderabad on 1st June to share a cab with me and split the cost. All the buses and trains were running packed (thousands of weddings were scheduled for 1st June that season). After a painful sleepless journey (from the tooth and blaring obscene film music in that cab) seated next to the driver, I reached Hyderabad; with another painful exercise convincing (pleading with) them to drop me home while they said I should take an auto-rickshaw from a bus-stand, lest they miss the *muhurtam*[5] (of their boss' son's wedding). And I ended up paying more than my share. I did not know these people. Nor did I know Raju, at the hotel. It takes all kinds to make this world!

In that moment, I thought of all the people who ever fed me or gave me shelter in the villages. And in that moment, I fell in love with my home in Hyderabad all over again; and longed for my bed and the tree by the window and my dog. I now have a memento of my own from the Godavari journeys, something nobody can take away from me – the chipped tooth (second in rank after one from a childhood story)!

*****

5. The moment of the betrothal, which is considered the most important in Telugu Hindu weddings.

The Colonial Monument - Kunavaram Forest Rest House

The Rest House, built in 1911; a painted memory of the Height of Flood Level (HFL) - 78.75 feet on 16.08.1986

Rubbles and stones and a stream - the road to Kondamodalu

Trying to move the jeep out of the stream

Frozen in the lens - the lamp in the hut at Lankapaka (post midnight)

Reflections...

Reflecting....from a Distance

An Explanatory Note: Year 2011, Some Distance

After all the physical journeys[1], the support of two academic fellowships (in Kolkata and Chennai)[2] helped me further place the journeys in perspective, in the sequence that they had happened. Those were also important political moments – the elections in West Bengal (which saw change of guard) when I was in Kolkata; and the Land Acquisition and R&R Bill 2011 was placed in Parliament, when I was in Chennai. Debate on land continues to be important and so also, the nature and form of economic reforms, understanding of 'development'. There is no question that the depth and truth of people's own knowledge and life experiences (reflected in their voices I have tried my best to record) say more than can be said. I record here – as a matter of sharing, as another historical truth essentially – my own personal attempt to understand things seen on the ground as also from my lived experience. There are no linear or sequential links to the issues discussed in this section. Thinking and writing on them happened at different intervals in time and they are not based on any order of importance.

***

OF LANDSCAPES...

I have been thinking about the perception of superiority of one landscape over the other: thus the perception of superiority of riverine (delta) tracts over the drylands, of hill and forests, and the caste here over the tribe there. In the ancient Tamil context, I had at one point, worked on the concept of *tinai*, the five eco-zones, represented by a corresponding idiom, emotion, represented in Cankam[3] poetry, each landscape and corresponding emotion having their own place in the scheme. Even if this was a mere idea, yet it was a beautiful one. Brahminical religion in many ways managed to crush these ideas of diversity, both in its sacred literature and hagiography in certain ways. The temple structure, too, added to the universality of resource-perception and distribution. Tanks (water bodies) and their presence within the Brahmin settlements (*agraharas* and *brahmadeyas*, in south India, as royal/state grants) have been associated with certain castes and dominance and control over these resources has continued through time. Some forms of power and prestige have no 'expiry dates'. I would see some concepts as similar though far removed in time and space – one example was the introduction of the Water Users Association model in Andhra Pradesh which was meant to be a democratic exercise, radically altering the idea of 'ownership' of resources, but ended up with people from higher castes and with more clout becoming 'Presidents' of these WUAs (Water Users' Association). WUAs incidentally, became negotiators in the irrigation structure of the state at one level, but ultimately the instruments of state control over these structures. Rivers, even in these ways, and riverine tracts, have continued to feel the dominance of the dominant caste structure, which modernity has only reinstated and re-formulated to its convenience, so that, control of river and its direction towards a particular segment (may not be in terms of hierarchy of caste, but that of economy, industry or large scale agriculture, owned by big farmers, again from higher castes). The riverine, as a landscape, has dominated economic discourse in AP; where other landscapes are not relevant for power politics. The brahminical religion has its role to play in this, as well. As it fitted in very well even within the colonial structure, as its status was not threatened just as with other dominant castes (in AP, the Kammas, Reddys, Kapus, Rajus, and several like them). The Buddhist and Jaina religion in regions that are today in Andhra Pradesh and Telangana states are important in the diversity of landscapes that were allowed in those religions – through the idea of non-killing, ahimsa, and the movement of their monks and nuns through the rocky terrains, hills and forests, which moved into their texts for matters of ideology as well as praxis; and gained for them followers from other sections of society. In the Buddhist case, looking at the spread of the religion, in Andhra Pradesh, for instance, the riverscapes and tanks are not seem in control and dominance sense as the religion itself respects nature. In the Jaina case, it spread essentially across Telangana region, relatively dryland, and its sphere of influence did not spread all that much in coastal delta region where the Saiva and Vaisnava religions seem to have dominated. The idiom of granting lands, for sacral sanction, and along with it, resources used to build tanks within agrarian settlements and agrarian expansion of a certain kind essentially emerged from (in what is today Andhra Pradesh) the Brahminical religion. And the perception of nature, natural resources, different social groups, too, changed accordingly. Thus, the importance of a forest and hill and dry and scrub land, is not perceived in dominant religious tradition of brahminism and its offshoots. Although its sacred literature may refer to the forests and pastoral lands but when it came to settlements, the preference was always the deltaic, riverine tracts. Mythology emerged to correspond to these ideas. Major Vedic gods or goddesses are usually shown subduing the god or goddess of the hill tribes and those living in the forests. In Puranic mythology it takes the sober form of 'co-opting' or 'adapting' what cannot be violently subdued. Of course, it is another matter that you do not see much of the Jaina or Buddhist ethic in operation when it comes to questions of power and commerce, even among followers of these religions and hence there can be no generalisation here. But at the level of philosophy and thought, one is trying to locate diversity of landscapes in dominant religious traditions, because at some point, these do seep into politics and perceptions of the superior and inferior. I am not speaking here of the *political choice* of being Buddhist, initiated by Babasaheb Ambedkar.

***

MOST BEAUTIFUL PLACE ON EARTH

But why not about landscapes? Why not about beauty? Each season has its own kind of dawn; each dawn is broken by competing calls of the roosters. The villages between the hills and Godavari, the mist and the clouds adorning the hills, and the Papikondalu, the Lapwings showing the way, playing with me, as they did one of those dawns? Are beauty and feelings of awe and wonder the preserves of the elite, or the ones that have made 'beauty' and 'nature' their leisurely get-away? Will water overwhelm these hills and fields and the mist and the clouds and will these become the homogenous landscape of development? Shouldn't diverse landscapes – with their own reasons to be in the realm of the universe and scheme of the earth – just be what they were meant to be, diverse? Just as communities are meant to be? And should be allowed to be? Shouldn't these landscapes remain beautiful in their proximity (in a non-possession kind of way) to the people today being marginalised and sidelined? At the other end, should beauty or nature (not spoken of in some circles) be something to feel defensive or shameful about? Is there a need to revisit the utilitarian, which has today raised paradoxes of productivity and capital, bereft of the soul? Is there need to revisit the idea of nature in the context of the marginalised? Are they unmindful of it? Are they so oblivious to it that specialists from the literate elite teach them all over again to conserve and preserve, in a vacuum, as 'preserves' or 'reserves', again oblivious to all their other (politically imposed) limitations?

***

A NOTE ON DISPLACEMENT

Displacement essentially makes you a lesser human, aching to belong in a context that does not let you belong unless you are willing to adapt to or adopt a set of rituals of the system, giving up your own innate identity to be called someone they think you should be, should have been, before reaching that state of belongingness (again, belongingness donated to you by their charity). You lose a lot more than you gain, in the end, and with a question – was the gain worth that losing, in the first place? Has it helped me evolve or has it destroyed my essence?

***

WOMEN IN A SUBMERGENCE ZONE: A DISCOURSE IN DISMEMBERMENT[4]

I use the term 'dismemberment' as an analogy of a separate, disjointed, amputated limb-site, cut off from the nerve-centre of the country's developmental process/democracy and also as a part that holds no interest to the larger body politic/body-economic. Submergence for me is both a physical and a metaphorical idea. Physical in terms of the dam-building exercise (including land acquisition) and the resultant havoc; and metaphorical in terms of drowning of voice and expression, and a certain way, of perceiving nature-human interdependence, in the flood of a dominant river "utilisation" discourse. As for my own location in this, while I speak of the women here as outside of me, physically, socially, economically, or historically, I still speak of them as me, in a sense, from within my own helplessness, anger or resistance. We are dismembered in our respective battles, but usually we fail to see a structural edifice that governs our lives and in fact intrudes into our lives – through media, political decisions and use of language. Sociologically, dismembering may be seen as transforming markers of 'tribe', 'caste' groups into a class category in an alienating, routed-towards-the urban economy context. If getting rid of identity markers of caste may be seen as a redemptive feature in an otherwise caste-ridden society, provided (and that is important) the lower castes can truly be allowed space to transcend their caste status in an equal space (and not in a non-problematised free-market canvas), at the same time, for a tribal society, it may not at all be redemptive to be rid of a cohesive context of the tribal social context. In the Polavaram case, some of the tribal hamlets have been shown different R&R colonies in different villages which many of them find unacceptable. The communities feel this will break their cohesiveness. The hierarchy of class subsumes markers of the tribe and related deeply embedded identity (self, vis-à-vis the other) question, but not towards a more egalitarian, idealistically democratic co-existence, where other selves, too, have relegated or consciously rejected their respective social markers. Does the loss of 'tribal' identity lead to a more redemptive, affirmative, assertive citizen identity in a Rights context? It does not seem to be the case; rather it seems to be a move towards a more ambivalent, rather negative universal space. Wallerstein (although in a different context) has said, that "...by a sort of impeccable logic, particularisms of any kind whatsoever are said to be incompatible with the logic of capitalist system, or at least an obstacle to its optimal operation. It would follow that within a capitalist system it is imperative to assert or carry out a universalist ideology as an essential element in the endless pursuit of the accumulation of capital. Thus it is that we talk of capitalist social relations as being a 'universal solvent', working to reduce everything to a homogenous commodity from denoted by a single measure of money."[5] Race and gender inequalities in this context, he notes, are on the rise.[6]

Dismemberment is a deeply loaded term and not about mere physical displacement or dislocation. It is built upon the acute alienation of the individual, or community via loss of memory – rooted in social-cultural, geographical-spatial historical contexts - in the face of larger political processes they have no control, nor autonomy, over. Seeing it through a gendered lens means to look at aspects of physical, forced, uprooting from a historically- lived context and a largely imbalanced negotiation they need to make with the powerful and their mediators. There is a sense of powerlessness, helplessness, loss of identity within the community and while the common spaces they inhabited as historical actors gave a centrality to their being and role, dislocated, they enter an ambivalent, and hence problematic, non-historically located space. The position of the SCs is very different but in their case, their caste identity, in spite of the class context, does not leave them and in fact is aggravated in the case of the poor among the SCs. The fisher communities, too, have a cohesive community context within different *jati* identitites (Vaddi, Jalar and Yanadi being predominant in the Godavari region). Loss of river for them is also a loss of their community socio-economic identity, as one found out. They also have notional fishing areas within the same river. In some cases, where they were enlisted for NREG works, they found it very difficult to assume the role of manual workers on agricultural fields they had never tilled in their lives. Somewhere, economic reforms of the kind we see have in their own way, added to fragmented spaces and a certain middle class monotone existence/or even existentialist angst, at times, seems to have developed around this economy and because of it. If at my urban level I have to tread cautiously, even battle, against the existential angst as a woman, in the lives of Godavari's women the daily battles without a collective female space mean a different kind of alienation. Is it possible to retrieve a feminist, feminine, if need be, ecological perspective of life-affirming life-systems and rivers (about men and women from the disempowered communities, reduced to the margins of history)? At the same time is 'women's voice' homogenous? I didn't find it so, in my experience. Yet, I could find some amount of collective expression at least among the poorer, landless, tribal and dalit women, which can be called a feminist, or women's perspective. The deeper, human, emotional levels of lived experience of women are absent in official and non-official dialogues and even in 'modern' citizenship terms. More than 2.5 lakh years of lived human settlement history seems to make no heads stop and turn and even wonder, at the very least. In our cities of today, we can count perhaps up to a few years of being housed in one colony, or tenement, as a luxury. Attachment is seen as a problem – emotional or physical. And (in the urban context) if the attachment might be to a tree by the window[7], they look askance, urging you to get your priorities right. But in those villages, these attachments might just 'sit around', somewhere in the vicinity, as an amiable presence, and that too, as a part of life, and in fact celebrated once a year in ritual expressions. If the river could tell a story it would narrate stories of 'want' and 'need' (a distinction women were very astute about) – which may be seen as reflected in the tourist launches versus passenger boats. How are natural phenomena perceived in the mainstream language? People are conditioned to look at, report, 'address' many natural phenomena as 'disasters' that always must be controlled, as if it were an epidemic. Tales of woe and sorrow and helplessness are streamed into people's living rooms on such occasions. So, heavy rains during the monsoons (so far as media is concerned, usually only in Mumbai) are always a 'problem to commuters' (how dare it rain this way!). But aren't the monsoons about rain? Would people like Mallada Narasimhamurthy and his family, or Vijayalakshmi, by the Godavari estuary and the joy on their faces as Godavari rushed down in full flow after an exuberant coming, ever find place in such depictions?

In a study on people's adjustment to floods in the Yamuna floodplain in Bangladesh, Bimal Kanti Paul notes, "...All respondents of the study villages refer to annual flooding either as barsha or bonna. The former is a normal inundation, which is crucial for production of aus and aman rice. This type of flood is perceived...as a benevolent agent providing sustenance to the farmers and, thus...an accepted and much-anticipated event. ...Floods that rise 8 feet above the broadcast aman fields but do not overtop the village mounds or homestead land are called barsha."[8] Further, "flooding is an intimate part of rural life in the villages of Bangladesh, and it is deeply imbedded in their culture."[9]

The idea of growing food versus buying is an idea women in the villages held strong views about. It is the difference between self-sustenance and dependence (home, as in the environment they somewhere feel powerful enough over, and the market). The articulation of women is strongest among the tribal and dalit women of the Godavari region in the submergence zone. Among the tribal communities it is especially marked in those villages which have had a long history of peasant-tribal resistance and a strong sense of self and identity. In the case of the dalits (SCs, BCs) the awareness of a long period of exploitation and a certain amount of fearlessness is evident in their being vocal about their protest and grievances. If there is a sense of "adequate" it is in connection with food and shelter (a dwelling built in camaraderie) than it is about cash that it has come to mean for some activists,

NGOs, government and industry. The strong sense of value for nature and natural resources (which they perceive almost as an urban middle class person would perceive a home with modern contraptions) was marked in almost all the tribal hamlets I visited. But over the years, the influence of the cable network with its string of popular Telugu TV serials highlighting only a certain regional culture within Andhra Pradesh and certain class and castes, cannot be underestimated. It will remain to be seen if this affects their value (valuation – even in its economic sense) of nature and natural resources and their farm land. There was a politically charged sense of rights and *not* a victimhood notion among those women. Since women were most articulate in their disapproval of the dam and selling their land to the government, the government officials approached the men and tended to isolate the women. It was invariably the men these officials called to the Revenue office, and their attitude was one of casually chatting them up and gradually broaching the subject of land over 'bonhomie', or if that did not work, threaten that if they let this opportunity go, they would lose both land and money. But women's voice is neither important to the government nor the mainstream media. And the women have a strongly political voice, no less political than the voice of the urban, educated icons of middle class activism, in fact much more so. Mutchika Suramma, whom I met at different times between 2006 and 2008, incidentally, passed away in 2010 after battling with Cancer. And the cases slapped against her, including Sedition, hadn't been cleared by the time she died. She and a few others kept running between Rajahmundry jail and their village since 2005. Till her last breath she did not sign the paper of consent to give her land for the project. When I first met Kuncha Varalakshmi, Mandal Parishad president (Pydipaka) in 2006, she had spoken to me as a Mala (dalit) Panchayat representative expressing her views as a dalit. Between August 2006 and December 2007 much had changed in local dynamics and Polavaram project as well. By 2007 her understanding was in tune with that of the administration. Her Mala (SC) and woman identity was subsumed under the overpowering state and upper caste logic. And being (life) partner to a non-tribal upper caste landlord of Pydipaka who had sold land to the government for money, made this an intriguing situation. She had moved to Polavaram by 2008, while I had first met her outside her home in Pydipaka in 2006. By 2008 the disempowerment and loss of livelihood of the SC agricultural labourers (working on this landlord's fields) of Pydipaka was complete. For the administration/government, Kuncha Varalakshmi was a dalit who had given consent to the Polavaram project without representing the cause of fellow dalits. For the dalits, she transformed as an agent of the upper caste and the state. Women losing their land to the dam, especially, the Scheduled Tribe and (landless) Scheduled Caste, invariably mentioned the loss of *gaali* (air), *vatavaranam* (this environment – of their villages by the river and forested hills) while men, barring few from the tribal communities, while not willing to leave their homes, were yet concerned about the R&R package (*"RR packagi"*). And there was a very crucial difference – the non-tribal upper caste women (from the Kamma, Raju communities) who were the wives (never owners of land) of medium and large landholders (owning tribal land in contravention of the 1 of 70 Act in the V Schedule Areas) almost always let their men do the talking and it was these men who were the first to accept the government's offer to legalise their illegal landholdings by selling them. However, in private conversations, where possible, these women would also lament the loss of this kind of environment, though adding, at the same time, that they would not be able to buy any alternate land elsewhere for this kind of money that the government was giving them. The tribal and dalit women, on the other hand, saw land as the feeding bowl. The economics as these women see it, is very much part of the environment – including flora and fauna – they live in. They do not dissociate the two.

Dams as Temples? Sure! Jawaharlal Nehru may not have been off the mark when he called dams "temples of modern India" which should actually be read today as exemplifying and signifying the politics of Hindu temples in Indian history: as redistributive spaces with a set of priestly mediators between the devotee and the deity within the sanctum sanctorum, built with money and meant to convey the power and prestige of the king/chief, etc who got it built (in today's times we may place the state and the industrialists and the engineers in these positions). Temples necessarily became alienating spaces to a large extent in the Hindu context and in fact creating wider and wider distance between the worshipper and the god/goddess within the edifice (which can be seen today as well, and in fact in many temples this distance continues to grow wider by the year, in spite of democracy, in the form of the steel railings, trappings of modernity; richer the temple, longer the distance between the deity and the worshipper); sufficiently brahminised/Sanskritised. Dams, similarly, once built, grant the power of mediation between people and the river to the priests – here the government engineers in charge of the gates, its opening and closing closely monitored, timed, just as the temple gates and the most important door of the sanctum sanctorum are. Strangely enough, almost all the engineers in charge of the dam gates (that one has come across in the dams observed) are male. Even those calculating the rise and fall of the dammed river waters, onset of floods, release of water to the farmers, etc, happen to be male in most number of cases, and this may not just be coincidental. Through the priests, the state (just as rulers in the past) wields the power. The intimacy of the people and the river (just as the intimacy between a personal god/goddess and the human, if one were to accept the existence of a god concept) is broken, and sufficiently distanced and neutralised by the priests of the dam. Dams indeed work as temples in a modern nation promising the best to the richest, in the volume of water and direction of water flow released. The marginalised – castes and tribal groups – have no centre-space in this establishment nor are they important to it. They happen to be incidental 'beneficiaries' of the crumbs from this whole status play, no matter how far our democracy may have come today. Polavaram and river waters in Andhra Pradesh exemplifies the dams-as-temples metaphor. A nation, especially a democratic nation, built upon its relationships and linkages with others in this system means that we can no longer look at displacement except as a systematic elimination of citizenship and participation of people whose historically lived spaces are today the contested spots for maintaining global linkages. There are no singular dispossession zones, anymore. In the same way, the disadvantage for women from marginalised communities can not be measured in terms of numbers alone, but has to be seen with a political engagement with these larger processes.

The transition of a tribal society to a class society is never smooth. Negotiations here are based on losing self-hood. For women, in these contexts, the process of loss to recovery of self and identity is a long, tedious process when there has been no movement into an empowering context in a monetised market colonialism (where rules are set by a global market) situation. In many cases the identity re-accessed if not recovered is not the identity the woman may have desired, but is forced to take on. It could be a purely economic identity such as 'construction worker', 'wage earner'. For most tribal communities hitherto displaced move/fall to a deskilled physical labour, where their knowledge of important socio-economic processes within their communities, such as agriculture, traditional medicinal practices, knowledge of forest species, etc, is of no consequence to the literacy-centred market economy. The larger world political economy and geo-politics is creating circumstances that necessitate the build-up of technologies, systems, governance, that seeks easily available physical labour provided by these communities, to keep the system going, increasing the number of such daily wage earners in rural and urban areas. It is difficult to engage with this critically when it is couched in terms of Right to Work. In the Godavari region – East, West Godavari, especially which was part of the Madras Presidency – there were two parallel social contexts. A tribal area with its own resistances and struggles against the zamindars, muttadars and the British colonialists, and a Godavari anicut-fed delta with its commercial cropping with linkages to the world market thanks to the colonial rule that extracted from them every penny spent on the anicut. Tobacco, sugarcane and paddy made these landlords intensively water-dependent and hungry for more land. The delta farmers continued to depend on water for their commercial crops post-Independence and continued to usurp land in the tribal areas in spite of protective Acts in place. Most of the delta capitalist farmers who later became industrialists and political rulers of Andhra Pradesh have continued this tendency. Even prior to colonial rule this area was a network of canals and tanks built by kings and chieftains but the access to these networks would have certainly been caste-based and the lands belonged to temples, Brahmins, and the other upper castes. In this sense the upper castes of the delta have had a smooth transition to a class context and are today enjoying the benefits of globalisation. They cannot think beyond land and water, both of which were commodified for them for centuries. The dam discourse is male – in terms of construction, engineering 'marvels' (engineering dams as being marvels of technology; engineers deified, as Arthur Cotton in AP, etc); and at the same time, the anti-dam discourse is also male in the case of Polavaram dam, where discussions, debates, from the opposite sides are put forward by men, and women have usually not been consulted on the same. Terminology such as 'cost-benefit analysis' etc which have been put forth as arguments come with an assumption that women do not know the technicalities, and these are meant for dialogue with the state, and for 'policy advocacy' which is again a male exercise. Women exist only in numbers as displaced, or in names, as one of the many who oppose on largely universal grounds of the displacement. Women's groups, few and far between, who once opposed the dam, again, looked at numbers of women displaced, again in R&R terms. Robin Tennant-Wood, discussing the Snowy River of Australia and the politics of dam building and the engagement of women in the discourse, comments on the nature of writings that came out on the same. So, "The male histories of the project…with chapters under such titles as 'Politics of Water'…and 'Tunnels of Blood'…focus almost entirely on the physical hardships and technical aspects of creating a system of dams to divert an alpine river in a climate of post-war economic renewal or the issues and problems of a large ethnically plural labour force, referring only women as a separate category of Snowy life…"[10]

Further, "The regulation of a river's flow can also be perceived as a male initiated end of the river's female function: the cessation of the flow that heralds the end of fertility…The natural seasonal flow of rivers as opposed to their function as economic agents in the global theme and it is this that underpins the male and female relationships to rivers…The loss of place, to women, is part of a greater loss of self."[11]

The damming of Godavari is not to be seen in a vacuum, but as part of this hegemonic control over river via politics that is not merely patriarchal, but also casteist, in favour of the dominant castes of Reddys, Kammas, Kapus, etc who have incidentally held important positions in the state governments right from the 1950s. The formation of capitalist farmer class has been documented in many studies in the region which also urges one to look beyond mere displacement from a dam discourse. In the late 1960s, early '70s, Daniel Thorner studied the capitalist farming class of coastal delta region. The nature of development over the years can be gauged by reference to one particular detail. For instance, "The agricultural future of the Andhra coast would be assured, he felt, by the building of a series of dams to control the Godavari River. Just as I had heard in Visakhapatnam, he declared that 95 per cent of the Godavari water rolled *uselessly* into the sea. Only 5 per cent was tapped. If the river were dammed, the agricultural output of the Godavari Valley would go up three or four times. What is more, there would be enough water to meet the requirements of a steel plant at Visakhapatnam."[12] This was an agricultural officer speaking to Thorner who was visiting the coastal farming communities. Thorner had gone there in 1959 and then returned back in 1966, when he was comparing coastal Andhra to the 'deficit' areas such as Bihar, lower Bengal and inland Orissa. And he says, "I found coastal Andhra to be a completely different world. Practically everyone had the equivalent of two-and-a-half to three meals a day. This was true even in non-irrigated villages, though in these the meals of the poorest villagers might be dull and repetitious – ragi[13] gruel and ragi porridge alternating with ragi porridge and ragi gruel. Best-off were the delta districts, such as East and West Godavari. There even the families of agricultural labourers who were utterly landless earned enough to eat three meals per day, supplemented by vegetables, spices, chillies, occasionally dry fish and tea, coffee or toddy. Compared with the rest of India, this was, in relative terms, an affluent area. The deltas abounded in signs of change, development, progress. It was no accident that when, later in the year, old fashioned and obscurantist elements in northern India rioted against cow-slaughter, coastal Andhra demonstrated in force for a new steel mill."[14]

To this may be added the upper caste, class of coastal Andhra whose entire outlook of development was directed towards capitalist formation and market economy, a link induced by the colonial regime through the anicut which made the delta farmers access the river water towards a thriving agrarian economy which diversified into other economic activities, as is shown by Thorner during his visit. Thus, he says, "The best-off families in the villages were benefiting in ample measure from Government help. They were getting Japanese, Polish and other foreign power-cultivators, tractors, etc through government loans at low interest. The really large-scale farmers – in effect, millionaire cultivators – were investing their gains from agriculture in trade, transport, money lending, and industry. *The main complaints of the cultivators along the coast were about irrigation and drainage. They wanted more water to grow more crops and they badly needed modern drainage systems to get the water out of the fields.* These questions of water supply were long-term problems involving large expenditure."[15]

It is not hard to draw the connection between the capitalist farmers' aspirations that early, a decade into the creation of Andhra Pradesh State and nature of river water discourse (especially Godavari water discourse) that ensued from these aspirations. Significantly, Thorner never consulted women of these farming families, or in general to record their views. But he does mention women agricultural labourers when he notes the agricultural wages paid to them in those days. The big farmer landlords he met are male and the dialogue on coastal Andhra 'affluent society' remains confined within male circles in terms of the people he meets, including the agricultural officer, etc. There was a certain construction of agrarian and thereby gender (delinked from agriculture) discourse with expansion of land under cultivation of commercial crops post-anicut on Godavari in late 19th century. This impacted the tribe-caste conflict; also where in tribal society in East and West Godavari district women played an equal role in agriculture, gradual socialisation into agrarian context controlled and dominated by upper caste non-tribal landlords changed that too in most cases. Carol Upadhya has noted that "the modern history of this region has been determined in large part by the construction, in the late nineteenth century, of major irrigation systems off the Krishna and Godavari rivers. These canal irrigation schemes virtually obliterated subsistence agriculture and created a near mono-crop economy in the deltas, transforming them from areas of frequent famine to ones of intensive wet rice cultivation." [16] Further, "by the early twentieth century a new stratum of rich peasants had emerged in the deltas that was rapidly developing into an entrepreneurial commercial farmer-capitalist class. In addition to agricultural growth, there were other changes in the late nineteenth and early twentieth centuries that contributed to the development of this class: rural-urban migration, the spread of education, and the development of caste consciousness…"[17] She also traces to this region, an entrepreneurial class which managed to keep the land intact while diversifying into business – small and large – and a class that made the most of agrarian reforms (post-Green Revolution). Compare this with the situation of the tribal communities in the same region vis-à-vis their hold over land. Writing in 1998, for instance, Jaswantha Rao points out, "According to a report by the department of tribal welfare, out of 18,48,000 acres of land in the agency areas, non-tribals hold 7,53,435 acres, i.e., about 48 per cent. The non-tribal holding is 52 per cent in Khammam district (4,07,368 acres out of 7,71,604 acres)… The situation is more or less the same in West Godavari district. All the 29 villages in Jeelugumilli mandal, the 53 villages in Buttayagudem mandal and 19 out of 23 villages in Polavaram mandal are notified Scheduled villages…More than half of the land has been cornered by non-tribals…"[18]

In another study, it has been noted that in the Devipatnam mandal, one of the biggest mandals facing the threat of submergence and displacement of tribal communities in East Godavari district, 735 non-tribals (upper castes) were occupying tribal lands in 72 villages to the extent of 4,689.60 acres, as in 1982.[19] In spite of several regulations in the years, 1917, 1959 and 1970 to prevent tribal land transfers. The most blatant disregard for tribal communities' right to their land was shown by the state government which bought thousands of acres of land from these very non-tribal upper caste landowners – as was seen during the journeys. Writing in 1998, Rao had noted that "About 2,000 girijans have been arrested during the past three years in 146 cases and, invariably, all the cases were registered under non-bailable offences to put the people behind bars for months together. Most of 176 cases registered against non-tribal landlords fell into bailable category and no one was detained in the jail and most were set free immediately after their arrest."[20]

Late human rights activist-advocate K. Balagopal, too, wrote, that "out of the 72,000 cases decided under the LTR (Land Transfer Regulation) till September 30, 2005, 33,319 were decided in favour of the non-tribals and 33,078 against them. Of the 3,21,683 acres of land involved in these cases, 1,62,989 acres were confirmed in favour of non-tribals and 1,33,636 against them."[21]

What is women's place in these political intrigues linked to land, essentially ruled by patriarchy? How is economic value generated for the communities? A tribal family is valued at Rs. 1,50,000 (or a slightly higher amount, now with the new Act) as an all-time settlement for permanent dispossession. Women's work does not even figure in these calculations, or trees of each household, such as toddy, tamarind. The forests are 'evaluated' in terms of umber of 'productive trees' (alone) and the 'loss' has been 'compensated for' in cash to the Forest Department of Andhra Pradesh. In a sense, both nature and women are perceived in a similar manner. The value is in terms of economic 'productivity' and women's economic productivity is marginal to this exercise, since it cannot be measured. For trees, again, there is a value based on the market value of a certain species – timber, use in paper industry, furniture for the rich in the city – not in terms of the number of birds, insects and animal species that may be visiting a particular species of trees. To a large extent, the nature of anti-dam discourse has almost discounted or totally negated these 'valuations', steeped as they are in the dominant understanding of displacement. There is no effort to see the deep inter-linkages and interdependence of human societies with the natural environs. On the other hand, the reductionist environmentalists discount the human element. Closeness to nature as a notion is increasingly significant in times when 'Nature' is being sold to the rich in the cities either as a brief 'escape' to a 'pristine' (non-human habitational) ambience, or as detached, decontextualised pockets of green in their very apartment complexes or homes. Forests and river have been compartmentalised in commercial terms to construct a vast blackhole with a linear, universal time and an organising principle called 'development' subsuming the diverse historical-cultural space that comprised the flowing river, with its diverse fish and aquatic species, a rich forest with its animals, birds, amphibians and insects and a diverse human (non-consumptional) context within. Dismemberment of women means drowning them in the perpetual waters of an irretrievable, irreversible history of doom; submerging them in a cultural, historical, physical and spatial worldview that gave them all but 60 years (from 1947) of the notion of independence and democracy (to be celebrated annually) without giving them the choice of the nature of freedom they might have preferred inhering the dignity of human life. Within a matter of years, we shall have a sea of controlled, dead water for which women will have paid the highest price, on par with nature.

***

EXCLUSIONS, IDEA OF FORESTS, ETC

I speak here only in the context of the adivasi/tribal communities in the Polavaram submergence zone and not about all the Scheduled Tribes in India, in general. "According to the present records the composition of the forest is more or less the same as what it was when the Forest Department came into existence more than a century ago. However, it can be stated that the species of economic value such as teak and rosewood, which had a high demand in the past and are still having, have now become rare. Similarly, due to high demand for bamboo, the percentage of bamboo has also gone down. To improve the percentage of valuable timbers, the Forest Department has also taken up plantations of teak and bamboo over extensive areas." [22]

"In AD 1877 Dr. Brandis who was advising the Government on its forest policy drew attention to the grazing, fire, indiscriminate cutting and shifting cultivation practised by the hill tribes, who were thus ruining the forests. He therefore recommended that the Government should introduce a Legislation and sanction the reservation of large compact blocks of forest area. Consequently, the Madras Forest Act came into force on the first of January 1883, and several of the forest areas in this district were constituted as reserves between AD 1891 and 1900."[23]

The closest relationship fraught with deepest mistrust and irony of existence existed between the forest department and the tribal groups in these parts. This has remained. Fraught with dangers that familiarity breeds. The foresters rule over the terrain they survey and have kept the dependence of the tribal groups on the forest department so complete that if it is not the revenue officials, the forest department can do them in, as and when they wish. I have observed some activists appreciating the British for their 'special provisions' for the tribal communities through the Agency Areas demarcation, without looking at the politics behind such a rule. I just cite two references to bring home the attitude behind that 'provision' which is truly problematic. The former Godavari District of Madras Presidency was made up of ten taluks and two deputy tahsildar's divisions. Of these, notes the Gazetteer, "Yellavaram, Chodavaram, Polavaram and Bhadrachalam are tracts covered with hill and jungle and inhabited by uncivilised tribes to whom it is inexpedient to apply the whole of the ordinary law of the land. Under the Scheduled Districts Act of 1874 these have been formed into an Agency in which civil justice is administered under special rules and the Collector has special powers in his capacity of Government Agent. They are consequently always known as 'the Agency' or 'the Agency tracts'."[24]

Again, "In the Agency...the deputy tahsildar's divisions of Polavaram, Yellavaram and Chodavaram and the taluk of Bhadrachalam, all of which are remote tracts covered with hill and jungle, sparsely provided with communication, shunned by the dwellers in the plains and inhabited by backward tribes who are most illiterate and ignorant of the ways of the world, and yet ready to go out on the warpath if once any of their many peculiar susceptibilities are wounded. In country, and to people, such as these, much of the ordinary law of the land is unsuited and a special system has consequently been introduced."[25]

It must be noted that the intervention and in fact interference with forests of Rampa (not as much, on account of the tribal groups being far more revolutionary in this region), Bhadrachalam, Rekapally (Rekapalle of the colonial records) divisions as well as in the mangroves had started with the colonial system and not all of what they left behind was as pristine a forest type as it was when they found it before their rule. They were selecting which species to plant in areas which were in their control. This continued post-1947 and continues till date. The idea of selling contracts for tree felling, for sale, and exporting started with the British and continues till date. Hence, introduction of species and forest intervention and a protracted tribal-forest officers contestation are aspects negated in the discourse that looks at tribal communities as an 'unsullied', 'homogenous' entity, meant to be 'protected' almost like a species to be conserved. It kills the very real and dynamic tribal history as a group that has had the most precarious yet sustained struggle-prone relationship with the empire and the state or any establishment that has sought to usurp their territory. More importantly, if there were parts of these forests that were left unreserved, it was thanks largely to the active resistance of the tribal communities against the British administrators were largely unfamiliar with guerrilla tactics and the fear of disease was overwhelming. In the Polavaram context, some of the Kondareddis, Koyas and others consider their land and forest as more valuable than money and there are those who feel otherwise. There are the ones who accept mainstream ideas of development and others who don't. Though we look at the majority opinion to reflect the truth, but gradually, there comes a time in democracy when even a sense of 'majority' is a construct on the one hand or on the other, that which is majority opinion (at the commencement of a project that will destroy that majority) soon gets pushed down to a minority opinion, fighting as it is, against a system far too ruthless. It is important to look at the tribal communities as having their own internal contradictions. Even as it is to look at the fact that they were forcibly kept away from participating in democracy as complete and fully aware citizens, precisely on account of the 'special provisions' and 'Agency areas' they were designated to, by the colonialists, thus appointing an Agent of the colonial state to directly control them which later, in a free nation, got a different nomenclature. The real challenge is not about saving the few tribal groups as mere names within our constitutional apparatus but undoing the very idea of politics and economics that rules over the communities who lived with pride as their local histories reveal are today rendered as some poor cousins needing protection. They do not exist in any significant numbers in our public institutions, or our government offices. Did they choose for themselves this state of permanent exclusion and 'special' provisions in the name of further exclusion? On the other hand, if at all they must be 'assimilated', should it not be from the point of view of their cultural difference being accepted as a respected and respectable part of the democratic rubric and not a patronisingly donated charity to the lesser mortals? How many tribal festivals are assimilated in our list of 'public holidays'? Does any forester get trained under a Kondareddi or a Koya or other tribal person? Is staying in their villages, *with* them (not as overlords or officers and certainly not as the only protectors of the forest) part of any Forest officer's training? At the other end, I have not seen a single Kondareddi or Koya person (from their villages) being invited to deliver a lecture in the universities in Hyderabad, even for the token celebration of International Day of the Indigenous Peoples, in spite of new departments for studies in exclusion and inclusive policy. In fact, they are not even invited as 'special guests' to speak of their own displacement from their own experience.[26] While it is important to accept the reality of tribal/adivasi parents aspiring for a certain life for their children (including the worldly goods that the mainstream society has now popularised, including TV, CD players, clothes, and education as I have seen happening in the villages I visited), at the same time it is important to question the universally imposed worldly aspirations through media and films, school text books and radio and governance. Unless the worldly aspirations are seen as varied, and accepted in varying degrees and forms, or unless they undergo transformation to include different ideas of living, being, nature, and happiness, and accepting different expressions of human view of life, as an artist, poet, farmer, potter, fisherman – all being equally important – there is little sense in token representations. When tribal communities' sense of pride, very rooted in their identity, is not acknowledged (as also with the dalits, and other marginalised castes), there is no redemption in the numbers game that is played each time their homes and lands are targetted to make way for development which seems to exist in a vacuum, 'sanitised' of these very communities and the lands of their toil. And even activists ultimately join the game to argue about what could be the most ideal compensation for this sanitisation. In the name of giving them a 'fair price' for their labours and their identities which have since been Scheduled in the Indian Constitution devoid of the respectability that is supposed to be part and parcel of the Schedule. Institutions such as INTACH have hardly ever sought heritage status for tribal languages and locales nor have institutions or politicians ever fought for 'classical language' status for any of the tribal or other marginalised groups' languages. There is more to the tribal communities than an essentialised identity imposed in documents without understanding these associated larger aspects under a universal idiom of wealth, 'good living', which is of little use or no use to the discourse against Polavaram dam or any other project directly affecting tribal villages anywhere in the country. Displacement needs to be questioned against what is going to be built in its place, and not simply written off with a cheque of compensation decided from these problematic paradigms.

Incidentally, in the course of my assignment on the Kurnul-Cuddapah Canal (noted earlier in the book, in the journey of 2010), I came across a piece of documented history with regard to both the Kondamodalu tribal peasant agitation and the Telangana Armed Struggle.[27] I had the chance to meet Comrade 'N.S.R', Nandyala Srinivasa Reddy, one of the many active participants of the Telangana Armed Struggle and former CPI MLA from Nalgonda, and his son, Comrade M. Krupakar, who was associated with the CPM, Nalgonda. In the course of our conversation, and learning of my Godavari journeys, I was handed over this volume, titled *Historical and Polemical Documents of the Communist Movement of India, Vol. II (1964-1972)*, published by the Tarimela Nagi Reddy Memorial Trust, Vijayawada, 2008. It contains reports on Kondamodalu. Strange for me to find this more than a year after I had been told the story by Kondla Gangaraju and Illa Rami Reddi in 2008! I did not know there was a written account of the same from the point of view of the revolutionary party that had spearheaded or guided the movement along with the Kondareddis of Kondamodalu. I quickly noted some of the important developments recorded here, as a continuation of the story narrated by Rami Reddi in 2008. I share here some parts of the said report, verbatim, as it is the 'official' record of the Party. The report, in brief, deals with differences within the Communist movement, in general and among the revolutionaries, in particular. The methods used by the Srikakulam girijan movement were not used in the pockets of East and West Godavari. There seems to have emerged differences within the cadres (outside of the tribal communities) as to the methods to be implemented. The report in the book mentioned above does not mention leadership from among the tribal communities. It is a centralised Party looking at the movement that originated with the Telangana Armed Struggle, to break away later from the united CPI and seek new base in the tribal settlements in the hills of Kondamodalu. Essentially, leaders from outside the social context of the tribal community initiated this movement, post-1950s. Gradually, over the years, it seems to have taken its own shape and form adapting to the tribal social-cultural context. I also observed that while the Communist Party of India (later CPM, mainly) established a strong base in the Khammam and to some extent, West Godavari, the CPI (ML) seems to have found its base in East Godavari district (of erstwhile Madras Presidency and Andhra Pradesh). Earlier, these parts saw the rise of the guerilla forces of Alluri Sitaramaraju. What is the reason behind this geographical distribution is hard to say without studying it. But mapping the political history of the tribal communities and the Left movements in the Godavari region should be a fruitful exercise. Besides the book mentioned above, we have the earlier seminal volume by P. Sundarayya, *Telangana People's Struggle and its Lessons* (first published in 1972 with several reprint editions). But as yet there is no comprehensive history of the different forms of Left politics in these parts.

"A Report of Kondamodalu Tribal Peasant Movement"

"We started our work in Kondamodalu area in January 1969. It was in March 1969, the local police camp was set up in Kondamodalu. Reserve police camp came in April...Tilling of lands has begun in June 1969. The people seized back crops grabbed by the landlords earlier. Lands which were not yet tilled were tilled and sown. The people moved into action in a big way...In October 1969, the girijans stood organised to put up resistance against the landlords when they sought to destroy the crops raised by the girijans...The building of East Godavari Girijana Sangham has begun at a time when the class struggle in the girijan movement of Parvathipuram Agency Area of Srikakulam district was advancing in the thick of serious class battles. The activities of the Sangham had taken a militant form in the forest areas of Y. Ramavaram and Maredumilli...The activities in Kondamodalu area were started in Jan'1969 only after the State Committee has taken the steps with a comprehensive view and as part of building the revolutionary movement. By this time, the District Committee has concretely decided then Comrade Simhadri Subba Reddy who was sent to this district as the Organiser for this pocket of Kondamodalu. The report we are publishing here was written by him when he was in Secunderabad Jail in November-December, 1970."[28]

"Harassment by the forest officials, the exploitation by the landlords and money lenders were the main problems in this area...Girijans who were hitherto getting frightened even at the sight of forest officials and praying them offering hens or money not to foist cases against them, have shown the courage with the consciousness gained from the Girjana Sangham and warned the forest officials..."[29]

"Karnam and mansab together were collecting the revenue tax from the girijans for bringing the lands under the shifting cultivation. On one side, the forest officials were harassing the girijans calling the shifting cultivation as illegal. On the other side, the revenue officials were harassing the people by demanding the payment of tax as per law...These were areas where comrades had taken up the programmes against the atrocities of forest officials, the plunder and domination of Muthadars. Together with these problems the injustices caused by the paper mill managers to the coolies while calculating the wage for the bamboo cutting have also come to our notice. These injustices included taking 35 bamboos more for a load of each bullock cart on the plea of not being upto the standard..."[30]

Pages from Telangana Armed Struggle: The Godavari Forest Region

Meanwhile, Sundarayya's book focusses on the Telangana Armed Struggle against the Razakars of the Nizam government and later the Indian state, involving direct action on land reclamation by the marginalised sections, among other issues. He writes, of the movement in Palavancha forest area (Kothagudem), "Even during the Razakar days, the base of operations had been the Godavari forest area. This had become the main base after the Indian Union army intervention. In the later days, the Party and the squads extended beyond River Godavari of the Bhadrachalam area, a part of old Setharamaraju's Koya revolt area. The Government concentrated its efforts to evacuate us from the forest areas...(by) evacuating the koya tribal people to the outskirts of forests. It burned down their hamlets and resorted to mass beatings and mass murders..."[31]

"The hard core that remained at the end, just before the withdrawal, were 150 Party members and many sympathisers in the plains, and 50 koya Party members and 500 Koya militants in the forest area. But this was one area with which the State Centre could not keep any contact for nearly 2 years till the armed struggle was withdrawn...Questions about the wisdom of conducting armed struggle became more insistent...The leadership heard the radio announcing the decision of withdrawal of armed struggle on October 21, 1951. By the end of February 1949, after thousands of members of village squads and village panchayats had either been caught, and even those who had resigned and surrendered were thrown behind prison bars by the Military Government, the remaining retreated to the forests while some others reached the towns..."[32]

"The exploitation of these innocent people was horrifying, landlords of the plains were the village authorities (patels and patwaris) for these tribal people who collected ten to hundred times more than the land revenue payable to the Government. They secured pattas on these lands for themselves, keeping the tribals as temporary occupants...Forest officials were devils incarnate vis-à-vis these tribal people. They exploited the labour of these people in a thousand and one ways for official and personal work..."[33]

In a sense, in both cases the Left leaders understood the conditions of the tribal communities similarly and the idea of setting up base in this region also followed a similar trajectory. The point of difference at that point emerged around the continuation of armed struggle. Gradually, other polemical differences arose. On the whole, the Godavari forest region, the home to the tribal communities, has had a long and continuous history of standing up to the state in protecting their rights. The centralised leadership from outside could not have sustained itself had it not been for the these communities whose own ideas of community-owned land, community living and sharing of resources, culturally given, even before the revolutionary leaders appeared in their midst (though Sundarayya paints a picture of the innocent tribes, held down by blind superstition and "antiquated social customs"[34]), helped lay the foundation for the movements.

***

TWO GRAND DESIGNS AND A LINEAR WATER NARRATIVE[35]

Deifying Arthur Cotton has been a specific delta trait in Andhra Pradesh State. You can spot Arthur Cotton's statues all over the coastal region even today. When one of his descendants visited the State in 2007, he was accorded the honour of a "State Guest" and taken around town, and understandably, to Rajahmundry as well. He may not have expected the kind of reception he would receive from engineers and ministers in the State. A similar kind of deification happened in the times of the late Y.S. Rajasekhara Reddy, former Chief Minister of AP. In one instance, one of his Party (Congress) colleagues, J. C. Diwakar Reddy, in a public meeting in Parigi in AP, "declared that Y.S. Rajashekara Reddy was Sir Arthur Cotton in a previous birth. In the present birth he is none other than the sage Bhagiratha, bringing irrigation waters to parched lands in AP."[37]

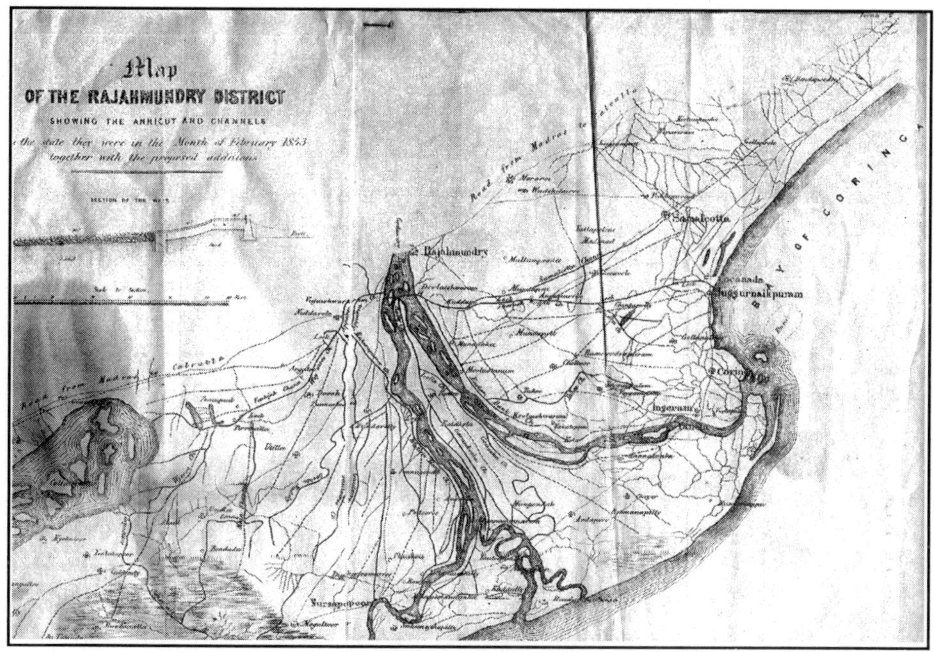

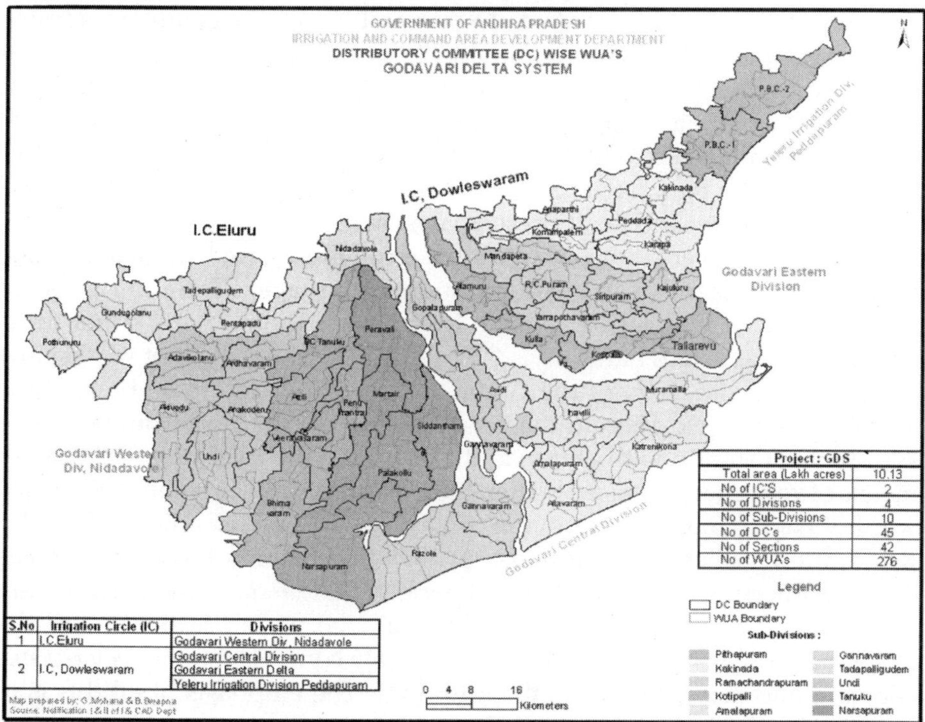

Project : GDS	
Total area (Lakh acres)	10.13
No of IC'S	2
No of Divisions	4
No of Sub-Divisions	10
No of DC's	45
No of Sections	42
No of WUA's	276

S.No	Irrigation Circle (IC)	Divisions
1	I.C.Eluru	Godavari Western Div, Nidadavole
2	I.C, Dowleswaram	Godavari Central Division
		Godavari Eastern Delta
		Yeleru Irrigation Division Peddapuram

Top – Map of Rajahmundry District showing anicut and channels in the month of February 1853; Below – Map of I &CAD, Government of Andhra Pradesh showing the delta system[36]

Another campaign feature used to be screened prior to a feature film in movie theatres (single screen theatres, especially) in Hyderabad in the years 2006-07. I had seen the ad-film (or campaign film) for the first time in year 2007. It enthused me so much I noted down the text of the same in the theatre itself. It never failed to amuse me – the creativity of the person who made the film. It was produced by the AP Television and Film Development Corporation for the Government of AP. It went like this:

Kubera rushes to Vaikuntam, the abode of the god Venkateswara (or Vishnu)[38], the presiding deity of Tirumala-Tirupati temple and says, "What is the meaning of all this? You have repaid all my loans! What is to happen to my livelihood now…?"

Venkateswara replies to Kubera, "What can I do? I have managed to get the maximum amount towards loan repayment from one just one Indian State, Andhra Pradesh, which has become Swarandhra Pradesh (lit. golden Andhra Pradesh)! With a ruling party so generous, and with its jalayagam, the farmers have become so that they have been donating to me generously. In fact, I am now indebted to the farmers of AP. Thanks to the jalayagam they have added to my hundi collections."

Kubera remarks, "So the farmers of AP are the present-day Kuberas!"

This film was made in the context of YSR's ambitious Jalayagam programme comprising more than 40 irrigation projects (several at that time without clearances) when they had commenced in 2006. But the story of Jalayagam does not start at Jalayagam. Its origins and the language of the state take from the original source, the first barrage built at Dowlaishwaram, by Sir Arthur Cotton in the year 1852. The narrative, then, is indeed linear in many senses as I believe, and shall try to analyse here.

The Anicut over Godavari (Arthur Cotton Barrage at Dowlaishwaram, near Rajahmundry)[39] - In Arthur Cotton's language, "The river must be restrained from wandering, which, from its having no hard strata in its course, it always does naturally…"[40] And again, "the system of works now in progress in the Delta of the Godavery…are intended to embrace these four objects, viz – to restrain the river; to preserve the land from floods; to supply it constantly with water; and to pervade the tract thoroughly with means of very cheap transit."[41] Then there was also his Christian ethic which was invoked in many places in the context of building the anicut. Thus, "In Rajahmundry they have now learnt that Englishmen have some ideas beyond collecting revenue. They can perfectly understand the wisdom and benevolence of Government in executing such works as these, and the difference between a Government that merely compels them to *pay revenue* (sic) and one that also *enables* them to do it. The emphasis of all this is to give them entirely new ideas of what a Christian Government is, and thus to prepare them to receive Christianity…"[42]

The construction itself did not begin nor end with the anicut. The embankments had breached several times (records state in 1886, 1892 and 1900, may be even later) leading to inundations. It was a case of agriculturists becoming far more dependent on the structure thus established. Increase in revenue was the intended reason for the irrigation works, with water being provided to grow certain crops and pay stipulated water cess, rather than (what is falsely propagated by the engineering establishment in AP, till date, that it was meant) to protect the region from constant floods and famine. Thus, in 1823-24 (when the anicut was not built), the revenue from the Godavari Districts is shown as Rs. 21,84,747; in 1843-44, it is Rs. 17,25,841. It jumps in 1894-95 to Rs. 88,21,322 (highest) and by 1902-03, is Rs. 32,28,810[43] (this is also the time when many ryots are reportedly refusing water from the irrigation system and its networks on account of the high water rates, etc. Constant irrigation was seen as the means to generate revenue, even if it changed the traditional cropping patterns and methods of irrigation. Cotton's statistics post-anicut, in the Godavari Districts, proved the benefits of these works in economic terms to the Empire. "The revenue of the Delta including that part that is in Masulipatnam, has increased about £ 60,000...(and) the amount of money re-circulated in the district had increased to £ 100,000, above the average in years preceding the works; the internal traffic is now estimated at 180,000 tons carried thirty miles..."[44]

Irrigation works also gave the colonial powers the advantage of bringing into its records land that may have hitherto been out of its sight. Every ryot using the water from the works was liable to be taxed. For his part, Cotton foresaw a larger role for private enterprise on the Godavari in the immediate future. As he said, "The Government *must* therefore take the lead; but every acre they irrigate, and every mile of cheap transit they provide will, assuredly, open the way for, and lead to the extension of private enterprise..."[45]

The empire flows again – I couldn't but help recall the Delhi Press Conference that Dr. YSR (late) addressed with regard to attracting investments in the coastal corridor, and the hard sell of the idea of advantages of the region with Godavari being a star selling point (2006). The problem is the uncritical acceptance of the logic of dividends. "The canals which drew their supply from the Godavari have converted the delta...in to one vast expanse of paddy fields broken by gardens of fruit trees. It is one of the richest tracts in Andhra Pradesh. Failure of crops is practically unknown and the main canals furnish an excellent means of transport. The system now comprises 500 miles of main and branch canals and 2,000 miles of distributaries; it irrigated close upon a million acres in 1919-20, and yielded a return of nearly 25 per cent on a total capital outlay of Rs. 154 lacks." [46]

The colonial anicut created an edifice around the irrigation system which has continued post-Independence: of bureaucratic control and dominance of authorities in terms of releasing water (how much, when, etc, decided by the edifice). There has been a long history in these regions of internalising a certain language of river waters and what to do with them, in several time periods in history even prior to the colonial rule. The first thing that any Chief Minister did after coming to power in Andhra Pradesh since the formation of the state in 1956 was to express a kind of despotic (benevolent or otherwise) idiom of power over river waters and 'bringing the river waters to the people'. Brahminical mythology has also played a role in this kind of idiom (after several myths associated with Ganga, Godavari, Kaveri and so forth, being brought down or humbled by the sages). The discourse that globalisation (of the 1990s) led to the nature of developments in this region (delta, coastal Andhra) – the idea of dams, and all the rest, including SEZs – needs to be re-visited in the context of another historical truth of this region. In the case of the Godavari, changes were already being brought about through state intervention on varieties of local crops, ways of cropping, etc, and introducing varieties for a global imperial market which one has already mentioned earlier. So, after a brief 'halt' to the process of private capital accumulation, post-Independence, which policies such as land reforms, etc. affected (not really having impacted these rich landlords so much, but revising the idea of land ownership, perhaps) there is as yet a longer historical lineage that must be acknowledged. Especially since it is this very group of enterprise-loving communities who hold the political reins of Andhra Pradesh as it is today. It will remain open to question whether the newer formation of Seemandhra and Telangana will break from these legacies of the deltaic context of land and river control. A resurgence of world market capitalism for Godavari region and its upper class is hardly a new phenomenon. But the difference lies perhaps in the bargaining capacity, in favour (relatively speaking) of the capitalists, when compared to the Socialist past, of which India has not had a long running period, starting 1947.

***

READING WATER IN THE PAPERS OF THE GODAVARI DISTRICT ASSOCIATION

I found this Association an interesting element in the delta history for its avowed aims and the nature of its discourse with the colonial government especially on matters relating to agriculture and the problems dealing with – by and large – the farming classes of the District (with some exceptions at times), though they may have interacted consistently with the government on other issues. The Godavari District Association (GDA, henceforth) was established in 1895 ("Under the Act XXI of 1860"). The members (in rotating positions of President, Vice President and Secretary) included, among others: N.Subba Row Pantulu, Ganjam Venkataratnam Pantulu, K. Perraju Pantulu, C.V. Perayya Sastri, K.R.V. Krishnarow Bahadur and so on. Among the "objects of the Association" were – "to diffuse knowledge among the people by circulating pamphlets, leaflets, tracts or otherwise & by providing for the delivery and holding of lectures, exhibitions, public meetings...& thereby to foster among the people a healthy public opinion; to ascertain the actual wants and grievances of the District, and adopt constitutional and loyal measures in bringing them to the notice of the authorities concerned, and to co-operate with them in all matters calculated to promote the welfare of the people"[47]; and so on. The Association, as the language it uses reflects, was formed with ideals of 'cooperation' and 'loyalty' to the British colonial authorities and – as reflected in the proceedings – they did take up grievances, mostly of the ryots, regarding the water rates, breaches in canals, etc, through submissions to the Government. It becomes clear that the problems of the ryots, accessing the anicut system were not small. There seem to have been made a number of requests made by the ryots to the Association as is evident from the sample, which the Association seems to have pleaded with the Government to solve, but it appears they were not so successful in doing so. The GDA was formed by educated elite, from the upper castes – Brahmins, Kapus, Rajus, essentially, among others. This was a group of elite with immediate concerns around the agrarian pattern that the irrigation system constructed over an earlier one, and this happens at around the same time as a moderate nationalist as well as radical movements addressing, or opposing the colonial powers are ongoing in the same region. The 'instrument' that the GDA uses to address its immediate concerns, which it considers as the concerns of a larger mass of 'subjects of the empire', is one of petitioning in a language of submission and pleading to the powers for concessions. One does not know the extent of influence the farmers (here, of course the landed ryots) had on the GDA itself, or as to how much the GDA truly represented their interests but I found the exchanges between the GDA and the government (through responses, however un-yielding in content, of the requests made) more significant in the way the entire apparatus is built around the river Godavari in the form of a centralised bureaucracy. The river passes on from a free-flowing entity from the hands of the people to that of the state as an instrument of regulation and control; and from this end, 'concessions' are granted. While this logic was specific to a colonial state-owned river construct, with its own place in colonisation and imperialism, in a modern, post-1947, democratic nation, the logic does not seem to have been altered all that much. I share just a few relevant examples here of the GDA's interventions.

Problems with Drainage - A submission is made to "H.E. The Governor's Tour (At Cocanada)" by a deputation from the GDA, consisting of "Sri Rajah K. Ramachandra Raju Bahadur, Messers C. Seshagiri Rao Pantulu…and D.V. Suryaprakasa Rao…"[48] It notes, "Owing to the lack of proper drainage it was alleged that certain lands liable to submersion lost from 25 to 75 per cent of the normal outturn of crop but as the crop did not fall completely no remission could be allowed under the rules. Before the introduction of the pipe system the channels were deeper and their slope was greater, so that the water taken out of a channel at one point could be drained back again into the channel at a point below but with the raised channel beds of the present system such drainage was no longer possible. It should be remembered that the principal drains of the Delta were only water courses and were not constructed as drains by the Government. As this was so, the Government should hold themselves responsible for keeping them free of silt and clear of the dense grass which grows in their beds and arrest free flow of water..."

"H.E.'s Reply – His Excellency said that he was prepared to admit that the drainage system needed attention and also that it had no received sufficient attention in the past but that it was not from want of will to remedy such defects as existed that hitherto no remedy had been applied…"

Regarding Water Cess on Lanka Lands – There were grievances and increasing resentment among ryots about the water rates that were being charged. For instance, "Under the Hindu and Muhammadan rule river water taken for irrigation was not charged for and it had not been the intention of Government to make any charge for water taken in this way. It was not until 1900 that there was any provision empowering Government to impose water rate for the use of river water. It was pointed out that if river water destroyed the crop on dry lands Government accepted no responsibility and paid no compensation and yet if the same river water was used to irrigate a crop the Government considered themselves justified in charging the water rate." [49]

Where Public Works were not about the 'public' but profit - I found several submissions made by the GDA regarding the ryots' pleas for repairs of sluices that had been breached, which were a constant problem in the canal system of the anicut. The government either responded with Committee set up to look into the matter or if the repairs were done, it was done in keeping with the profit motive; if the said 'public works' did not seem to yield the kind of dividends on investment that the government expected, they were not undertaken. For instance, "There was a stream known as the Torrigedda which every year submerged the lands of some 20 villages. A sluice which was intended to prevent these floods had breached and had not been repaired. The villagers were able to grow only second crops owing to their lands being submerged during the first crop season. As the lands were classified as dry lands no remission was granted. The ryots had agreed to pay half the cost of the reconstruction of the sluice but the Government had not consented to repair the sluice as they said that the work would not be profitable. The request of the Deputation was that the sluice should be reconstructed. The Executive engineer replied that the sluice had been twice constructed and twice washed away and the Government had not refused to rebuild it because it would not be profitable but because the benefits accruing would not be proportionate to the expense involved."[50]

Our agrarian system was paying a heavy price for the supposed 'benefits' of the continuous supply of water rationed out to them from the top; with orders to grow certain kinds of crops if they were taking water from the irrigation canals.

Problems with Irrigation, Maintenance, Flood banks, Farmers at the mercy of the System (past and present) - There were problems with day-to-day functioning of the system created and the nature of pleas made for redressal; including an interesting case of requests to re-open a certain navigation canal used as a thoroughfare by farmers, traders and others. Incidentally, during the course of my journeys in the field, problems such as these, including breaches, etc, seem to have continued in the present day. In one instance, there was a breach in the Kovvada canal (extension of the canal network of the anicut) in Polavaram mandal that had breached during 2006 Godavari. Until reports last came in, the said breach had not been repaired in spite of several petitions by local farmers and others to the authorities. People used to link it with the construction of Polavaram project – since the Polavaram project works commenced the irrigation department was not showing the kind of concern (*sraddha choopinchadam ledu*) it should to these 'minor' repairs and people had been making requests to the authorities to get it repaired. The small farmers, especially, remain at the mercy of the structure developed around the Irrigation department.

"The Teki Drain – Letter no. 110 dated Cocanada 31st May 1914, From the Joint Secretary to the Godavari District Association, To the Collector, Godavari District, Cocanada. The attention of the Government was solicited by means of an interpellation in the Legislative Conncil in 1913 and by a Resolution passed in the Godavari District Conference held in March 1914 to the urgent necessity of a flood bank to the Teki Drain near Pallipalem in the Cocanada Taluk as in its absence great damage is caused annually to wet crops of the adjoining villages."[51]

***

THINKING ON THE NATIONAL LAND ACQUISITION AND R&R BILL 2011 (now the Right to Fair Compensation and Transparency in Land Acquisition, Rehabilitation and Resettlement Act, 2013)[52]

The value of the land is notional, in urban areas, if it is not measured by what that land was meant for, originally. In rural areas, agriculture, forest, pasture-land, grazing land defines the value of the land owned or collectively used. Value of land in hinterland urban areas goes up today in terms of the presence of apartments, roads, number of franchise supermarkets, and access to the airport (not the railway or bus stations anymore, to note). But yes to this list must be added the sale of the so-called 'green zone', where nature itself is the preserve of the rich. The poor and the lower middle classes almost never have access to greenery either in images presented of the same in advertisements or in real terms. I am not taking into account here the other kinds of property within the city that gains its 'monetary value' from being the oldest, or being closer to the power-centres, the colonial times, etc, as you find in Delhi, Chennai, Chandigarh and so on. Strangely enough in Hyderabad the areas that were closest to the power centre, the Nizam's, fell in appreciation when the new Jubilee Hills real estate boomed. That is another story. So we are living in a context of constructed notions – of wealth, class, etc. – which are evaluated through land. People buy land as a speculative investment (across cities through agents). At the other end you have those who buy land to become 'farmers' as a politically correct exercise. Again these are people with no organic connection with the land they buy or live on; or may not ever live on. Land or even notional land (as in apartment complex) has become an investment deal, and is not seen as something you built a certain quality of life on. Owning land is equal to owning cash. And this is important. One cannot critically address land reforms or the Bill without talking of these changes across India and the world. When all the options in urban spaces are getting exhausted, rural hinterlands have become sites of appropriation. Hyderabad has today enveloped and consumed parts of Mahbubnagar, Sangareddy, Siddipet and few other districts, which were essentially rural farmlands. The concept of notion again – so you can live notionally in Hyderabad but in reality, in what was a village in Mahbubnagar district. The idea is to construct a large urbanscape as the reality of the future. So, it is important to look at the development in the urban spaces in order to make sense of what is happening in the rural areas and why is there a need for a massive land acquisition with small dollops of charitable compensation. Urban spaces are more essential to world economy linked as they are to banks and the international debt industry; this is apart from other scientific experimentation carried out by multinational companies in agriculture, human and animal genetics, experiments on the human mind which have most of their consumers (or unsuspecting victims) in urban spaces. Television is yet another insidiously mind-controlling device in this canvas. Food growing rural villages where the farmer or peasant is the owner or tiller of her/his piece of land are quite inconsequential to the larger web. Land acquisition will only be complementing or constructing a huge mass of deskilled daily wage labourers, thoroughly dependent on either the private market players, world capital, or the government, which will look at 'convergence' and 'corporate social responsibility' to sponsor programmes of beneficiary-oriented schemes towards what is increasingly being called 'harnessing social capital'. It will also coincide with fluctuations on world capital market and geo-political developments. So in effect it does not begin, nor end, with land acquisition and compensation, per se. The AP government under YSR started Jalayagam for creating "98.41 lakh acres of new irrigation potential and stabilising 22.26 lakh acres by constructing a total number of 86 irrigation projects – which include 44 major, 30 medium projects, 4 flood banks and modernising 8 projects… since 2004-05…In fact, irrigation sector alone then accounted for 41.57% (amounting to Rs. 44,898 crores) of the total plan outlay during 2007-09 to 2009-10" as their official website revealed.[53]

The investment in Polavaram dam is big, so is the concept and design. Who will it feed, once it comes up, if it does? For one, the Andhra Pradesh Petroleum, Chemicals and Petrochemicals Investment Region – "a specifically delineated investment region with an area of around 603 square kilometers planned for the establishment of manufacturing & service facilities for domestic and export led production in petroleum, chemicals & petrochemicals. Andhra Pradesh is the first state to sign a MoA for PCPIR. A Memorandum of Agreement was signed between Department of Petrochemicals, Government of India and Government of Andhra Pradesh in New Delhi on the 1st of October, 2009. An investment of $ 3.95 billion is being made for external infrastructure of PCPIR in Andhra Pradesh. The investments are being made by the State Government of Andhra Pradesh (GoAP), the Central Government of India (GoI) and private players. Of the total investment made to set up the infrastructure for PCPIR, GoAP proposes to invest $ 450 million while GoI is assisting with an investment of $ 1.3 billion. The central government's assistance pertains mainly to up gradation of National Highways, improvement to link roads, airports, rail links etc."[54] Major Investors apart from Anchor Industries – Reliance, Eisai, Continental Carbon, Velankani, RCL, Naturol, ISPRL, SNF India, Air Liquide, Baker Hughes, Biocon, Phormozell.[55]

What did they assure for this region? 365 days assured industrial water supply through Visakhapatnam Industrial Water Supply Company Ltd (VIWSCO) – "India's first dedicated industrial water supply project completed at a cost of US $ 100 million Dedicated Water Supply scheme of 100 MLD at PCPIR commissioned. Proposed to be augmented by another 100 MLD."[56]

Where is the water coming from? According to the website of APPCPIR, water will be routed from the "Yeleru Reservoir (385 MLD volume of water); Pipeline form River Godavari (385 MLD); Samalcota Canal from River Godavari (220 MLD), Polavaram Left Main Canal from River Godavari –

under execution (1848 MLD)…95 MLD water supply scheme (through the Vishakhapatnam Industrial Water Supply Company Limited (VIWSCO) commissioned to supply water to APPCPIR Phase I; proposed 100 MGD pipeline from River Godavari for Kakinada cluster of PCPIR."[57]

With the formation of Telangana State on 2nd June, 2014, there is expected to resume another round of debate on the utilisation of Godavari waters for the coastal corridor, excluding genuine agrarian needs of Telangana.

With regard to irrigation projects displacing people, the Act assures – "a minimum of one acre of land in the command area of the project for which the land is acquired: Provided that in every project those persons losing land and belonging to the Scheduled Castes or the Scheduled Tribes will be provided land equivalent to land acquired or two and a one-half acres, whichever is lower."[58] But the question is, if the water from the project is already being streamlined to feed into these industries downstream, where exactly in the 'command area' is the government to find one acre land each for the displaced, which amounts to thousands of acres, which in AP case, have already been apportioned? Thus, "62 industries with an investment of at least Rs. 10 crore each in East Godavari, 18 industries with an investment of Rs. 5-10 crore, 62 small industries with an investment of Rs. 192.16 crore…"[59]

More than ever before, each land lost in the village today ends up as a structure in the city. Finally, to end with this note – the AP Information Commission website has this quote from Gandhi prominently displayed – "Real Swaraj Will come not by acquisition of authority by a few but by the acquisition of capacity by all to resist authority when abused." Was it pun-intended? Should we instead say, 'real Democracy, unmediated, will come not by acquisition of land by a few but by acquisition of capacity by all to resist the authority of a few to acquire land'?

***

ON POPULAR MOVEMENT

Popular resistance and struggle against the dam among the tribal communities, continues in its own steam at the local level. Hence the idea of 'movement' and the way we would like to look at it, needs to be revisited. Everyday battles that people continue to have with the government authorities over matters of the compensation, or not accepting the terms that the authorities have laid out, are also movements of everyday in ways that are not always visible. The Kondareddis and Koyas and other tribal groups of these parts question the authorities, fight their legal battles in spite of not being literate most of the times. But if there is no mass resistance as one voice, the reason is the play of several multiple, multi-locational debilitating factors which stem any wave of organised protest, though efforts were made at different times. Debilitating factors such as everyday struggle for sustenance – food, education, health, fight to protect their lands and forests each day – are their constant battlegrounds not seen as an extraneous, 'outside' force in the visage of a common enemy. There is no luxury of distance for them to see all these as a process set to work by a government. In case of Polavaram the state, in fact understood these locational complexities and contradictions. The state in fact was the first mobiliser using a ruptured social structure: caste. The non-tribal upper castes in the Polavaram submergence zone were its target for mobilisation. There was already a caste-tribe conflict; and the state never quite addressed the anomaly of the non-tribals holding so much land in the V Schedule (erstwhile Agency) Area. Once they purchased all this land from these upper-caste landlords, consolidating a legal wrong by making it 'right' in 'public good', it was easy to nip in the bud any possibility of a mass protest against the dam (which there was, in the initial moments, since the non-tribals were to lose all the lands they had been wrongly enjoying). Gramsci noted, long ago, and it rings true for the Polavaram context, that, "the history of subaltern groups is necessarily fragmented and episodic. There undoubtedly does exist a tendency to unification in the historical activity of these groups but this tendency is continually interrupted by the activity of ruling groups..."[60] Thus the solidarity of the dominant in the Polavaram submergence zone is that of men in industry, state, local landlords, local political leaders (including MLAs, MPs – one of whom has a contract for the dam works) and middlemen, more powerful than the solidarity of the affected marginalised. In tribal and rural areas, there are so many forces to struggle with everyday, with no autonomy or choice, for a processually strong, long-standing movement to emerge from the ground.

On the other hand, several processes in the economy have added to the fragmentation of people's movements, and the issues raised are equally fragmented, locational and locale specific. One could locate these locale-specific, fragmented sites and issues within the global development discourse. Ludden discusses this development discourse as being part of a global phenomenon, with nations no longer being truly independent in a global economic paradigm. He has noted that, "Inside outwardly oriented development regimes, national territories fragment spatially into a collection of potentially profitable sites for business investment, at the same time as national problems and populations fragment into global 'target' groups, interests, and issues. Women, poor people, indigenous people, the environment, health, microcredit, and governance: the list goes on and on of specific topics of global development specialisation, each with its own experts and leading institutions, each focused on some particular feature of national space. The World Bank's *World Development Report* represents an annual compilation of leading global issues, to be tackled in each country separately under the discipline of the global regime. Meanwhile, in the world market economy, a repeat of the core trend of the first globalisation – the creation of specialised sites for capital investment and labour control –is well underway...India is now a collection of regional development histories."[61] I would add further to say that river politics as is happening in Andhra Pradesh (both erstwhile Andhra Pradesh and the present one, with Telangana in the contentious battle since June 2014), and displacement as a result of it, are aspects of creating these 'sites'.

The phenomenon of NGOs in this context makes popular resistance further impossible; by the time a movement picks up, funded projects are lined up to disband the same. These are not far from the 'specialisation' mode that Ludden mentions. Thus, there are numerous organisations with their own expertise and modus operandi which may not always converge into a mass movement, which ends up being a good thing for this "governance conundrum" that Ludden is talking about. This happened to a large extent in the case of Polavaram dam discourse. Here, few NGOs became the self-appointed or government-appointed negotiators on behalf of the tribal communities for adequate compensation. At the other end, during the movement for creation of Telangana State, some groups mobilised an anti-dam movement on the premise that it submerges the largest number of villages in Khammam district, in Telangana region (now State). This premise, too, did not look at destruction of entire tribal villages (on either banks of the river) as a problem, but only focussed on the villages that happened to be in Telangana. Even in this way, the movement remained fragmentary and disjointed. If there were occasions of likely joint movements, they seemed to break away every other moment. And most times, survival is the longest (political) 'movement' of all in these parts. NGOs, here, exist in an ambivalent space, within the world political economy of charity-funded exercise, with their own forms of mobilising towards one-off, event-based, project-centered activities, incapable of (or afraid of) deeper political engagement with these paradoxes[62].

In Years 2005 - 2006

There was a larger movement[63], alright, but not in terms of a grand spectacular event, barring a few rallies in the Polavaram area and intermittent *padayatras* (long marches) by Left political parties, such as the CPM and CPI (ML), groups such as Agency Girijana Sangham, NGO representatives and a few human rights groups such as the Human Rights Forum, besides the inevitable big names addressing a meeting or two at some central location. In years 2005 and 2006, radical groups under Koya and Kondareddi leadership had staged dharnas at the dam spillway site. It was multi-locational as it was multi-dimensional. It was the kind of movement that saw involvement of various kinds of people involved at several levels to press the AP Government for a re-think and to halt the progress of the project on the ground. Briefly, of course, there was a stay on works on the spillway site, but as has been pointed out earlier, the stay was only on paper. Moreover, on the ground, a more important work was ongoing, non-stop, through 2005 and 2006 (and of course later) which none of these other movements (in cyber space and judiciary and elsewhere) managed to bring a halt to: land acquisition. How legally valid and justified was the acquisition itself violating environmental and PESA regulations and of course the inter-state conflict with Odisha and Chhattisgarh? The state continued with land acquisition reflecting despotism of a patriarch in the form of the then Chief Minister Y.S. Rajasekhara Reddy, negating every argument, plea, suggestions made by political organisations, Parties, engineers, human rights activists, academics, former bureaucrats, etc. Carrying on with Polavaram had taken the magnitude of an 'ego-trip', hard to bend or satiate. I cite here some of the more pertinent points of the anti-Polavaram discourse – on the ground and in cyberspace – as being part of this historical process. Even in late 2005 and early 2006 the nature of debate on the Polavaram dam was far from being a singular, uniform voice. Many people came out in solidarity against the Polavaram project, each from their own experiences and expertise, and some with a background of work in the area. Many organisations brought forth alternative designs and proposals involving no, or very minimal submergence, hoping the Government would consider them. There were also Gram Sabha resolutions passed by tribal villages opposing the project.

People's Protests, Petitions, What People Said to the CEC (July 2006)

I see these as very important part of the movement, for many memorandums, letters, submissions came from people working on the ground, and the CEC took cognizance of some of these in submitting its Order for the Government of AP to consider. I mention some of the most important from copies NGO representatives, political groups and individuals shared, when they submitted these petitions at Bhadrachalam to the CEC. [I decided to retain the original language of the petitions.]

The tribal leader Sonde Veeraiah of the **Adivasi Samkshema Parishad,** in his memorandum, dated 29th July 2006, to the Chairperson of the CEC, writes (I quote verbatim), "lion share of our aboriginal tribal sects are dwelling in dense forest of Dandakaranya in AP adjacent to the river bank of Godavari since time immemorial...Governments, under the shadow of development, themselves violating all Agency laws which were intending to welfare of aboriginals..., consequently, under gross violation of Agency laws urbanisation was taken place rapidly in Scheduled Area of Andhra Pradesh, due to that the influxion (sic) of non-tribal became aggressive with the result the existence of aboriginal tribals drawn into peril and development of tribals became dormant...After construction of the said

Dam whither the Tribal entity?...In this connection we totally and strongly opposing the construction of Dam as the Agency Area will be inundated and not only ancient culture of aboriginal will disappear but entity of tribe will come into end…The Hon'ble Chairman may be pleased to convey our grief and grievances to the Apex Court and recommend to stop the construction of Polavaram dam."

In their "**Memorandum to the Central Empowerment Committee**" the **Agency Girijana Sangham, Rythu Cooolie Sangham,** and some members of the **Kondamodalu** Gram Panchayat, submitted. They pleaded, "Kondareddy and Koyadora tribes are traditional sons of the soil having a history of thousands of years. This part of the Godavari valley, consists of fertile land and hence not only Girijans but considerable number of non-tribals migrated to this area. In comparison to many other scheduled areas this part of Godavari valley (East Godavari, West Godavari & Khammam) is densely populated. Some of the non-tribals who migrated to the agency area, taking advantage of the honesty and innocence of tribals, took possession of thousands of their agriculture land…Even though tribal land alienation (prohibition) Acts are in vogue since 1917, our lands were being snatched away by hook or crook by some of the migrated non-tribal people…However in 1969 the tribals of this area…united under Agency Girijan sangham (and reclaimed the land)…Unfortunately the government has not legalised our defacto ownership of the land…Now we are living in our land with self respect, peacefully and hopeful of a better future for our children. But alas! Now a bolt from the blue is falling on us! It is nothing but the Polavaram project…We the girijans are not rich individually but we are joint owners of highly rich land and forest; while rehabilitating the loss of this land the government is not taking into consideration this aspect of community ownership…We earnestly hope the Central Empowerment Committee will recommend the cancellation of the disastrous Polavaram Project and…to save the lives of 2 1akh people, their culture, their habitat and the natural forest which can not be replicated else where by spending any amount of money."

Girijana Sangham members Lakshmana Rao and Sriram Naik, in their memorandum to the CEC point out that "The National Tribal Policy draft in para 8.6 (released by GOI on 6th July 2006) states that there shall be a threshold of displacement, viz. the maximum number of persons that can be displaced in one project. Project involving displacement of more than a fixed number, say 50,000 would not be considered, if the majority are STs, or would be subjected to more stringent appraisal norms. The proposed Indira Sagar displaces 1,80,000 of which more than one lakh are Tribes. So, construction of Indira Sagar would be a greater violation to the National Tribal Policy (draft)…"

Meanwhile, in a "**Requisition of Thotapally Village Displaced People (Bhadrachalam Mandal) to CEC"** I found a different take on the project. They write, "The four villages – Bhadrachalam, Kunavaram, Chinturu, V.R. Puram, are to be submerged fully. The package announced by the government is good for the tribal communities but hopeless for the non-tribals. The land value decided by the government, at Rs. 1,15,000, if it is increased to Rs. 2,00,000 we wish to state that the government may take our lands without any obstacles. And we wish to request that along with the tribals, even non-tribals may be provided alternate R&R housing sites. We have no objection to the construction of Polavaram dam."

In a similar vein was this letter from Bellamkonda Ramarao, President, **Mandal Congress Committee, Burgampahad, Khammam District (AP),** of course on expected lines, saying "We the agriculture ryots irrespective of caste (ST, SC, BC, OC) etc, of Burgampahad mandal are willing for the construction of "Polavaram Project". But due to the project construction the agriculture lands, houses, and vacant sites are submerging and request for payment of compensation to all of them irrespective of caste (i.e. ST, SC, BC, OC)." Incidentally the people who made these requests (and handed over the copy to me) were non-tribal upper caste landed farmers of the region and the latter were of course Congress Party Workers, who were seen waving Congress flags at the Punnami Guest House on 29th July when the CEC was flooded with memorandums from a huge gathering of political activists, NGOs and others.

In a "Letter to CEC Chairperson, P. Vijay Krishnan", Gandhi Babu, the Convenor of ASDS (V.R. Puram) made six points about problems with the Polavaram project. Among these were – "The displaced people will migrate to upper reaches (according to a study by Prof. Thimmareddy, social scientist, these people will again resort to *podu* there when displaced from their original settlements)...In the Reserved Forests they will have no rights on forest land; It will cause permanent damage to the delta, according to the former engineer, K. Ramakrishnaya who had looked at the water resources and utilisation in AP in 2004...The eastern canal of the (Dowlaishwaram Cotton anicut) has already caused damages, 250 tmc from Polavaram (after the project is built) in the kharif season will mean worse damage; a study of the Singareni collieries by the Godavari shows that Bhaskar colony there has graphite ores in excess of 250 tonne. The Polavaram dam submergence will lead to graphite dust killing organisms and aquatic life in Godavari where it mixes in water." He also presented to the CEC a copy of his letter (duly signed) to the Chief Minister of Andhra Pradesh, which says, among other things, "Public hearings were held on festival days so people could not attend. Tribal representatives were not given a hearing. People should be told in their own language about the project, but this was not done. They did not even beat the *dandora* (drum) in the villages about the public hearing…We oppose the carrying forward of the project without discussing these in public hearings."

Mainstream Political Parties and their Stance

The **Telugu Desam Party** has always been in favour of the dam, since the surveys for the Polavaram dam had been ongoing during the TDP regime. I present here some of the views of the other Parties in the Journey 2006.

Communist Party of India (Marxist) - B.V. Raghavulu, State Secretary of CPM, AP, and Politburo member had issued several statements in media regarding the issue since late 2005. In one of the books on Polavaram he wrote, "Let me make it clear that our party is not against the construction of multi-purpose projects. Our demand is that it should be done without disturbing the livelihood of tribal people...The State government has furnished inaccurate information to the Government of India on various aspects of the project. For instance, the number of displaced people as per the site clearance (Sept. 19th, 2005) was 1,17,034. But according to an environmental clearance (October 24, 2005), it was 1,93,357. Likewise, the land to be acquired for the project according to the G.O. (Government Order) No. 93 (dated May 24, 2005), was 75,177 acres. But under an environment management plan (EMP) submitted to the Government of India later, the area was shown as 1,55,182 acres…According to a GO No. 93, issued in May 2005, the ayacut was shown as 23 lakh acres. In the environmental clearance it was shown as 7.2 lakh acres. But again in an affidavit submitted to the High Court, the figure was mentioned as 23 lakh acres...We have been urging the government to avoid, and if this is not possible, to minimise the displacement of people because of submersion. Uptil now, the State government has not heeded to our pleas....The CPI (M) strongly feels that it is mandatory for the Andhra Pradesh government to go for alternatives instead of causing this big displacement that is wholly unwarranted..."[64]

Telangana Rashtra Samiti – V. Prakash, former State General Secretary of TRS, wrote in the same publication (cited above): "This project would not benefit Telangana region in any way. Moreover, about 300 tribal habitations (according to government estimates) and around one lakh acres of crop area in Khammam district in Telangana will be submerged under the project."[65] He further states, "We have valid reasons to oppose the Polavaram project. This project intends to irrigate about 7.2 lakh acres in East Godavari and West Godavari district. The Bachawat award has made necessary water allocations to this effect. Of this, about four lakh acres are being irrigated by two lift irrigation schemes – Tadipudi and Pushkaram – on either side of Godavari. These two projects would be ready by the year-end. The same ayacut has been included in Polavaram project too....We at TRS, are of the view that the government that wants to divert Krishna waters to Rayalseema from Srisailam is constructing these three projects –

Pulichintala, Dummugudem and Polavaram – only to ward off opposition from coastal Andhra people. At the same time, we want to make it clear that irrigating parched lands in Rayalaseema is not an objectionable thing. But, diverting water through backdoor methods deserves to be questioned...."[66]

However, something changed later. In fact, there were charges of graft levelled against one of the TRS-supported businessmen on contracts for construction of the dam which were debated for long in Telugu news media (in years 2011-12). The National project was declared at the same time of declaration of Telangana state (year 2013-14). Yet, some members of the TRS remain opposed to the project since unequal utilisation of Godavari waters was the main plank on which TRS had built its initial discourse. In fact, TRS had organised a public rally and meeting against the dam in February 2006 in Bhadrachalam while the CPI (M) had organised a "Chalo Hyderabad" meeting on 9th March, 2006, and a large number of tribal communities from the three districts in the affected zone had participated.

CPI advocated a better and perfect rehabilitation package prior to the construction of the dam, initially, but gradually showed its allegiance to the tribal communities and the Telangana organisations opposing the submergence.

The CPI (ML), with its affiliated organisations, such as Rythu Coolie Sangham and Agency Girijana Sangham, although with a very small base in the villages has kept up their opposition to the dam till date. They have filed several petitions to this effect.

CPI (ML) New Democracy argued in favour of the dam but with the caveat of better rehabilitation initially but changed its stance by the year 2012 to 'stop construction of Polavaram dam'.

BJP[67], was not seen making strong statements against the dam. Interestingly, Adivasi Samkshema Parishad, though an affiliate of the BJP, I observed, had a rather strong tribal leadership at the grassroots with a tribal identity concern, and has not always represented the mainstream views of the BJP in local contexts.

The state-banned outfit of Maoists (in Andhra Pradesh and Chhattisgarh) were not (or so I observed) heard making too many statements (regarding Polavaram) in the villages that I traversed in the years 2006-2010, though there were statements on Telangana statehood. But I saw some wall posters against the dam in the border village of (Andhra-Chhattisgarh), Kunta at a bus-stand in year 2012. I did not have access to their statements sent out to selective Telugu media regarding the project. However, their views on the Polavaram dam may be seen as part of their larger political opposition to projects such as mining, etc. in tribal areas and against tribal displacement, and globalisation, in general.

Year 2005: Meetings in Hyderabad, Debate on Anti-Dam versus R&R and Adequate Compensation

Many meetings were organised on the Polavaram issue. I share some of these arguments in meetings and cyberspace reflecting the mood of the moments all of which are valuable historical documents of the nature of development-displacement discourse today.[68] The then Major Irrigation Minister, Mr. Ponnala Laxmaiah had invited Ms. Medha Patkar for a dialogue on the issue of the Polavaram rehabilitation package. A meeting took place in Hyderabad, with all the representatives of the tribal and dalit people in the movement. These groups had apprehensions that the movement will be led into a talk on "RR" or compensation when they were in no mood to budge from their anti-dam stance. Questions were raised as to why the Minister was not interested in a dialogue with the affected people instead of with Ms. Patkar, in spite of several petitions they had submitted. There seemed a clear division between 'national' activists and 'local' and a certain form of representational politics which, to the local people, side-lined several movements from the ground and where these local movements were not conveyed nationally, either by media or by the state. Some felt that there was a shift in gears from the time the Minister's meeting with the Ms. Patkar took place in Hyderabad, from a clear anti-Polavaram moment, to that of negotiations for the best 'package' (Ms. Medha Patkar, the leader of the National Alliance of People's Movements, denied it strongly); all the government communiqués reported that the Minister had accepted Ms. Patkar's suggestions for a better R&R package. This was reported by all the newspapers the very next day. It may be possible to second guess that the government had played a tactical game in this entire episode. So, at a meeting held on 4th December 2005 at Hyderabad[69] (organised to elicit opinions from participants for the meeting scheduled the same evening of Medha Patkar and the Minister), organised by the Solidarity Committee for Anti-Polavaram Project Struggle & NAPM, participants included tribal leaders and activists from across the State, members of NAPM and Telangana activists. Medha Patkar was a chief invitee at this meeting. Participants included National Front for Tribal Self-Rule, Yaksi, Samatha, Human Rights Forum, and individuals such as Prof. Kodandram Reddy who was then lecturer, and later became the leading figure of TRS Joint Action Committee and few concerned individuals. The participants had stated that compensation was not even a remote thought and that the NAPM leader should keep that in mind when she would speak to the Minister. A member of Adivasi Samkshema Parishad, A. Navin Kumar had said, "We have travelled 300 kms to come here, and we will not agree to R&R and reduction of height of dam. The question is that of survival of the Koya people and the Koya culture, as the Koyas are the majority group whose homes and lands will be submerged by the dam…If R&R and reduction of dam height is the subject of discussion today, then our organisation will leave the meeting..."

Another member of a Polavaram Vyetireka Adivasi Porata Committee (Committee of Adivasis Opposing Polavaram), Tellam Venkat Rao said, "The people of West Godavari are completely against the project...Some NGOs are arguing for reduction of height, asking for R&R, however we do not agree with this..." Illa Rami Reddi representing the Agency Girijana Sangham, said "Adivasi people are going to be the leaders of this movement against the dam. We tribal people have to stand together and fight together-no matter whose solidarity we get..." Kunjam Pandu Dora of the Adivasi Aikya Vedika said, "While there is a huge majority of us who oppose the dam, there continue to be some who argue for better rehabilitation or alternatives which will affect fewer people; however...you must unite and rally behind the will of the affected people who are struggling against the dam. They do not want the dam. We request you, Medha Patkar, to take this message of the people – that they are opposing this Dam and will not accept a good rehabilitation package." Kunjam Biksha Dora, a former MLA, from Khammam also spoke of "NGOs and intellectuals" being "ambivalent about the dam. They are indulging in double speak...I was an MLA for 10 years we consistently said "No" to the project. We do not agree with rehabilitation. Recently in Bhadrachalam we all met and unanimously agreed to oppose this dam collectively – so we must further the movement."

An interesting point was made by a person representing the Krishna delta region, quite unexpectedly. Erneni Nagendra said, "I have come from the Krishna Delta area and you must all be wondering why I am opposing the dam, which is supposedly being built to benefit us. They are going to submerge so many hectares of land in Krishna area. We said this in the public hearing to the CWC. They do not give us information. They tell us they will give us water from Polavaram – that too only in the rainy season. I express my solidarity with the struggling adivasis against the project Polavaram because, I think today it is them, tomorrow they will evict us."

Medha Patkar, in her response, made some important points, saying "I am one of you...You know very well the stands we are taking as Narmada Bachao Andolan and NAPM. Also the position we took when I was a member in the World Commission of Dams, where we studied many dams...There was only one dissenting note to the report and that was mine. Our stand is against the large centralised projects which exploit the resources on which communities survive. These resources are also to be considered to be intergenerational capital...when we visited the Polavaram area last month...we raised questions about the dam and many issues were brought forth...Within a few hours of that press conference, the CM called a press conference, and announced that he had obtained Environmental clearance from the MoE&F. Immediately we contacted the MoE&F, and they said that only conditional clearance had been given. They had still not obtained the clearance under the Forest Conservation Act 1980. The CWC also has not cleared the project...What has followed, is that the Irrigation Minister, Mr. Laxmaiah called (on phone) and said that I want to meet you and dialogue with you...Our organisations believe that a dialogue is necessary between the affected population and those who stand by them when there is repression or crisis. That does not mean that primacy does not rest with the people who are affected...The Minister wants to talk - lets us go and listen to what he wants to say... I said to him, we will listen to you, but... it is the people...in the affected Gram sabhas who will decide. Coming to the Narmada Bachao Andolan, 5 big dams have been stopped in the valley. The Sardar Sarovar dam was the only dam where rehabilitation has taken place, of which 10,000 people were given land. 40,000 are still in the area and fighting the dam... However... it is essential that when we are dealing with the state, we can have a dialogue without compromise...The game of the state is to corner the people. We should say we are all one...We should use the dialogue as effectively as possible...which may try to change, if not their hearts, at least some of their decisions..."

The printed record of the proceedings (which came into public domain as it was distributed) highlight the points the NAPM leader and others made to the Minister at the meeting, which included demands that no land be taken without agreement of Gram Sabha which apparently the Minister agreed to.

"The Government Agreed to the following:-

1. They have not got clearances from Environmental Impact, CWC, Planning Commission, etc.
2. They will provide all information regarding: Project costs, command and catchment areas and detailed areas to be benefitted/submerged etc.
3. Provide detailed information on land identified (giving districts, mandals, villages) or rehabilitation, and afforestation (compensatory forestry).
4. Will hold Gram Sabhas and take their consent before acquiring land or executing any further work on project.
5. The government will not use any police force.
6. The government will organise a State Seminar to discuss all issues pertaining to Water Projects and Water Policy in AP."[70]

Ultimately, the government's promises were never kept, especially on points 3, 4, 5 and 6 of the points mentioned here. Irrespective of local resistance in tribal villages, the dam work did not stop (in fact police force was also used, as has been noted). And one wondered if there was a thin line between a state being ruthlessly powerful and beguilingly co-opting. Several protests were held since. The Hindu, Hyderabad edition (dated 16 November), reported, "State CPI(M) threatens to stall work on Polavaram – The State committee of the Communist Party of India (Marxist) has threatened to launch direct action and stall the works on Polavaram (Indira Sagar) project by conducting indefinite picketing if the Government did not change its design...The CPI(M) is planning to lay siege to the Khammam directorate on November 21...B.V Raghavulu, State Secretary, said, "Once the project takes the inter-State proportions all the investments and efforts made thus far will be rendered waste." Again, on December 9, 2005, The Hindu reported (with a huge column space story with a photograph) - "CPI(M) sounds war cry against State's Polavaram Project" – (Araku, Khammam District) – The CPI (M) has decided to intensify its anti-Polavaram project struggle significantly taking support of all political parties, including the Telugu Desam and the TRS, beginning with a huge rally at the dam site... (Raghavulu) declared, "Be prepared for a do-or-die battle. Our goal is to make the Government stop all works relating to the project. There is no compromise on that count."

A workshop in Hyderabad conducted by the same Polavaram Solidarity Committee, titled, "Perspectives on a Struggle", at Hyderabad on 12th and 13th March, 2006[71] saw attendance of participants representing six affected Mandals (Devipatnam, Polavaram, Chintur, Velairpadu, V.R. Puram and Bhadrachalam). It was pointed out at the meeting that the state government was carrying out surveys of individual families, their landholdings, assets, trees on their land etc. In some areas these surveys are being conducted in the name of Indira Kranti Patakam (the former Velugu Programme), thus deviously collecting information by leading the villagers to believe that the survey is meant for some development loan which may be obtained under the Velugu programme. And the participants spoke of the end of all developmental works in the villages on the pretext of Polavaram project. Mr. K.G. Kannabiran, Senior Advocate, President PUCL (People's Union for Civil Liberties; he passed away in year 2010) stressed on the urgency of getting 274 Gram Sabhas of the 276 submergence villages, which lie in the Schedule V regions to pass Gram Sabha resolutions expressing their non-cooperation with the proposed project and their opposition to the same, as it violated their rights in Scheduled V regions. Mr. Nagaraju (the advocate we met earlier in my journeys) shared the information that the Government was purchasing land from non-tribals in Charla, Venkatapuram and Dummugudem mandals, at the rate of Rs. 1 lakh per acre, to rehabilitate dam-displaced people.

Gradually, the Committee itself became redundant. In the midst of all this there was an active cyberspace discourse (which happened between late 2005 and continued until late 2006 regularly, and intermittently thereafter until the end of 2008) on the Polavaram dam, initiated and moderated by intellectuals and activists in Hyderabad. The email alerts, news, opinions, etc were continuously flooding the inbox via a *polavaramgroup* started specifically for this purpose. This was a world discussing the more technical and highly literate issues of economics and cost-benefit and what could be done in terms of reports to be brought out, an information cell to be set up, etc. There was initially a frenzied phase of people seeking more information on the hard facts, and ways to go about organising a movement based on these facts. No popular movement emerged from these internet activities but those who protested on the field before and during these educated discussions continued to do so in a separate space, unrelated to the arguments and counter-arguments on cyber space, which is not to say that the discussions on their own were insignificant. Very important issues were raised. At the same time, many other parallel activities were ongoing – petitions filed with the Supreme Court, for instance, and perhaps a project or two of some of the NGOs and there were meetings organised in Hyderabad and elsewhere (apart from 'padayatra' – long marches). Interestingly, Illa Rami Reddi was the single Kondareddi from the submergence zone who found for himself a via-media – an email account perhaps opened on his behalf - to engage with the cyber-space in the midst of all these happenings, with updates from his end which didn't continue beyond a point. Starting in the first week of January 2006, there were a couple of email exchanges on where the money for Polavaram project was coming from. Then there were plans to start a "library of studies on Polavaram" which could be accessed by members on the list. Cyberspace ended up as its own 'library' of sorts for a short while. But there were also doubts even among the list members about the purpose of this activity itself. For instance, one list member wrote, "I am still not sure what the purpose of these studies or those that may be done in future really would be...If we don't have clarity on that, scholars will keep producing reports, NGOs will keep filing court cases and conducting workshops, political parties will keep up the rhetoric and the dam will get built and the people in the submergence will find their little local alternatives…"

There seemed too many fragmented alleys in the entire discourse. Perhaps singular and united voices are not possible in e-space. And probably that is a truth about democracy, too. These debates showed the reality of movement discourse, in general, and perhaps also in the urban politics of organising movements.

***

Notes

1. Until 2010, the records of which are here. But there was yet another journey that came my way, in year 2012, for a report on the status of the 'internally displaced' Gottikoya community (Muriyas of Chhattisgarh) settled now in

Khammam district of AP. In many ways they will now be thrice-displaced with the Polavaram project, as they had already left their homes in 'CG' (as they locally refer to their home State) for better pastures here. But theirs is not a simple displacement story as there are layers of complexity involved – not all of them who moved here have left the connection with their own lands and work in their home state. But there is another angle of how far back in history do you go to understand the movement of this community into what is now called Khammam district and AP? Is it necessarily a distress migration caused by the Maoist-State crossfire, or is there something with regard to the idea of boundaries and a region for tribal communities? Not all migration can be traced to Maoist violence alone. But the definite increase in Muriya/Gottikoya settlements in this district since 2007 onwards is a case for further probe.

2. At the Jadavpur University, Kolkata at the School of Women's Studies and IIT-Madras, School of Humanities and Social Sciences.
3. Ancient Tamil poetic corpus dated between 5th century B.C.E to 5th century C.E, with some fascinating translations by A.K. Ramanujan in his *Poems of Love and War*. Less poetic was George L. Hart's *Poems of Ancient Tamil, their Milieu and their Sanskrit Counterparts*.
4. Based largely on my paper, *A Discourse on Dismemberment: Women in Submergence Zone Case of the Indira Sagar Polavaram Dam on the Godavari River*, Occasional Paper, 11, School of Women's Studies, Jadavpur University, Kolkata, 2011. The space at SWS with its library, students and faculty was a huge contribution to this part of my understanding.
5. Emanuel Wallerstein, 1991, p. 31.
6. Ibid, p. 32.
7. I received a phone call, not so long ago, that finally even that tree by the window has fallen, ostensibly to a storm (while I was away from Hyderabad), but at a deeper level, for lack of a collective spirit that may have held on to it (as in the Chipko movement), embraced it and not hated its very existence, as some in my neighbourhood did (where cross-cutting cable wires are not seen as ugly). That the tree had to fall in my absence – well, what can be more painful? What kind of a house/home does one go back to? Without the 3 am Koel, the sore-throated Koel, the Green Bee-eater, the Oriole, the Common Babbler, the butterflies, and the yellow flowers, each with their spaces on and around that tree?
8. Bimal Kanti Paul, 1984, p. 9.
9. Ibid, p. 10.
10. Robin Tennant-Wood, 2006, p. 324. [About the efforts made by women to chronicle its submergence history and save it from death, under the leadership of Jo Garland and Dalgety and District Community Association.]
11. Ibid, p. 327.
12. Daniel Thorner, 1967, p. 252. Emphasis mine.
13. Finger millet gruel -ironically, the nutritional value of this millet is far superior to that of rice.
14. Ibid, p. 241.
15. Ibid, p. 241 Emphasis mine.
16. Carol Boyack Upadhya, p. 1376-77.
17. Ibid, p. 1377.

18. P. Jaswantha Rao, 1998, p. 81.
19. Trinadha Rao, *Policy in Wilderness: A Glimpse into Tribal Life and Governance in Agency Areas*, CARPED, Hyderabad, 1996, p. 30.
20. Jaswantha Rao, cf.cit, p. 82.
21. K. Balagopal, 2007, p. 4030.
22. AP District Gazetteer – East Godavari, 1979, p. 10.
23. Ibid, pp. 11-12.
24. F.R. Hemingway, 1915, p. 2. Emphasis mine.
25. Ibid, pp. 188-9.
26. I remember this because of a conversation in a university in Hyderabad with a Centre of this kind. When I mentioned to the young professor concerned that they might invite people to be displaced by Polavaram dam to give a lecture to the students, he replied, "We do not have a budget". I thought, what should be the cost for a train or bus ticket to Hyderabad and a day's stay at the University guest house for a man or woman from the Godavari?
27. I thank Mr. Reddy (the elder) and his son (and Amar, the grandson of the senior Mr. Reddy, because of whom I met them).
28. *Historical and Polemical Documents of the Communist Movement of India, Vol. II (1964-1972)*, p. 751.
29. Ibid, p. 754.
30. Ibid, pp. 754-5.
31. P. Sundarayya, *Telangana People's Struggle and its Lessons,* Foundation Books, Hyderabad (reprint, 2006), p. 182.
32. Ibid, p. 183.
33. Ibid, p. 184.
34. Ibid, p. 185.
35. This section owes much to a chance invite by Mr.Ramaswamy Iyer to speak on Living Rivers and Dying Rivers at the IIC, Delhi in 2012.
36. The Map (slightly damaged) above was procured at the British Library, London; it is singed by F.C. Cotton, Civil Engineer, at the Government Engineers Office on 28th February 1853. The map appears in *Profits upon British Capital Expended on Public Works in India as shown by the Results of Godavery Delta Works of Irrigation and Navigation,* London, 1856. The Map below was procured at the Irrigation and Command Area Development officer in Hyderabad in the course of my journeys (this is prior to the formation of the State of Telangana). I show both the maps to highlight the continuities in the administrative apparatus built on Godavari waters.
37. The news report was published in the Telugu paper Vaartha, in its Hyderabad edition, and sent over a group email on 25th April 2006. Bhagiratha is the sage who is credited with bringing Ganga to the earth through his penance form her otherwise 'secure' space in Siva's locks.
38. Kubera is considered the Hindu god of wealth. According to mythology associated with the Tirupati temple, Kubera is said to have lent money (on very high interest) to Venkateswara for his marriage to Padmavathi. It is believed that till date the god Venkateswara is only paying the interest on the loan and has not yet paid the principle! Which is why devotees, in return for his favours, donate money to the temple 'hundi' (donation receptacle) within the temple precincts, with the belief that they are helping Venkateswara (of

Balaji, or Venkataramana as he is called) repay his loans to Kubera. Tirupati is apparently one of the richest temples in the country.

39. One of my articles on Arthur Cotton was published in the context of the elections in AP in 2009, under the title, "The empire flows again" in http://www.indiatogether.org/2009/jun/env-cotton.htm (02 Jun 2009)
40. Cotton, op.cit, 1856, p. 8.
41. Ibid, p. 9.
42. Ibid, pp. 48-9. Incidentally, it is also mentioned in the records that Sir Arthur "heartily supported the Bible Society" in Henry Morris, *Sir Arthur Cotton: Philanthropist Engineer*, The Christian Literature Society, London, Madras, Colombo, 1907.
43. From Walch, author of the Engineering Works of the Godavery Delta – giving statistics for each decade from the Godavery district Manual, as quoted in Henry Morris, 1907, p. 21.
44. Cotton, 1856 (op.cit), p. 46.
45. Ibid, p. 60.
46. 'Irrigation from the Godavari', in A Century of Irrigation, Godavari-Krishna Delta: 1859-1959, Institution of Engineers, AP, Hyderabad, 25th August, 1959 - a commemorative volume compiled by Dildar Husain, p. 18.
47. *Selections from the Proceedings of the Godavari District Association,* In 1912 and 1913, Published by Godavari District Association, Sujanaranjani Printing Works, Cocanada, 1914. (Price of the same is mentioned as 6 *annas*). I chanced upon this reference at the India Office Collections at the British Library, London.
48. Ibid, pp. 72-3.
49. Ibid, pp. 74-5.
50. Ibid, p. 75. Emphasis added.
51. Ibid, pp. 34-5.
52. These thoughts owe to an invitation I received to speak at the Madras Institute of Development Studies, Chennai on 26th September 2011, following publication of my op-ed piece, "Facilitating Displacement? The Draft National LA&R&R Bill 2011" in The Hindu, September 13, 2011. I also use some part of that article.
53. www.apgov.nic.in . Check 'irrigation projects'.
54. See www.appcpir.com . This MoA was signed under the previous UPA-II regime with the Congress in power in Andhra Pradesh state.
55. Ibid.
56. Ibid.
57. Ibid. Under Section "Water". See my Journey 2010 for related discussion.
58. See Section Land for Land in the said Act, which is in public domain.
59. APIIC supplement as cited.
60. Gramsci, *Selections from Prison Notebooks; Notes On Italian History*, pp. 54-5.
61. David Ludden, 2005, p. 4050.
62. Of course, the situation today is increasingly debilitating even for the NGO as a construct that started very much within world capitalism and ideas of philanthropy not altogether unproblematic. But with increasing surveillance of and cracking down on NGO funding carried out by governments across the world (especially the 'liberal 'governments), it is a question as to how long this system will survive.
63. I mention only the groups (political or otherwise) that I encountered in the

course of my engagement with the issue in the year mentioned here – 2005-06; not the ones who may have been involved in the anti-dam discourse but I did not know of or come across.

64. B.V. Raghavulu, "A Thorough Review Needed", in Biksham Gujja, et.al., 2006, (pp. 45-9); pp. 46-9.
65. V. Prakash, "Danger to Telangana", Ibid, p. 59.
66. Ibid, pp. 60-61.
67. BJP-led NDA has now formed Government following the elections in 2013. Their perspective on water includes debatable interlinking of rivers; it will be interesting to watch their response to the Polavaram issue.
68. Needless to say, some of the apprehensions of the local tribal and dalit groups regarding the 'compensation lobby' came true as witnessed in later years of my journeys.
69. This is based on the documented proceedings of the meeting, a copy of which was shared by a human rights activist in Hyderabad.
70. From the above source.
71. A research scholar from a University abroad had compiled the proceedings of the same. I took the liberty to extract the most important points raised in that workshop from the proceedings (which are meant to be open source documents), to highlight some aspects of the nature of the debate.

Epilogue/Postscript...

Was I outside of it all, witnessing and aching to belong inside as I continued my journeys? Don't know.

In the years that I travelled and wrote, some of these questions in my head I did not dwell upon in this book. With my limitations and nature of resources, I could not cover the entire submergence zone villages – I managed just a minuscule part of it, though I wish I had seen them all. I do hope though that what people said, among those I did meet (some of whom who could not be accommodated in the book for want of space), represent the larger issue in all its diversity and complexity.

A point about the Telugu word of 'modernity' (wherever displacements happen): *'RR package'*; it is not difficult to see why some peasants/farmers are accepting compensation. For decades now their survival has been one filled with everyday struggles against money lenders, exploitation by non-tribal upper caste men gaining pattas on tribal land through connivance with the Revenue officials, working as agricultural workers without any rights; unscrupulous traders, etc. News reports on all these issues abound in the district editions of Telugu newspapers since decades, too. And funnily enough, these issues just move into the sphere of local legend and things are said in casual conversations as an internalised reality to contend with. The role of the Polavaram project in this midst was also to dig up some of that muck all over again so older people would revisit some of these classic land-related legends and younger people could put names to faces or faces to names (and revisit the old land records), and for those long-lost names and faces (through their descendants) to make a beeline to the Revenue officials and prove their 'bonafides' and sell lands they had illegally owned. In the course of the journeys I realised (also from my own experience) that when faced by debilitating trauma, where your immediate survival is a question, you are caught between the very real dilemma of taking care of yourself or fighting the system. Many lose out for want of supportive mechanisms. Organisations and groups that come from outside your own context (in this case the tribals, and the Scheduled Castes, or even the 'BC' fisherfolk) can see the problem from the advantage of distance, and make fine calculations and schemes as to what might work. They seem to get the answers right (when seeing the situation at face value, that is) but those who face the situation upfront, cannot see beyond a day, two days, or a month and all their energies are fixated on getting out of that immediate crisis. Yet within these situations, truly heroic are those that continue to resist for it must take a huge effort to rise beyond immediate and self-trampling situations and look at the larger. I have met both these kinds of people and seen both ends to realise that a disempowering situation can either suffocate or force one to resign to one's fate, or give rare strength and courage to plunge neck-deep, irrespective of consequences, into an ultimate act of resistance. From the point of view of the state (as it has come to be), it always seeks to work from amidst the majority who suffocate or resign and seek easier ways out, before it can come to the stronger ones who resist, by either charging them for crimes they did not commit (pushing them into a tedious legal process, which takes its own time and can test their patience) or plugging out their life-blood in other ways. The former works creditably well, as it also comes with sops doled out at intervals to these communities. Such as an employment guarantee scheme, or forest cooperatives, seeing them as 'beneficiaries'. The majority cases are of those who give up or give in, against force and power. The rest either go to the other extreme of a war-like situation, and there are those who join a path of negotiating with the state or act as mediators from both ends. Whether or not the suffering communities become active members of – or assert agency in – either of these sides, is an issue that must be dug into some more.

Where is the Media in this canvas? Does it belong at all or does it remain satisfied watching from a distance and getting others to watch or read?

Through the journeys I also understood that by the time the communities got over one hurdle of that moment, after many protests and requisitions to the MLAs, or government officials, with media reporting on the same, a new scheme or an Act would be ready to tease them and realisation would dawn that this new Act was, in fact, far more problematic than the issue they had been fighting against, which may have nearly got resolved: they got the state to acknowledge their Rights to the Forest (through the FRA) but didn't envisage the National Park would do them in. There are other examples – Right to Education and 'Merging' of government schools[1] in the Scheduled Areas would happen almost together. So, yes, you have a right to demand education, but your child will have to walk a few more miles to reach the new 'merged' school, so might as well give up, as many have done, and do. The RR envisages schools, but the few colonies already built did not construct school buildings; tribal communities who had to shift there found sending their children to school a greater problem. The state gave them a health plan (Arogyasri), ostensibly 'free' health care; but merged the plan with private hospitals in the larger towns and cities. So the fishermen and tribal people would be seen in hospitals in Rajahmundry Khammam town for the 'first referral' and then be sent to Hyderabad (all the way) for surgeries and major illnesses. So you travel for days and hours to reach health care which was otherwise possible within the former Primary Health Centres, had they functioned. So, do you fight the 'idea' of Arogyasri, or the RTE, or just make do? For, the Acts and Schemes are implemented ostensibly after years of 'demanding' these rights, so 'you asked for it', why do you complain? At the level of 'higher' economy, extremely significant MOUs or MOAs have been signed, which may even be discussed in cities and even outside India by groups and organisations as "policy" matters (or advocacy); while those discussions continue, in the meanwhile, landless dalits or tribal peasants would still remain focussed on the (by now 'smaller') immediate issue of R&R. 'Smaller', because of what has happened at the magnificent level of deciding the fate of millions precedes, and sometimes sprints, way ahead for none but the 'specialist' group to make sense of. In the meantime, officials have already started implementing the process on the ground, hitting where it hurts. Sometimes, one wondered – the struggle of a dalit, landless agricultural worker or a tribal peasant, or a fishworker continues at her/his level just as my returning home to 'writing the story' from 'the field', 'truthfully recording history', as it were, continue. But by the time the story has got its space, the person I spoke to has lost her/his battle or is contending with an all-new problem. The next time we meet, it will be the same old story, but an altogether different story, and there we start again. In the stress laid on 'speed' there is no time to understand the problem, you just report them; there is no time to think and accept or reject a proposition, you just take the money – like Seeta had said *mundu occhinavaalake ticketlu* (first people in the queue get the 'tickets') – but to which show?

I ended up – thanks to the journeys – with too many questions of my own condition beyond the people in this displacement process. Which money is 'good' money (in academics, or media, or NGO world)? Is there money which is 'good'/'safe', 'non-problematic'? How does independent journalism or independent academics survive? Do they? Is a salaried job with a 'part-time indulgence' in activism ('managing' both that some people do) the only way to make sense of the whole thing and be relatively 'secure'? But do security and activism go together? Is it possible to believe that even today, people can, and do, plunge into an issue that needs to be addressed as a matter of personal politics, or as a basic responsibility towards the identity they assume for themselves in that given time, viz. 'journalist', 'academic', 'photographer', 'chronicler', 'teacher', 'Rights activist', or most simply, a Fellow Human?

In My Diary

17th December 2011 - Phone call from Singanapally. One Ranjit Kumar, son of the Sarpanch of Singanapally, a Koya, called. He said, "There is fraud in the name of Polavaram package. Some people were given pattas on land, but no money. Initially they showed land to the Tuthigunta and Sivagiri people in Gunjavaram, where Mamidigundi people are building homes. Some people received cheques in 2007-08 but have not been shown any land, till date. Tuthigunta and Sivagiri people have been given pattas. Kundrukota people gave consent to the government to take their lands on the promise of land worth 100 acres in Buttaigudem two years ago, through the involvement of the former MLA. But people didn't know what they were signing on. They have not seen that land…Till date people haven't been settled on that land they were promised. Packages are being doled out in bits and pieces…The officials are giving few people either money or land, in order to convince other people to give up their lands. The situation is not at all good here for tribal people. At Chegondapally, 100 cheques of just half the amount were given to people... We do not know if this dam is going to be good or bad, but so far as package is concerned, nothing is going right."

I met Raju (Ballaparaju), fisherman from Singanapally for the second time (after May 2010, but where?!) He called me, out of the blue, to inform me that his uncle is admitted at the Nizam's Institute of Medical Sciences (Hyderabad) under the Arogyasri scheme[2]. He had a clot in his brain, he said. What a long way to come to get a cure! Singanapally to Hyderabad? Apparently the doctors in a Rajahmundry hospital referred the case to NIMS. Ballaparaju said "the compensation situation isn't too good." He wanted to know if I had written about them, since I had interviewed them. Of course I did. I showed him the article published in an English daily, which had his picture…

14th-20th July 2012 – A Godavari Journey came my way (To Khammam District and Kunta, Chhattisgarh)

I got yet another chance to visit Godavari in July 2012[3] (with its own 'accident and memento', which is another story for a different time[4]).

14th July in My Diary, again – Godavari again! Came to Godavari last in 2010. Bhadrachalam again! This time round, staying in a place called Sivananda Asramam behind the bus stand, thanks to an old Post Office acquaintance (of one of my Bhadrachalam visits) for very cheap! Quiet in its own way, a quaint old temple within its precincts…The Ramalayam (at Bhadrachalam) has changed phenomenally since I last saw; they are aping Tirumala (Tirupati)…Met Adinarayana (member of Human Rights Forum), Sriramamurthy (who spoke of Manye Seema the last time I met him 2006/07) and Sonde Veeraiah (Girijana Samkshema Parishad), who informed me of a meeting on Polavaram to be held (15th)… Am glad to be here by coincidence (and to have got to know…).

15th July (in my diary) - Reflecting – The meeting on Polavaram was organised at the MDO office in Bhadrachalam town in the context of re-opening of tenders the next day (16th July) for Polavaram spillway construction. Human Rights Forum, Tudumdebba, Adivasi Samkshema Parishad, Telangana Praja Front, some tribal student groups, few academics and activists from Hyderabad were present...I noticed there was no representation from East and West Godavari districts and assumed this must be a meeting of the Telangana groups. While some spoke of (again) 'adequate R&R' others were speaking against the Polavaram dam... Some of them mentioned a National Forum against Polavaram Dam (which exists, which was news to me). There was analysis of what had happened so far and misgivings of the role of NGOs in the entire movement with questions on their integrity and involvement. An economist present at the meeting (Dr. Revathi)[5] pointed out the difference in perception of adivasi women in the field who were now in the "compensation mode" which was a shift ever since the Polavaram dam commenced (according to her analysis based on one of her studies in a few villages in Khammam district in another context). Others mentioned the significance of the forthcoming local (Panchayat) elections and need for building a united voice against the dam. I notice newer people (and)...the absence of dalit organisations in the meeting.

16th July – In My Diary – The government boys hostel just across the road from here interests me in its daily activities – seeing young boys of all ages (must be between 8 and 13 years of age) come out, play, get in, emerge again at set times of the day (yesterday being Sunday). Today Mr. N.S.C. Bose (!) who is part time caretaker of this asramam (and has a tiny shop selling water, condiments, and a STD booth) fuelled my curiosity further, speaking of it, and of an old Collector in the past who had done a lot for it (built a compound wall, put up some swings for children, etc) and spruced up the building. Today the structure leaks at places, the compound wall has been pulled down, a hospital dumps its waste on the compound of the hostel, and there is an overflowing garbage bin just by the hostel. He spoke of how Bhadrachalam had been encroached upon by several small traders who had set up shop, and even by government offices...Then suddenly he said, 'why don't you go see the hostel; they must be feeding the children now' (it was evening time).

I did. Some children were busy eating (excited that they had been served egg today!) their supper. The cook was an elderly woman who rattled out their routine – *idli* and *pongal* or *pulihora* for breakfast, an early supper after they return from school. By 7 pm they sit down to study with a tuition master; he was about to arrive any moment and the children were all curious about me (a new visitor) in their midst. I have an appointment with them tomorrow at 7 am! All of them are from poor families; few are orphans, some have been abandoned by their families – one of them had the most angelic face, large doe eyes, cherubic face (must be 7 years old)...The children come from Venkatapuram, Khammam, and V.R. Puram. Apparently one woman pays a visit regularly from Khammam town and acts as guardian to two of the orphaned children..."

The Lost Farmer - Varaka Muthaiah

Today I went to a village near Gangireddycheruvu by Bhuvanagiri near Kukunuru to meet the Gottikoya people there. And at the bus stand, I met a man who has by now exhausted all his compensation money from the R&R for Polavaram. He is a Mala (SC) farmer, Varaka Muthaiah, from Valairpadu. Suddenly, that man asked me, "Polavaram *ostundantaara madam*?" Polavaram has become a refrain and the issue is raised inevitably even when I am not asking people about it...The number of policemen asking questions has increased in this area. Today a Forest Beat Officer, too, followed and stopped us (I was with a farmer who was guiding me to a Gottikoya settlement) to enquire as to why we were entering a Protected Area without permission or without giving prior information to the Forest Range Officer... The SC farmer, by the bus stop, was going blind in one eye. His village is Nalavaram in Vaḷairpadu mandal. We were waiting for a bus, when he asked me where I came from and for what purpose. And suddenly began to narrate his story. He and his brothers apparently owned 15 acres and they received compensation of Rs. 16, 50,000 and each of them got around Rs. 4 lakh. He said, "in our village alone, more than 130 acres come under the *barragi*; in Valairpadu mandal at least 1000 acres or more...Some farmers have 60–80 acres of land. I paid off debts (with the money), since we are four of us, we shared that money. Some money I spent on my daughter's marriage for which I gave Rs. 1 lakh in dowry alone. I have four daughters and one son and all are married. I came here (Bhuvanagiri) to take a loan from someone. My eyesight is failing. I worked for 17 years in Hyderabad as a gatekeeper, at the residence of a filmmaker in Filmnagar colony. I used to give my land on lease (*kowulu*). At first they (government officials) said they would give us houses. We all opposed the project initially. But when nobody listened to us, many resigned to their fates and gave away their lands. So we did, too. Aren't there so many people who protest? It is of no use. Is the government listening? From the four lakhs that I got, not a penny is left now. Since the dam has not come, we have given the land (that was given to the government) on lease and I get Rs. 10,000 per acre for that.[6] I make up to Rs. 40,000 in a year. I wish I had never been so greedy about the compensation, I have lost the only land we had and the money, too..." And then, again, he added, he has no money left neither for the bus fare back home nor for his treatment (of the eye problem). And if he failed to find the man who was to give him the loan that afternoon it would be hell. *"Dabbulu kalavakapote naa pani Godaari."*[7]

Meanwhile, the last phone call I received from him (in August 2012) from a telephone booth, he asked me about an eye doctor whose name I gave him. I don't know if he came to Hyderabad, ultimately.[8]

17th July'12 – At the Ananda Nilayam Hostel (for boys) at Bhadrachalam

In My Diary - That boy with a cherubic face is named Sainatha; he studies in sixth grade, he told me. His father is no more, but his mother is alive and works as agricultural labour; he has a younger sister and brother and since his mother cannot afford to educate him, he has been sent to this school. When I asked him what he dreams of doing when older, he said he would think about it and let me know the next day! Ananda Nilayam Social Welfare (Department) Hostel for boys is a neatly maintained place but the building is coming apart, it seems. The roof has cracks at various places. The children do not complain, though. They come from economically poor backgrounds, are aged between 6 years and 13, studying in I to VIII grade, or up to the IX grade, after which they have to leave the facility. A tuition master stays with them – an engineering student by the day. He tutors the senior boys for an hour or two each morning and once in the evening. It was lovely to meet the children. And I thought all these places, spaces, have been possible due to at least a semblance of welfare state India started with. So many school children who cannot otherwise afford schooling find a place, a shelter, food clothing and education for free. The children seem so happy; their eyes don't lie. The cook has been working here since 25 years, she says – an old woman with a benign smile. Clean rooms. Just one tin box each is all they have as their small lives' belongings. They wake up at 5 am and follow a strict regimen; wash their own clothes and plates and glasses...; no gadgets, but there is a TV set they hardly seem to get time to watch. 6 to 7 am is study time, then off for a bath and ready for school after a quick breakfast; return at 5 pm, early supper and by 7 pm ready for 'study time' for an hour and off to sleep. Met 6 year-old Borla Pullaiah of Eruvandi (near Sarpaka, close to Bhadrachalam) who lost his mother; his father grazes cattle, and his uncle admitted him in this hostel as there wasn't anyone to look after him at home. Satram Sathish in the sixth grade doesn't have parents; his grandmother brought him up; he came here three years ago. His sister is stays in a similar facility for girls and is in the fourth grade. He is from Kothagudem and wants to become a 'policeman to catch thieves' when he grows up. The children here get a 'pocket money' of Rs. 62 a month for buying soap, shoes, bag , towels, etc. and they are given four sets of clothes, school uniforms, a tin box, notebooks and have a small library with story books and other educational material. They get two meals in this hostel and get mid-day meals at school. This particular hostel is a 'mixed hostel' while there are separate hostels of a similar kind for SC, ST and BC children.

This hostel, I learnt (from the warden, Mr. Prasada Rao), was built in 1962. Once they leave this hostel, some return home, while others continue schooling staying in hostels in Khammam (up to tenth grade). He says many children who had studied here are in good jobs today and some of them return, to donate books and help in other ways. Donations are exempt from Income Tax. A tribal boy named Vetti Deva who is in the ninth grade was left here ten years ago by someone; nobody comes to look him up. The warden's friend in the USA has offered to support his schooling; a private school principal comes by to celebrate his birthdays with cake and new clothes. Some donations come from Clubs and associations. Some lawyers provided them with dictionaries for the library. In spite of the leaking roofs, hospital dump on their compound (which the children clean up) and other problems, the children are happy about being able to go to school. The warden asks me to write about the school's condition. The children ask me to visit them again, and write to them, or call them (on their tuition master, Manoj Kumar's mobile), each one giving me an individual hand shake.[9]

Today I also saw the ITDA Nutritional Rehabilitation Centre at Bhadrachalam, a facility meant for pregnant and lactating mothers and for undernourished children for emergency treatments, with a residential facility…A rather well-maintained, clean place. A woman had just been admitted with her one-year-old child who weighed, shockingly, 3.16 kgs! Looked like an infant. The child would be here until its weight and vital statistics improved…

If Polavaram Dam Waters Gush in?

But…some good things do not last. This school, as well as the ITDA centre that I saw, will likely get submerged under the Polavaram dam if one were to go by the water levels of the 1986 Godavari (*enabhai-aaru Godavari).* Importantly, the officials till date do not count Bhadrachalam as a potentially submerged area, though few experts had warned of its high possibility from the inundation of the previous years. By a freak chance I got the proof of the likely submergence of the Sita-Rama temple today. Had to buy a charger for my mobile phone and entered into conversation with the owner of the kiosk close to the Asramam, who also happened to be a stringer once. He shared with me some photographs of Bhadrachalam and the temple under water during the *enabhai aaru* Godavari and wondered if that will be the state of affairs once the dam waters gush out. The stringer had been among those who had taken a few photographs at that time. He shared with me some (not all his) which had appeared in the newspapers. What coincidence to find this man, and see these pictures. One of the images is of the elephant of the temple (which had been tied up in the precincts) which could not get out when the waters enveloped the place!

I got the pictures copied on to my mobile phone for want of any other device.

18th July – In My Diary

Finally, saw the historic Chintur *santa*, the Wednesday market I always missed (in earlier journeys). People from Chhattisgarh, Odisha and Andhra Pradesh converge at this weekly fair which is a riot of colours. And went to a village in V.R. Puram mandal – Darbalanka, named after a stream locally referred to as Darbalanka – settled by Gottikoya people originally from Chhattisgarh. The submergence zone is indeed a story in diversity as it is in historical complexity, all of which will be dissolved in one mammoth project.

16th July - Seventy-year-old Giga Thammayya of Kondapally (a Koya whose land comes under the Polavaram submergence) adds, "Around hundred and one people came from Chhattisgarh, cleared forests here and started cultivating. Old fields will drown under Polavaram dam not the new ones (on which the Chhattisgarhi Koyas have settled down). For the Polavaram R&R we went and saw the lands that the government showed us. We do not like those lands. Many of us here occupied and reclaimed lands here from the 'pedda rythus' (non-tribal upper caste landlords). We want those lands to be regularised in our names. The forest department officials filed cases against some of us (under the Wildlife Protection Act). We got out on bail. We did not get pattas under the *adavi chattam* (Forest Rights Act). The ones who did get pattas are not cultivating those lands. Nobody in our village has got any compensation so far for the Polavaram project. We have no issues with the Gottikoyas from Chhattisgarh who have settled down here, at least in our village. The government is not recognising them as girijans (tribals). Many of them do not have voting cards; almost none has patta on the lands they are cultivating. There are ten families of Gottikoyas in Kondapally...The forest department is planting Eucalyptus and Bamboo for BPL (ITC)[10] on the lands that the tribal people are cultivating."

Rama Kartam is a Koya from Singanaguda in Odisha in Podiya Panchayat of Malkangiri district. He says, "I came here ten years ago. There was not enough water in our village (in Odisha) for crops. Our people had between two and three acres of land there. Wages were not adequate. They used to make us work for two to three months and pay us Rs. 50 per day in wages (the big farmers); sometimes we would not get any wages for a whole day's work. We used to work mainly under the Muslim liquor contractors and Telugu Kapus from Valairpadu (from Khammam district) on their chilly farms." Their migratory route was Singanaguda to Malkangiri through Dornapada, Kunta, Bhadrachalam to Kukunuru. He says, "Our land (their original land in Odisha) will also submerge under Polavaram dam. It is close to Motu but it is not in the list of submergence villages. Kalmela, Podiya and Sukma will also probably go under the dam waters. My younger brothers have some land there – around five acres or so, but no pattas.

***

19th July on a Rainy Day

At Kunta you can see the confluence of rivers Sabari, Sileru, Kalleru. Entire Kunta panchayat will be submerged under Polavaram. Madakam Krishna, former Sarpanch (whose wife is now Sarpanch) of Dondra panchayat informs that the Chhattisgarh government has as yet not issued any public notification about the Polavaram dam and as to how many villages will actually be submerged under it. All the farmers here take the rivers coming in normal times as the likely mark for what can happen later due to a forced release of waters. The place, intriguingly, has a good number of Bangla residents, and a Kalibadi, a Kali temple in the midst of a colony settled by the former Bangladeshis. Though Kunta itself, and the larger panchayat will come under complete submergence under the dam, yet one perceives in this place none of the fear, resentment, anger, anxiety, or pain of loss among the people, as most of them are as yet unaware of the extent of dislocation that is to visit them at some point, and one of the reasons is also the lack of information from either the government or other sources. Incidentally, all my instruments of photography had been used up to their limits by then[11] so I could not capture on camera of a poster stuck on the wall of the bus-stand regarding Polavaram dam.

There is an entirely different story of Polavaram dam in Chhattisgarh with its own problems. In the overwhelming story of Maoist-State crossfire Polavaram dam or submergence is least of their concerns – the Chhattisgarh state's or even of those who come in regularly for their 'fact-finding' and documentation of this particular issue once every few months. It seemed so, at the glance, in my first visit to that part of the region. It would perhaps require another journey in another time to record.

***

Postscript: Year 2014[12]

By July 2013, the Telangana State became a Reality. We now have the Polavaram issue contested between two new States, Andhra Pradesh and Telangana. Polavaram has been declared a National Project. At stake are Godavari and investments in Hyderabad. Special packages for Andhra Pradesh (which includes the Rayalaseema region) include waiver of excise duty for new industries in this State and fund flow of Rs. 50,000 crore from the Centre for next three years to this residual state. Increasingly, the debate on Polavaram will be the debate on industrial versus agricultural needs as it will be on the status of tribal communities in both these new states.

January 2014: Newer Resistances

In the last two years – some new associations are in place as also a new spirit of resistance seems to have affected the older organisations of the tribal communities in this area. A new anti-Polavaram dam discourse is beginning to make itself felt though in a small way through interesting forms of protests including street dharnas. For instance, one of the protests included preventing the AP Tourism boats from taking off for their Papikonda cruise down Godavari. Local level organisations such as Adivasi Samkshema Parishad, Adivasi Vidyrathi Samkshema Parishad, Girijana Samkshema Parishad, Adivasi Kondareddula Sangam, etc are re-activating the anti-Polavaram movement. Then there is a new organisation I hear of, the Kondareddula Sangham. Things are different in East and West Godavari within a new Andhra context.

February'14 – Electoral Discourse

Fascinatingly, political parties, including Congress, TDP, CPM, etc, have entered the ground afresh, with their own ideas on Polavaram dam. It is again election time. Polavaram dam will now figure in these electoral debates. The Congress (at a dharna at V.R. Puram recently) has raised a few demands, which include giving Rs. 25 lakhs per acre as compensation for tribal land; CPM and TDP too are advocating increase in compensation to the tribals. Interestingly, the local cadres of CPM are opposed to the Polavaram dam itself, while its leadership has been advocating change in design to affect minimum submergence. CPI (ML), CPI-ML (New Democracy), etc. are among the smaller (in terms of sphere of influence) parties that have again iterated their opposition to the dam. There are talks about merging some of the villages of erstwhile Khammam district with Andhra Pradesh, especially ones that face submergence in the proposed Polavaram dam. On February 9, 2014, The Hindu (National Daily) reported – "Scores of tribal people owing allegiance to Adivasi Samkshema Parishad (AVSP), Girijana Samkshema Parishad (GSP), Adivasi Students JAC and Adivasi Kondareddi Sangham took out a huge rally in the temple town on Saturday denouncing the Centre's reported "merger move." The protesters set ablaze the effigies of the political parties including the Congress, TRS, BJP, TDP and other organisations in protest against the alleged failure of the parties to safeguard the interests of Adivasis. It is mandatory on the part of the government to take the consent of the gram sabhas to make changes of any in the boundaries of the agency areas under the Fifth Schedule as per the PESA Act, said Sode Chalapathi, district president, GSP."

People believe that this idea in fact pre-empts inter-state riparian and submergence issues, with associated legal battles between the new States and aims to remove these hassles so that the dam works can proceed without a glitch, since too much money has been pumped in already.

29th May, 2014 - An Extraordinary Gazette on Polavaram

The President of India, Mr. Pranab Mukherjee, promulgates The Andhra Pradesh Reorganisation (Amendment) Ordinance, 2014 (No. 4 of 2014) - released by the Ministry of Law and Justice (Legislative Department), dated "New Delhi, the 29th May, 2014/Jyaistha 8, 1936 (Saka)"

"WHEREAS Parliament is not in session and the President is satisfied that the circumstances exist which render it necessary for him to take immediate action;

NOW, THEREFORE, in exercise of the powers conferred by clause (1) of article 123 of the Constitution, the President is pleased to promulgate the following Ordinance:—

1. (*1*) This Ordinance may be called the Andhra Pradesh Reorganisation (Amendment) Ordinance, 2014.

2. Amendment of section 3 of Act 6 of 2014.

In the Andhra Pradesh Reorganisation Act, 2014, in section 3, for the words, brackets, letters and figures "Khammam (but excluding the revenue villages in the Mandals specified in G.O.Ms. No. 111 Irrigation & CAD (LA IV R&R-I) Department, dated the 27th June, 2005 and the revenue villages of Bhurgampadu, Seetharamanagaram and Kondreka in Bhurgumpadu Mandal)", the words and brackets "Khammam (but excluding the Mandals of Kukunoor, Velairpadu and Bhurgampadu but not including its revenue villages of Pinapaka, Morampalli Banzar, Bhurgampad, Nagineniprolu, Krishnasagar, Tekula, Sarapaka, Iravendi, Mothepattinagar, Uppusaka, Sompalli and Nakripeta under the Palvancha Revenue Division, and the Mandals of Chintoor, Kunavaram, Vararamachandrapuram and Bhadrachalam but not including the revenue village of Bhadrachalam under the Bhadrachalam Revenue Division)" shall be substituted."

***

Finally, while the larger games are played, I put an end to the chronicle thus far. I leave with a wish for Godavari to come like she does, stay like she does, take herself back, be the way she is meant to be – with her anger, love, pain and joy intact in the seasons of their relative being – living with the people who may be who they wish to be, in their respective freedoms. And as I leave, just a thought. One cannot predict the future, but one of these days, I want to go back to Godavari and this time, if I get to stay at the Kunavaram Forest Rest House – if still not abandoned forever – I want to make my peace with the lizards, that tree frog on that door and even the bandicoots on that false ceiling.

*****

A Chance Journey - Bhadrachalam upwards (2012)

Bhadrachalam Temple steps leading to Sabari-Godavari (or vice versa)

A poster of the meeting held at Bhadrachalam on Polavaram

The garbage bin at the entrance of the Ananda Nilayam Residential Hostel for Boys at Bhadrachalam

The dilapidated building of the residential hostel

The Lost Farmer - SC farmer Varaka Muthaiah of Valairpadu at the bus-stop near Bhuvanagiri (Kukunur)

The Children of the Ananda Nilayam Hostel - Sainatha, Satish and others

All the children gather for a group photo outside their hostel (a building likely to go under the dam waters, if it is built)

The Bhadrachalam stringer's collection- a newspaper clip of the temple elephant struggling in the waters from the enabhai-aaru Godavari (1986 Godavari - this and next 2 from the stringer's collection)

The Rama temple's 'gopuram' under water - 1986 Godavari (the stringer's collection)

Godavari waters reach up to the bridge across Sabari-Godavari

The colourful weekly Chintur santa (a fair visited by people from both sides of the AP/Telangana-Chhatisgarh border)

Fruits from nearby villages at the fair

Everyday essentials at the santa

Confluence of rivers Sileru-Kolleru-Sabari at Kunta, Chhattisgarh

Tribal children at the Tribal Welfare school at Kunta (which is in the Polavaram dam submergence zone)

Notes

1. This was a grand plan of the AP government to curb unnecessary expenditure on government schools with fewer (they claim) children; merged schools according to this plan would have better infrastructure and more teachers and more children and make it easier for the government to disburse benefits like mid-day meal and other schemes.
2. The Government's health scheme which gives free consultation and bed but medicines, tests, have to be done with their own money, without any concession.
3. Got a chance assignment, thanks to Prof. Dandekar, then at IRMA, to look at the issue of 'internal displacement' with specific focus on the Gottikoyas (or the Muriyas, some of whom I had met back in 2007) now settled on the then Andhra Pradesh-Chhattisgarh border. I used the opportunity to record any new developments on the Polavaram dam issue. I got a chance to also look at the blurred lines of Sabari-Godavari at the bordering village Kunta. The timing of my visit meant some extra suspicious looks and stares as to whosoever was entering this zone. Journalists and political activists are especially regarded with suspicion.
4. Accident within a bus, with its own lessons in immobility, for two months, and with a memento!
5. A senior economist and professor at Center for Economic and Social Sciences, Hyderabad.
6. Some farmers' lands have been leased back to the government who then leases it out to contractors for the dam works, either for dumping debris or some lease out land to temporary cultivators. It seems like a strange situation.
7. *Naa pani Godaari* as an idiom is used by some communities in a situation of hopelessness, likened to drowning in Godavari waters, perhaps.
8. He still calls, once in a while, to tell me his story, as also to simply enquire after me. In spite of the fact that our meeting that afternoon must have lasted barely fifteen minutes. His son works as a Security staff in a private company in Hyderabad; Muthaiah's eye is still a problem; no money for surgery.
9. Which I did not end up doing; the phone gave way, the mobile number got lost; life took other turns and the story on the social welfare hostel never got written; adding to all other of life's guilts. I still remember Sainatha, Saikrishna, Deva, Sathish, and all others on that laughter-filled, joyous evening. And the next morning of bidding goodbye, hoping the hostel and the concept, will survive.
10. What was formerly the Bhadrachalam Paper Board (BPL) was taken over by the ITC Company several years ago, but local people still refer to it as 'BPL'.
11. Later on, many recorded moments of Kunta trip – names, places – also disappeared with the loyal mobile phone developing its own snag!
12. I had a chance to revisit Bhadrachalam and surrounding villages in early 2014 as well (to write about the destruction (by the Forest department) of some Gottikoya houses in the village Chimilivagu on which a petition had been filed in the State Human Rights Commission (Andhra Pradesh), since even a government school had been destroyed. I stayed again at the Sivananda Asramam, gave pictures to Vijayalakshmi, the attendant-supervisor and her young son; met the children of the Hostel by Sivananda Asramam. Their old

hostel is being repaired and they have been housed in a temporary house nearby. I gave them prints of their pictures; made them very happy. They took a picture of mine as well, this time. Mr. NSC Bose was not there this time round; his job is now undertaken by one Mr. Krishnamurthy. And I could not meet the stringer at the kiosk, whose name I had forgotten – which I usually never do.

Some Numbers of Those to be Dislocated, Lost

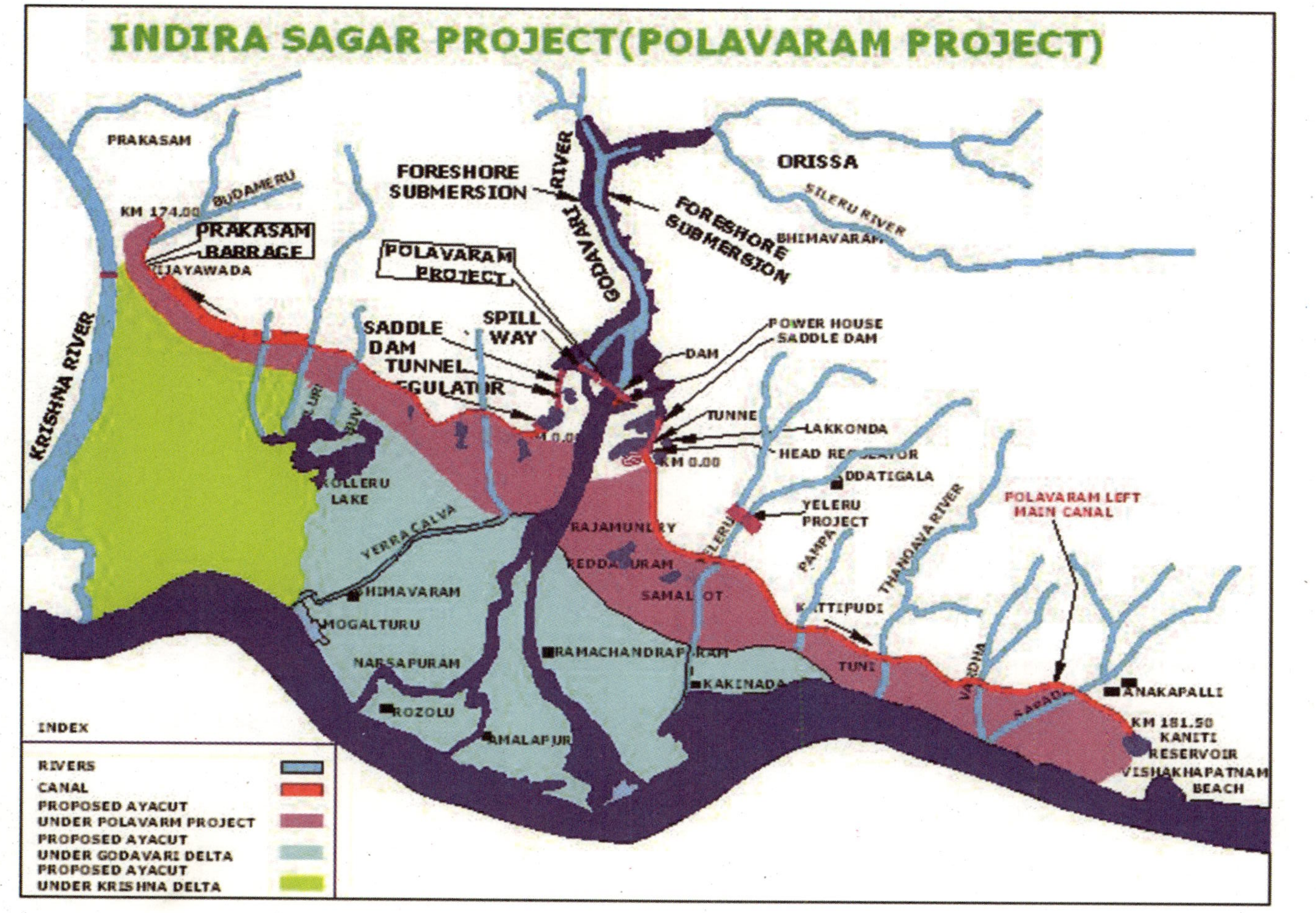

INDIRA SAGAR PROJECT(POLAVARAM PROJECT)
PRAKASAM
BUDAMERU
KM 174.00
PRAKASAM BARRAGE
FORESHORE SUBMERSION
GODAVARI RIVER
ORISSA
SILERU RIVER
BHIMAVARAM
FORESHORE SUBMERSION
POLAVARAM PROJECT
KRISHNA RIVER
SADDLE DAM
SPILL WAY
TUNNEL
DAM
POWER HOUSE
SADDLE DAM
LAKKONDA
KM 0.00
KOLLERU LAKE
YERRACALVA
RAJAMUNDRY
SAMALKOT
YELERU PROJECT
PAMPA
THANDAVA RIVER
POLAVARAM LEFT MAIN CANAL
KATTIPUDI
MOGALTURU
NARSAPURAM
KAKINADA
TUNI
ANAKAPALLI
KM 181.50
KANITI RESERVOIR
VISHAKHAPATNAM BEACH
ROZOLU
INDEX
RIVERS
CANAL
PROPOSED AYACUT UNDER POLAVARM PROJECT
PROPOSED AYACUT UNDER GODAVARI DELTA
PROPOSED AYACUT UNDER KRISHNA DELTA

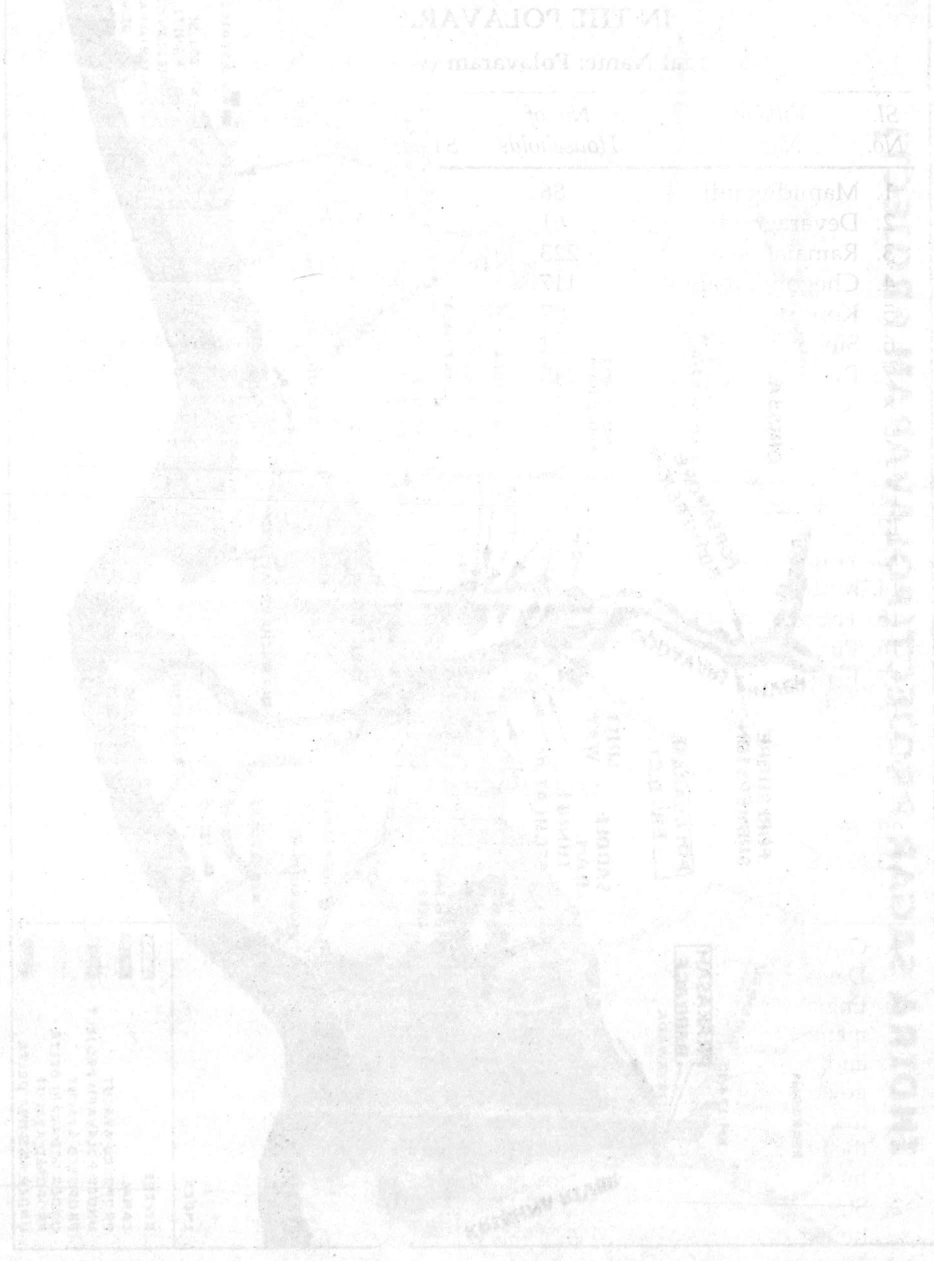

1. "MANDAL WISE LIST OF SETTLEMENTS"[1] TO BE SUBMERGED IN THE POLAVARAM PROJECT

Mandal Name: Polavaram (West Godavari District)

Sl. No.	*Village Name*	*No. of Households*	*No. of ST Households*[2]	*Total Population*	*ST Population*
1.	Mamidigondi	86	84	353	342
2.	Devaragondi	61	61	213	213
3.	Ramaiahpeta*	223	5	994	20
4.	Chegondapally	117	115	450	448
5.	Kondrukota*	87	46	321	165
6.	Singanapally*	94	74	363	272
7.	Pydipaka*	243	16	1033	58
8.	Kothuru*	104	0	411	0
9.	Kothamamidigondi	32	32	108	108
10.	Madhavapuram	42	26	146	85
11.	Thallavaram*	129	124	498	478
12.	Gajulagondi	64	64	208	208
13.	Wadapally	77	63	270	216
14.	Mulakalagudem	67	63	234	222
15.	Thutigunta*	131	92	474	348
16.	Pudakalagondi	3	3	15	15
17.	Erravaram	54	52	212	206
18.	Pallapuru	35	34	132	128
19.	Paidikulamamidi	32	32	123	123
20.	Sarugudu	24	24	83	83
21.	Tekuru*	78	34	296	132
22.	Bornagondi	11	11	47	47
23.	Cheeduru*	54	49	215	199

1. Government of Andhra Pradesh, I&CAD (Irrigation and Command Area Development) Department, Note on Indira Sagar (Polavaram) Project, Chief Engineer, Indira Sagar (Polavaram) Project, Dowlaishwaram. Some of the Districts mentioned here may change post-formation of two separate States of Telangana and Andhra Pradesh. The information here is from the report circulated by the government in years 2005-06. I retain the district names as they are within the mandals as given in the said report. But details of which Mandals will be part of the new Andhra Pradesh state, are already given in the President's Promulgation mentioned in the Epilogue.
2. Strangely enough this list does not provide numbers of SC, BC and other Caste households.

* Marked by me. These villages in all these tables have a substantial number of non-tribal upper caste households, and BCs, mainly from the fish-hunting communities (or fishworkers), and many SCs. That should explain the numbers (few STs) in Pydipaka, Ramaiahpeta, etc.

24. Sivagiri*	95	84	330	288
25. Koruturu*	62	60	216	210
26. Sirivaka	39	39	138	138
27. Sriramavaram	16	16	49	49
28. Thotagondi	39	39	137	137
29. Kamavaram	3	A (?)[3]	12	A
	2102	**1342**	**8081**	**4938**
		(64) (?)[4]		**(61) (?)[5]**

3. Not sure what is meant by 'A'. No explanation provided in the report.
4. Not sure what does (64) mean. No explanation is given for it.
5. Same question as above.

Mandal Name: Devipatnam (East Godavari District)

Sl. No.	Village Name	No. of Households	No. of ST Households	Total Population	ST Population
1.	Anguluru	70	30	277	111
2.	Pudipalli**	180	1	776	2
3.	Paragasanipalli[6]	43	32	196	153
4.	Gonduru	130	80	510	295
5.	A.Veeravaram	29	5	128	17
6.	Veeravaram lanka[7]	57	0	223	0
7.	Nagalapamu	72	57	245	197
8.	Dandangi	74	32	309	131
9.	Gangampalem	63	62	251	240
10.	Bodigudem	16	16	76	76
11.	Ravilanka	43	38	174	145
12.	Devipatnam	340	12	1398	34
13.	CH.Ramayyapeta	133	36	562	148
14.	Koyala Veeravaram	29	27	114	104
15.	Gubbalampadu	21	21	79	79
16.	Seetharam[8]	39	33	148	130
17.	Damanapalli	61	49	272	209
18.	Ganugulagondi	15	15	60	60
19.	Yenugulagudem	50	50	202	202
20.	Lingavaram	15	15	52	52
21.	Toyyeru **	253	8	1151	29
22.	Agraharam	31	31	121	121
23.	Mulapadu	43	38	179	153
24.	Manturu	137	81	510	287
25.	Madipalli	65	47	246	174
26.	Penikilapadu	38	35	154	142
27.	Mulametta	41	41	150	150
28.	Mettaveedhi	60	59	219	218
29.	Kachuluru	58	56	223	217
30.	Suddakonda	4	4	21	21
31.	Kothagudem	31	31	128	128
32.	Gonduru	31	26	110	100
33.	Kondamodalu	153	71	583	267
34.	Talluru	70	65	277	261
35.	Teliperu	50	44	196	170
36.	Nadipudi	28	26	106	98

** Number of non-tribal upper castes and other castes more than STs here.

6. In some records, and in popular use, it is Paragasanupadu.
7. As the name suggests, this is a 'lanka' (islands formed from alluvial deposits) land, which produces rich crop of tobacco (usually that is the crop grown) post Godavari's coming.
8. Popularly referred to as Seetharamam.

37. Somalapadu	37	37	153	153
38. Kethanapadu[9]	37	37	128	128
39. Kokkirigudem[10]	53	53	204	204
40. Metlagudem	57	56	178	174
41. Tadivada	44	44	169	169
42. Peddagudem	65	65	218	218
	2866	**1566 (54)**	**11476**	**5967 (52)**

Mandal Name: Bhadrachalam (Khammam District)

Sl. No.	*Village Name*	*No. of Households*	*No. of ST Households*	*Total Population*	*ST Population*
1.	Gogubaka	105	57	441	252
2.	Gommukoyagudem	110	109	472	464
3.	Kapavaram	91	0	401	0
4.	Gollagudem	56	2	244	13
5.	Tripuranthaveedu	55	2	305	21
6.	Sitapuram	124	1	603	6
7.	Rachagompalli	25	7	114	29
8.	Gouridevipeta	66	13	274	45
9.	Nandigama	311	2	1320	7
10.	Murumuru	125	90	593	444
11.	Gommu Murumuru	49	0	247	0
12.	Kothagudem	76	74	353	343
13.	Nandigamapadu	66	0	309	0
		1259	**357(28)**	**5676**	**1624(28)**

9. Also referred to as Kathanapalli in popular reference, and in some other government records.
10. Or, Kokkarigudem.

Mandal Name: V.R. Puram (Khammam District)

Sl. No.	*Village Name*	*No. of Households*	*No. of ST Households*	*Total Population*	*ST Population*
1.	Choppalle	128	77	535	320
2.	Ramavaram	86	71	393	310
3.	Somulugudem	67	64	260	250
4.	Ramavarampadu	55	43	210	168
5.	Koppalle	41	41	163	163
6.	Ravigudem	35	24	136	91
7.	Borigudem	11	11	51	51
8.	A. Venkannagudem	99	96	393	381
9.	Prathipaka	60	60	260	260
10.	Tustivarigudem	44	44	211	211
11.	Gundugudem	28	16	106	60
12.	Chintaregupally	202	50	878	211
13.	Kannayyagudem	31	31	125	125
14.	Sunnamvarigudem	51	48	199	185
15.	Nuthigudem	46	45	241	237
16.	Mettagudem	4	4	11	11
17.	Gurrampeta	92	92	425	425
18.	Ummadivaram	174	168	678	648
19.	Annavaram	103	101	474	469
20.	Rekapally	284	108	1145	439
21.	Vaddegudem	406	19	1727	85
22.	Dharmatallavaram	75	75	328	328
23.	V.R.Puram	587	63	2427	236
24.	Thotapalli	106	105	442	438
25.	Rajupeta	29	0	118	0
26.	Rajupeta Col.[11]	238	1	1001	6
27.	Seetampeta	90	2	412	10
28.	Sri Rama Giri	118	36	499	135
29.	Chokkannapally	98	93	361	342
30.	Kotturu	6	6	20	20
31.	Kalthunuru[12]	63	59	230	214
32.	Jeediguppa	100	77	428	334
33.	Isunuru	24	14	70	43
34.	Ravigudem	39	37	151	142
35.	Mulakalapally	66	36	267	149
36.	Mutyalamma Gandi	17	17	65	65
37.	Bhimavaram (Khammam)	2	2	9	9
38.	Ippuru	37	32	175	159
39.	Kotaragommu	53	53	209	203
40.	Pochavaram	79	71	281	253
41.	Tummileru	75	74	296	292

11. Colony.
12. Also referred to as Kalthanuru.

42. Kondepudi	17	16	65	61
43. Kolluru	30	29	116	111
44. Gonduru	25	25	103	103
45. Narsingpeta	75	66	331	303
	4096	**2202 (53)**	**17025**	**9062 (53)**

Mandal Name: Kunavaram (Khammam District)

Sl. No. *Village Name*	*No. of Households*	*No. of ST Households*	*Total Population*	*ST Population*
1. China Polipaka+	64	0	298	0
2. Pedda Polipaka+	31	1	144	3
3. Gunduvarigudem	24	24	99	99
4. Pochavaram	104	40	383	131
5. Pochavaram Colony+	119	37	510	160
6. Kachavaram	221	0	960	0
7. Dungutta	67	66	281	277
8. Kollapadu	80	77	351	340
9. Tekubaka+	150	22	632	90
10. Venkataypalem	94	52	434	252
11. Potlavai+	68	67	235	234
12. Gommu	34	0	143	0
13. Ayyavarigudem+	50	42	213	177
14. Pedda Narasingapeta	54	49	248	225
15. Chinna Narasingapeta	73	53	293	218
16. Karakagudem+	59	57	240	232
17. Ayyavarigudem+	44	25	189	114
18. Kondayagudem	108	0	405	0
19. Gommugudem	26	0	102	0
20. Kumaraswamygudem	46	0	174	0
21. Marrigudem+	113	74	480	294
22. Palluru	199	66	846	271
23. Jaggavaram	42	34	139	103
24. Jaggavaram Colony+	201	65	841	255
25. Suchirevulagudem+	49	0	194	0
26. Suchirevula	726	11	3075	40
27. Kunavaram+	449	85	1820	350
28. Tekulaboru+	41	38	164	149
29. Kondrajupeta	62	62	263	263
30. Pusugudem	146	141	665	638
31. Jinnelagudem	26	26	98	98
32. Bandarugudem	157	133	581	483
33. Pedda Arkuru	194	188	755	730
34. Repaka+	5	0	14	0

+ These villages have a large number of SC and BC communities; SCs are mostly landless agricultural workers.

35. Musaragudem	95	92	393	382
36. Bhagavanpuram	143	75	540	307
37. Mulluru	118	98	471	391
38. Kuturu	158	144	635	575
39. Abhicherla	103	101	418	410
40. Lingapuram	16	12	90	72
41. Koderu	26	26	112	112
42. Tallagudem	96	63	387	268
43. Bajjavarigudem	12	12	46	46
44. Ravigudem	144	144	607	609
45. Regulapadu	74	74	305	305
46. Vekannagudem	160	160	619	619
47. Pandirajupally	44	44	201	201
48. Wolfordpeta	145	141	657	644
49. Sabari Kothagudem++				
	5260	**2721** **(51)**	**21753**	**11165** **(51)**

Mandal Name: Chintoor (Khammam District)

Sl. No.	*Village Name*	*No. of Households*	*No. of ST Households*	*Total Population*	*ST Population*
1.	Kalleru	130	116	537	478
2.	Chidumuru	183	165	863	790
3.	Kuyuguru	131	88	595	394
4.	Chatti	354	270	1490	1121
5.	Veerapuram	122	122	525	525
6.	Chintoor	561	134	2442	578
7.	Bandarugudem	52	49	203	195
8.	Kummuru	228	225	980	970
9.	Gorrelagudem	13	13	68	68
10.	Mamillagudem	108	107	488	482
11.	Chuturu	143	142	635	632
12.	Mukunuru	118	101	444	382
13.	Tummarigudem	42	42	180	180
14.	Agraharapu Koduru	226	211	944	875
15.	Jallivarigudem	37	37	162	162
16.	Ulumuru	116	96	486	395
17.	Mallithota	145	140	612	591
		2709	**2058** **(76)**	**11654**	**8818** **(75)**

++ The original government record shows this village but does not give data. I retain it as it is.

Mandal Name: Burgampadu (Khammam District)

Sl. No.	Village Name	No. of Households	No. of ST Households	Total Population	ST Population
1.	Sridhara	133	2	672	6
2.	Veleru	168	44	788	193
3.	Rayeegudem	104	0	525	0
4.	Venkatapuram	175	6	719	31
5.	Gumpanpally	36	0	175	0
6.	Buvangiri	43	40	172	161
7.	Alligudem	89	45	436	196
8.	Ganapavaram	87	0	454	0
9.	Ibrahimpeta	120	0	547	0
		955	**137 (14)**	**4488**	**587 (13)**

Mandal Name: Kukunur (Khammam District)

Sl. No.	Village Name	No. of Households	No. of ST Households	Total Population	ST Population
1.	Tondipaka	115	15	599	76
2.	Mittagudem	131	0	596	0
3.	Banjargudem	149	104	682	463
4.	Amaravaram	427	118	1841	543
5.	Komatlagudem	126	70	529	279
6.	Upperu	248	25	1215	117
7.	Koyagudem	53	52	228	225
8.	Reddygudem	29	29	136	136
9.	Damaracharla	273	1	1197	4
10.	Yellappagudem	31	31	108	108
11.	Cheeravally	188	0	856	0
12.	Kothur	29	29	120	120
13.	Marripadu	56	24	246	101
14.	Madharam	168	62	707	250
15.	Kowdinyamukti	35	17	167	81
16.	Vinjaram	185	0	865	0
17.	Mutyalammapadu	102	0	445	0
18.	Kondapalli	230	26	1075	127
19.	Koyagudem[13]	35	35	147	147
20.	Maredubaka	214	85	933	333
21.	Kiwaka	219	0	896	0
22.	Kummarigudem	61	60	242	238
23.	Kukkunoor	608	13	2379	33
24.	Ramasingaram	418	5	1723	17
25.	Kistaram	64	59	295	279
26.	Kurlapadu	44	42	165	156
27.	Lankalapally	23	22	113	107

13. Mentioned twice, with different statistics.

28. Isakapadu	59	43	246	185
29. Uppara Maddigatla	70	9	321	41
30. Sita Ramachandrapuram	88	0	355	49
31. Gommugudem	90	0	419	0
32. Dacharam	211	83	1004	413
33. Kothur	35	0	144	0
34. Bestagudem	184	14	861	61
	4998	**1082 (21)**	**21855**	**4689 (21)**

Mandal Name: Velairpadu (Khammam District)

Sl. No.	*Village Name*	*No. of Households*	*No. of ST Households*	*Total Population*	*ST Population*
1.	Rudramakota	266	2	1139	6
2.	Pochirevula[14]	143	83	664	340
3.	Patha Pochirala	63	62	303	301
4.	Lachigudem	71	60	309	269
5.	Pepallagommu	238	50	1054	225
6.	Nadimigommu	50	0	183	0
7.	Maddikotla	81	75	379	361
8.	Velerupadu[15]	314	28	1328	116
9.	Nagalagudem	107	49	420	189
10.	Tatiluragommu	36	3	159	18
11.	Bhudevipeta	88	0	376	0
12.	Sriramapuram	20	10	87	46
13.	Jagannadhapuram	271	83	1134	339
14.	Sagarpally	113	110	486	478
15.	Korrajulagudem	33	33	145	145
16.	Tirumalapuram	34	13	133	47
17.	Kannaisutta	183	168	716	648
18.	Nallavaram (old city)	57	0	214	0
19.	Nallavaram Colony	79	2	318	11
20.	Kothuru	79	76	300	288
21.	Chigurumamidi	143	63	655	275
22.	Bollapally	68	67	234	232
23.	Yadavally	66	66	296	296
24.	Borradigudem	29	28	103	98
25.	Katkoor	120	52	480	207
26.	Tekuru	75	74	299	295
27.	Kasaram	71	71	316	316
28.	Khoida[16]	132	106	549	428
29.	Tallagondi	11	11	44	44
30.	Pusagondi	22	22	94	94

14. Also referred to as Puchirevula.
15. Also spelt Valairpadu.
16. Also referred to as Koida/Koyda.

31. Tekupally	47	42	183	169
32. Perantalapally	35	35	1541	151
33. Chittipeddypalem	44	44	185	185
34. Siddaram	37	37	126	126
35. Chintalapadu	7	7	22	22
36. Padamatimetta	68	66	332	321
37. Burrathogu	149	137	686	626
38. Turpumetta	40	40	205	205
39. Kakisanoor	63	61	219	216
	3553	**1936 (54)**	**15026**	**8133 (54)**

2. LIST OF FLORA AND FAUNA IN BHADRACHALAM SOUTH (AND NORTH) AND PALONCHA DIVISIONS

Most of the Polavaram dam submergence will be in Bhadrachalam South Division of the Forest (including parts of Papikonda National Park). The Forest area is managed under three divisions – Bhadrachalam North (Venkatapuram and Bhadrachalam Forest Range); Bhadrachalam South (V.R. Puram and Kunavaram Range) and Paloncha (Kukunur and Palavancha Range). Some forest area under Kukunur and Bhadrachalam range will also be submerged.[17]

[Note: This list is *not* the list of 'affected' species, but I make the assumption (or I contend) that since these species are officially part of the larger forest habitat that will be submerged by the dam, they are likely to be either destroyed or impacted in one way or the other due to the project.]

List of Flora in Bhadrachalam (South) Division[18]

Trees

	Scientific Name	*Local name*
1.	Acacia nilotica	Nalla tumma
2.	Acacia leucophloe	Tella tumma

17. Personal communication, DFO Bhadrachalam South, July 17th, 2012.
18. Source: Bhadrachalam (South) DFO Office (17th July 2012); I have retained the list in its original form. Some local names did seem dicey, but I retain them since this is as an official record. I finally got information I was looking for, thanks to a new DFO-Bhadrachalam South Division (who had joined a few days before), Mr. S. Kishan Das. The list of Fish species though comes from a "Baseline Environmental Status Prepared by SPB-PC &VIMTA Lab (as applicable for Bhadrachalam South, Paloncha and Bhadrachalam North Divisions) for the ITC Limited-PSPD Expansion Plan; since the forests are contiguous with the Polavaram dam submergence area, I take the liberty to share this list which concerns the same expanded forest habitat. I thank Mr. Subhani Syed of the organisation Asha, Chintur (Khammam district) who secured the ITC EIA through RTI, for sharing the same with me (in 2010).

3.	Acacia sundra	Sundra
4.	*Adina cordifolia*	Bandaru
5.	*Aegle marmelos*	Maredu
6.	*Ailanthes malabarica*	Mattipala
7.	*Alangium lamarkii*	Udgu
8.	*Albizia amara*	Nalla regu
9.	*Albizia lebbeck*	Dirisinam
10.	*Albizia procera*	Tella chindugu
11.	*Alstonia scholaris*	Edukula pala
12.	*Anogeissus latifolia*	Chirumanu yelama
13.	*Antidesma ghaesembilica*	Tella pulusu, Jonna pulusu
14.	*Azadirachta indica*	Vepa
15.	*Bassia latifolia (Madhuca indica)*	Ippa
16.	*Bauhinia racemosa*	Ari
17.	*Bauhinia variegata*	Mandari-devakanchanamu
18.	*Bombax ceiba*	Buruga
19.	*Boswellia serrata*	Anduga
20.	*Bridelia montana*	*pantuga-ponthangi*
21.	*Bridelia retusa*	Koramaddi-koramanu
22.	*Buchanania angustifolia*	Sarapappu-murali
23.	*Butea monosperma*	*Moduga*
24.	*Caryota urens*	Ithas-jeelugu
25.	*Cassia fistula*	Rela
26.	*Chloroxylon sweitenia*	Billugu
27.	*Cleistanthes collinus*	Bankakkairi
28.	*Dalbergia latifolia*	Virrigudu-jutiga
29.	*Dalbergia paniculata*	Pacharai
30.	*Dillinei pentagyna*	Revadi
31.	*Diospyros melanoxylon*	Tumki
32.	*Emblica officinalis*	Usirikai
33.	*Erythroxylon monogynum*	Devadaru
34.	*Feronia elephantum*	Velaga
35.	*Ficus benghalensis*	Marri
36.	*Ficus hispida*	Bodamarari
37.	*Flacourtia cataphrata*	Kanregu
38.	*Flacourtiia ramonchi*	Puthkithada
39.	*Gardienia gummifera*	Manchi dikki karingua
40.	*Gardienia latifolia*	*pepda bikki karingua*
41.	*Gardenia lucida*	Yerra bikki
42.	*Gardenia turgida*	*kukkaelka*
43.	*Garuga pinnata*	Garga-garugu
44.	*Gmelina arborea*	Gummadi
45.	Grewia tiliaefolia	Thada
46.	*Hardwickia binata*	*Yepa*
47.	*Holarrhena antidycenterica*	Pala
48.	*Holoptelea integrefolia*	Nauli
49.	*Hymenodictylon excelsum*	Dudidda chepida
50.	*Kydia calcina*	Konda pothari

51.	*Lagerestroemia parviflora*	Chennangi
52.	*Limonia acidissima*	Torralelaga
53.	*Mangifera indica*	Mamida
54.	*Melia azadirachrta*	Neem- thurakavepa
55.	*Memyceylon edule*	Alli
56.	. *Mimusops hexandra*	Pala
57.	*Mitrgyna parviflora*	Battaganapu
58.	*Morinda tinctoria*	Thogaru
59.	*Nyctanthes arborstristis*	Pogada
60.	*Odina wodier*	Gumpena
61.	*Ougenia dalbergiodes*	Chikkudu
62.	*Polyalthia cerasoides*	Chilaka duduka
63.	*Pongamia glabra*	Kanuga
64.	*Premna tomentosa*	Navaru
65.	*Pterocarpus marsupium*	Yegisi
66.	*Phyllanthus emblica*	Usiri
67.	*Randia dumatorum*	Manga
68.	*Sapindus emerginatus*	Kunkudu
69.	*Scheleichera oleosa*	Pusuga
70.	*Screbera sweitenoides*	Mokkapu
71.	*Soymida fabrifuga*	Somi
72.	*Steruclia urens*	Yeuruponku thabasi
73.	*Strychnos nux-vomica*	Mushti
74.	*Strychnos potatorum*	Chilla
75.	*Sygygium cumini*	Neredu
76.	*Tamarindus indica*	Chintha
77.	*Tectona grandis*	Teku
78.	*Terminalia arjuna*	Tella-maddi
79.	*Terminalia bellerica*	Thani-thadra
80.	*Terminalia chebula*	Karaka
81.	*Terminalia tomentosa*	Maddi
82.	*Wrightia tinctoria*	Konda tangedu
83.	*Zyzyphus jujuba*	Regu
84.	*Zyzyphus xylopyrus*	Gotti-gotiki

Shrubs

1.	*Bridelia hamiltoniana*	Pachotkam
2.	*Cassia auriculata*	Thangedu
3.	*Celastrus senegalensis*	Danti
4.	*Crotalaria alata* —	
5.	*Dodonea viscosa*	Pulivalli- junglinar
6.	*Flacourtia separia* —	
7.	*Grewia hirsuta*	Chittijana
8.	*Gymniphorea montana*	Dhanti
9.	*Helictris isora*	Adavi -chamanthi
10.	*Indigofera pulchella*	Kondanthita
11.	*Indigofera tinctoria* —	

12.	*Pavetta indica*	Papidi
13.	*Phoenix faringefera*	Pasupu
14.	*Vitex negundo* —	
15.	*Woodfordia floribunda*	Jaji
16.	*Zizyphus numalarica*	Rhamnacex

Climbers and Creepers

1.	*Abrus precatorius*	Gulibinda
2.	*Acacia concina*	Sheekkakai
3.	*Atylosia albicana*	Adda teega
4.	*Bauhinia vahlii*	Theegamoduga-addatheega
5.	*Butea superba*	Belapalas
6.	*Calycopteris floribunda*	*pippinda theega*
7.	*Jasmium angustifolium*	Adavimalli
8.	*Jasmium rigidum* —	
9.	*Smilax aspida*	Kunmaari theega
10.	*Spetholobus roburghii*	Modugu theega
11	*Zyzyphus oenoplica*	Chirathalli

Herbs

1.	*Achyrantehs aspera*	Otthareni
2.	*Ageratusm conyzoides* —	
3.	*Euphorbia hirta* —	
4.	*Indigofera endosphylla*	Cheragadem
5.	*Vetveria zizanoides*	Vetiveru
6.	*Zingiber officinalis*	Allam

Bamboos

1.	*Bambusa arundanacea*	Mullam bangu
2.	*Dendrocalamus strictus*	Sadanam

Grasses

Saccharum spontaneum	Baru
Ischeaemum sp	Kache-kopri
Andropogon contortus	Yedagaddi
Andropogon schoenuathum	Chipurugaddi
Cyaodon dactylon	Darba gaddi

List of Fauna in Bhadrachalam (South) Division

Wild Animals

	Common Name	*Scientific Name*	*Local Name*
1	Rhesus Macacue	*Macace mulatta*	Kothi
2	Common Langur	*Presbytis entellus*	Kondamuchu
3	Tiger	*Punthera tigris*	Peddapuli
4	Panther	*Panther pardus*	Chiruthapuli
5	Leopard Cat	*Felis bengalensis*	Chiruthapuli

6	Jungle cat	*Felis chaus*	Jungupilli
7	Small Indian civet	*Viverricala indica*	Mullapandi
8	Palm civet	*Paradoxurue hermaphroditus*	Manupilli
9	Common mongoose	*Herpestes edwardsi*	Mungisa
10	Striped hyena	*Hyena hyena*	Kondresu
11	Jackal	*Canis aureus*	Nakka
12	Indian fox	*Vulpes bengalensis*	Gunta nakka
13	Wild dog	*Cuon alpinus*	Resukkuka
14	Sloth bear	*Melurus ursinus*	Yelugubanti
15	Common otter	*Lutra lutra*	Neerukukka
16	Smooth Indian otter	*Lutra presicillata*	Neerukukka
17	Honey badger	*Mellivora capensis*	Bigu khawar
18	Indian tree shrew	*Anathana ellioti*	Chunchu
19	Flying fox	*Pteropus giganteus*	Kabothi pakshi
20	Indian flying squirrel	*Peturista petaurista phillippensis*	Egire udhatha
21	Indian giant squirrel	*Ratafa indica*	Bettu udhataha
22	Palm squirrel	*Funambulus palmarum*	Udutha
23	Porcupine	*Hystrix indica*	Mullapandi
24	Hare	*Lepus nigricollis*	Chevulapilli
25	Gaur	*Bosgaurus*	Adavi dunna
26	Chinkara	*Gazella gazella*	Gaddi meka, Bhumajinka
27.	Black Buck	*Antilope carvicpra*	Krishna Jinka
28.	Chowasingha	*Tetracerus quadricornis*	Kondagorre
29.	Nilgai	*Boselaphus tragocamelus*	Manubothu
30.	Sambhar	*Cervus unicolor*	Kanusu
31.	Spotted deer	*Axis Podala*	duppi
32.	Barking deer	*Munticus munitjak*	Ariche duppi
33.	House Deer	*Tragalus meminna*	Mushika Duppi
34.	Wild boar	*Susscrofa*	Adavi pandi
35.	Indian Pangolin	*Manis crassis audata*	Adavi valuga
36.	Indian mugger crocodile	*Crocodilus palustris*	Musali
37.	Indian pond terrapin	*Melanochelys trijuga*	Neeti tabelu
38.	Starred Tortoise	*Geochelone elegaus*	Metta Tabelu
39.	Common garden lizard	*Calotes verricolor*	Tonda
40.	Frost calotes	*Calotte rouxi*	Tonda
41.	Fan throated lizard	*Sitana ponticeriana*	Tonda
42.	Common skink	*Mabuya charinata*	Nallikeesu
43.	Snake skink	*Riopa punctata*	Nallikeesu
44.	Chameleon	*Chameleon zeylanicus*	Usaravelli
45.	Monitor lizard	*Varanus benalensis*	Udumu
46.	Python	*Python molvru*	Kondachiluva
47.	Common rat snake	*Ptylas mucosus*	Jerripotoo
48.	Keel Back	*Exnochropis piscator*	Neeti Pamu
49.	Buffstuped keel back	*Anpliesma stolata*	Wanapamu
50.	Olivaceous keel back	*Atretium schestorum*	Nalla Wahlagille pamu

51.	Krait	*Bungarus cauruleus*	Katla pamu
52.	Banded krait	*Bungarus fascdciatus*	Bangarun pamu
53.	Cobra	*Najanaja*	Nagupamu
54.	Viper	*Viper russeli*	Katukarekula poda

Birds

	Common Name	***Scientific Name***	***Local Name***
1	Little grebe	*Podiceps ruficollis*	Buda bunga
2	Cormorant	*Phalacrocorax carbo*	Neeti kaki
3	Indian shag	*Phalacrocorax fusecollis*	Neeti kaki
4	Little Cormorant	*Phalacrocorax niger*	Neeti kaki
5	Darter	*Ashina rufa*	Pamutala neeti kaki
6	Giant Heron	*Ardea goliath*	Pedda budidarangu konga
7	Grey Heron	*Ardea cinera*	Nalla kalla konga
8	Pond heron	*Ardeola gravil*	Guddi konga
9	Cattle egret	*Babulcus ibis*	Tella konga
10	Open bill stork	*Anastomus oscitan*	Nathagottu konga
11	Night heron	*Nycticorax nycticorax*	NIsachara konga
12	White Necked stork	*Ciconia episcopus*	Tallameda konga
13	Black necked stork	*Ephippiorhyachus asiaticus*	Nellamada konga
14	Adjutant stork	*Leptonites javanicus*	Adjutant konga
15	White Ibis	*Threskiornis aethiopica*	Thella Venkaramukku Konga
16.	Black ibis	*Psuedibis papillors*	Nalupurangu vankaramukku konga
17.	Glossy ibis	*Plegadis falcinellis*	Merese vankatamukku konga
18.	Spoon bill	*Platalea leucordia*	Teddu (Chemcha) mukku konga
19.	Barheaded geese	*Aner indicus*	Charalatala bathu
20.	Lesser whistling teal	*Dendrocygna javanica*	Jilama
21.	Large whistling teal	*Dendrocygna bicoler*	Pedda Jilama
22.	Ruddy shell duck	*Tadorna ferruginea*	Brahmana bathu/ Chakravakam
23.	Pintail	*Anas acuta*	Sudimonatoka Jilama
24.	Common teal	*Anas cressa*	Neeti jilama
25.	Spot billed duck	*Anas pocillorhyneha*	Mukkumeeda machagala bathu
26.	Garganey	*Anas qnerquedula*	Jilema
27.	Shoveller	*Anas clypeata*	Paramukku jilama
28.	Red crested pochard	*Netta guffna*	Erratala bathu
29.	Common pochard	*Aythya ferina*	Bathu
30.	White eyed pochard	*Aythya nyeca*	Tellakannu bathu
31.	Tufted duck	*Aythya fuligula*	Juttu bathu
32.	Cotton teal	*Nettapus coroamandelianus*	Tella bathu
33.	Comb duck	*Sarkidionis melanotus*	Kommu mukku bathu
34.	Black winged kite	*Elanus caerulus*	Nalla rekkala dega

35. Honey buzzard	*Pernis ptilorhyucus*	Tenethatta gradda
36. Bhraminy kite	*Haliastur Indus*	Bhraminy dega
37. Shikra	*Accipter badius*	Dega
38. Crested goshawk	*Accipter tribirgatus*	Pitchukalanu vetade dega
39. White eyed buzzard eagle	*Butustue teesa*	Tella kannu gradda
40. Crested hawk eagle	*Spizaetus cirrhatus cirrhatus*	Pinchamu gradda
41. White bellied sea eagle	*Haliacetus leucogaster*	Tellarangu pottagala chepalu vetade gradda
42. Grey headed fishing eagle	*Ichltyophaga ichthyactus*	Budidavarnam talagala chepalu vetade gradda
43. Indian white backed eagle	*Gyps bengulensis*	Rabandu
44. Scavenger vulture	*Neophron percuepterus*	Peethiri gradda
45. Marsh harrier	*Circus aeruginosus*	Dega
46. Short toed eagle	*Circaetus gallicus*	Pottivella gradda
47. Peninusular crested serpent eagle	*Spiloruis chedamalanotis*	Pinchamugala garudapakshi
48. Osprey	*Pandion halaetus*	Chepalanupattu gradda
49. Kestrel	*Falco tinnunculus*	Dega
50. Grey patridge	*Francocolinus pondiceriacus*	Budidarangu Kanjupitta
51. Rain quail	*Coturnix comramandelica*	Puridu pitta
52. Jungle bush quail	*Peridu asiatica*	Puridu pitta
53. Grey jungle fowl	*Gallus soneratie*	Adavi kodi
54. Common pea fowl	*Pavo cristatus*	Nemali
55. Common crane	*Grus grus lilfordi*	Karunchu konga
56. White crested water hen	*Amaurimis phoenicurus*	Buta kodi
57. Water cock	*Callicrex cinerea*	Neeti kodi
58. Moor hen	*Gallinula chloropus*	Tumba kodi
59. Purple moor hen	*Porphyriia porphyria*	Nilabolli kodi
60. Coot	*Fulica atra*	Nalla bolli kodi
61. Pheasant tailed jacana	*Hydropahsinus chirurgus*	Podavuthoka jacana
62. Bronze winged jacuna	*Metopidius indicus*	Kanchu rekkala jacana
63. Black winged stilt	*Himanotopus himanotopus*	Ulava pitta
64. Avocet	*Recurvirostra avosetta*	Vankaramullu (Pieki); Ulavapitta
65. Stone curlew	*Burhinus oedicuemus*	Ralla pitta
66. Indian courser	*Cursorius coromandelicus*	Kalivi kodi
67. Colla Red Practincole	*Glareola pratincola pratincola*	Vadayam Practincole
68. Red Wattled Lapwing	*Vanellus indicus*	Errachenpaka Sethwa
69. Yellow Wattled Lapwing	*Vanellus malabaricus*	Pasupu Chenbala Sethwa
70. Golden plover	*Pluvialis apricaria*	Bangarma vanne plover
71. Large Sand Plover	*Pluvialis leschenaulti*	Teerapu Plover
72. Ringed Plover	*Charadrius hiaticula*	Valayam Plover
73. Little Ringed Plover	*Charadrius dubius*	Chinna Valayam Plover

74. Curlew	*Lumenius arquata*	Vankaramukku (krindiki) Pitta
75. Blacktailed Godwit	*Limosa limosa*	Nallathoka Godwit
76. Red shank	*Tringa totanus*	Eruputhoka upe pitta
77. Marsh Sandpiper	*Tringa stagnatilis*	Buradha Sandpiper
78. Green shank	*Tringa nebularis*	Akupachha thoka upe pitta
79. Green ssand piper	*Tringa ochropus*	Akupachha sand piper
80. Wood sand piper	*Tringa gloreola*	Sand piper
81. Little stint	*Calidris minuta*	Chinnathokaupe pitta
82. Brown headed gull	*Larus brunnicephalus*	Godhumavarnum thala samudrapu pavarum
83. Indian river tern	*Sterna auranita*	Naditeeram pavuram
84. Indian Sand Grouse	*Pterocles exustus*	Rathipalka
85. Painted Sand Grouse	*Pterocles indicus indicus*	Chitritha Rathipalka
86. Southern Imperial Green Pigeon	*Ducula aenea pusilla*	Poluga
87. Green pigeon	*Treron phoeuicoptera chlorigaster*	Akupacchha pavuram
88. Ringed dove	*Strentopelia decaocto decaocto*	Valayamgala guvva
89. Spotted dove	*Strentopelia shineusis suratensis*	Chukkala guvva
90. Large indian parakeet	*Psittacula eupatria eupatria*	Pedda ramachiluka
91. Rose ringed parakeet	*Psittacula krameri manillansis*	Gulabi varnam valayamgala ramachiluka
92. Blossom headed parakeet	*Psittacula cyenocephall*	Erupu varnumgala ramachiluka
93. Common hawk cuckoo	*Cuculus varius varius*	Dega Cuckoo pitta
94. Indian Cuckoo	*Cuculus microptronus agierapierus*	Cuckoo pitta
95. Plaintive Cuckoo	*Cacomintus passerinus*	Cuckoo pitta
96. Koel	*Sadya scolopacea*	Kokila
97. Crow Pheasant	*Contropus sinensis*	Jamudu kaki
98. Barn owl	*Type alba*	Tella gudlaguba
99. Scope owl	*Otus bakkamoena*	Chinna Gudlaguba
100. Great horned owl	*Bubo bubo*	Pedda gudlaguba
101. Brown Fish Owl	*Bubo zevloneusis*	Chapalaupatte Gudlaguba
102. Jungle owlet	*Glaucidue radiatum*	Adavi gudlaguba
103. Spotted owlet	*Anthena brama*	Chukkala gudlaguba
104. Common Indian night jar	*Caprimulgus asiaticus*	Nisachara pakshi
105. Franklin's night jar	*Caprimulgus asiaticus*	Franklin's Nisachara pakshi
106. House swift	*Apus affinis*	Swift
107. Palm swift	*Cypsiurus parvus*	Swift
108. Alpine swift	*Apus melba*	Himalayapu swift
109. Crested tree swift	*Hemi procne longipennis*	Juttugala swift
110. Small blue king fisher	*Alcede otthis*	Chinna chekumuki pitta

111. White breasted Kingfisher	*Haloven smvraensis fusca*	Tellarommu Chekumuki Pitta
112. Blue tailed bee eater	*Merops ieschenaulti*	Neelamthoka teneteegalu tine pitta
113. Green bee eater	*Meropsoerientalis orientalis*	Merapakai pitta
114. Indian roller	*Ceracias benghalensis indica*	Pala pitta
115. Hoopoe	*Upura aneps*	Hoopoe
116. Common Grey Hornbill	*Teekus birenstris*	Iberatha Pakshi
117. Malabar Pied Hornbill	*Anthracecores coronatus*	Malabar Telupu Nalupu, Iberatha Pakshi
118. Large green barbet	*Megaleamia zeylamica isernate*	Ukkumukku lakumuki pitta
119. Crimson breasted barbet (Copper Satta)	*Megalaima heamacephala*	Kempuvanne ukkumukku lakumuki pitta
120. Lesser golden backed woodpecker	*Dinoplum benghalense*	Bagaruvanne veepu vadrangi pitta
121. Maharata Woodpecker	*Deadrecepes maharetteonsis*	Pasupunethi Nalupu
122. Pigmy woodpecker	*Picodes namus*	Thelupu Wadrangi Pitta
123. Indian pitta	*Pitta brachvura*	Marugujju Wadla pitta
124. Ashy Crowned Finch	*Lark Budida Kireetam*	Jeenuvai Bharata Pakshi
125. Small Indian Skylark	*Alanda gulugula*	Chinna Bharta Pakshi
126. Red rumped swallow	*Hirundo danrica*	Errathouti swallow
127. Bay Becked Shrike	*Lanius vattatus*	Shrike
128. Golden Circle	*Orilous drilus kundeo*	Pasupumudha
129. Black headed Ortal	*Orlous mantheraus mantherus*	Nallathala Pasupumudha
130. Black drongo	*Dierurus adsimslia macrocorcus*	Bharadhwaja pakshi
131. White bellied drongo	*Dierurus caerulescone*	Tellapetta bharadhwaja pakshi
132. Green racket tailed drongo	*Dierurus paradiseus*	Rachet thoka bharadhwaja pakshi
133. Brahminy myna	*Sturnus pegadrum*	Brahmini goruvanka
134. Pied myna	*Sturnus contra*	Nalupu telugu goruvanka
135. Common myna	*Acidotheres tristis*	Goruvanka
136. Indian tree pie	*Dendracitta vagabunda*	Treepie
137. House crow	*Cervus solecdens*	Kaki
138. Jungle crow	*Corvus macrorhynchos*	Adavi kaki
139. Black headed cuckoo	*Coracina melanoptera*	Nallathala cuckoo shrike
140. Sirkir cuckoo	*Teccocva leschenavlti*	Cuckoo
141. Scarlet minivet	*Pericrocotus flammers*	Sindhuram
142. Small minivet	*Pericrocotus cinnamomous*	China pancharangula pakshi
143. Common Iora	*Aegithina tiphis*	Sukika
144. Golden fronted chloropsis	*Chloropsis aurifrons*	Nuduru Bangaram vanne harithavarna pitta
145. Golden Mantred chloropsis	*Chloropsis cochinchinensis*	Harithavarna pitta

146. Redvented bulbul	*Pycnonotus cafter*	Juttu pitta
147. White browed bulbul	*Pycnonotus luteolus*	Thellakannubomma juttu pitta
148. Common babbler	*Turdoides candatus*	Lambadi pitta
149. Jungier babbler	*Turdeides striatus*	Adavi lambadi pitta
150. White headed babbler	*Turdeides affieis*	Tellatala lambadi pitta
151. Tickell's blue fly catcher	*Muscieapa tickelliae*	Neelam dasari pitta
152. White browed fantailed fly catcher	*Rhipidura aereola*	Tella kanubommalu wisan-akarrathoka dasari pitta
153. White throated fantailed fly catcher	*Rhipidura albicellis*	Tella gonthu wisanakar-rathoka dasari
154. Paradise fly catcher	*Terpon phone paradisi*	Tella dasari pitta
155. Black necked fly catcher	*Monorcha azurea*	Nalupupacha dasari pitta
156. Grass hopper warbler	*Locustella naevia*	Midhuthala kampenaswa-ram pakshi
157. Tailor bird	*Orthotonus sutorius*	Darjee pitta
158. Magpie robin	*Copsychus saularis*	Nalupu telugu Robin pittaa
159. Pied Bushcat	*Saxicola caprita*	Nalapu Telugu Vsa Pitta
160. Indian robin	*Saricolordes fulicata*	Robin pitta
161. White throated ground thrush	*Zeethera cituina cvaactus*	Tella Gonthu Mohanranga Pede Pitta
162. Grey Tit	*Parus major*	Chinna Adavi Pichuka
163. Spotted Grey Creeper	_____	Chukkala Budida Varnam Chetlekke Pitta
164. Yellow wag tail	*Montacilla flara*	Pasupuvarnam thoka pitta
165. Large pied wagtail	*Montecilla macleraspatensis*	-
166. White wagtail	*Motacilla alba*	Bollithoka upe pitta
167. Tickell's Flower Pecker	*Dicacum crythrorhynches*	Poolu Tenethragu Pakshi
168. Purple rumped sunbird	*Neetarinia zeylonica*	Dhumavarnam thene-thragu pitta
169. Purple Sunbird	*Nectarinia aseatica*	Dhumravarnam Thene Pitta
170. White Eye	*Zorperops palpehrosa*	Tellakannu Pitta
171. Wry Neck	*Jvuse tornailla*	Vankarameda Pitta
172. House sparrow	*Passer domesticus*	Pichuka
173. Baya	*Poloceus ubilippinus*	Jeenurai
174. Black Throated	*Ploceus Benghalansis*	Nalla Genthu Gijigadu
175. Streaked weaver bird	*Ploceus manyar*	Geethala gijigadu
176. Red Munia	*Estrilda amandava*	Erra Pichuka
177. White throated munia	*Lonchura malabarica*	Tella gonthu pichuka
178. Spotted munia	*Lonchura perutuista*	Chukkala pichuka
179. Jungle green warbler	_____	Adavi kampanaswaram pakshi
180. Ashy green warbler	*Pricia racialis*	Budida varnam chinna kammana swarama pakshi

* * *

An additional Note

I wish to quote here some excerpts (with statistics) from a paper titled, '***Vulnerable endangered, threatened and rare species categories in the submergence area of Polavaram area***' by D Ramprasad Naik, S A Rahiman and Kaizar Hossain, from the Department of Geo-Engineering, College of Engineering, Andhra University, Visakhaptnam, AP, Department of Environmental Sciences, Andhra University, Visakhapatnam, AP and Department of Environmental Studies, GITAM University, Visakhapatnam, AP, respectively, published in the ***European Journal of Experimental Biology***, 2012, 2 (1):288-296, accessed through the Pelagia Research Library. The study gives a long and detailed list of "Distribution of Habitat and Status of the Mammals", "Birds as per species-wise", "Herpetile species", "Fishery resources", "Loss of Biotic Resources", "Loss of Mammals as per status", "Losses of migrated Bird species", etc in the submergence area of Polavaram dam. The list of species is nearly the same as the above list, but with the additional detail of their status as "rare", "endangered", "threatened" and so forth. The study points out that "about 13 species of mammals belonging to the vulnerable, endangered and threatened categories will be affected due to the disturbance of habitat conditions in the submergence study area" and "that a considerable extent of land resources including cropped areas and builtup areas floral and faunal resources including rare and endangered species particularly in flood plains and foot hills will face submergence owing to Polavaram Project..."

Some of the details are as follows

Fishery resources: The river Godavari has abundant fish resources which is the source of livelihood for a large population on the banks around the submergence area. The predominant species were common Carps – *Cirrhinun mrigala, Labeo calabasu,Catla catla*, and cat-fishes – *Mytos Seenghala, Mytus aor, Silonia childreni, Wallago attu, Pangasius pangasius,*

Bangarius bangarius, Chipeids – *Hilsa clisha*; Prawns – *Macrobrachium malcolhsoni.*

Loss of Mammals as per status in the submergence area of the Polavaram area.

1. Bonnet Macaque	Vulnerable
2. Common langur	Vulnerable
3. Indian giant squirrel	Endangered/Threatened
4. Sloth bear	Endangered/Threatened
5. Indian fox	Vulnerable
6. Jackal	Vulnerable
7. Jungle cat	Vulnerable
8. Wild boar	Rare

9. Hyena	Rare
10. Indian Tiger	Endangered/Threatened
11. Leopard	Endangered/Threatened
12. Sambar deer	Vulnerable
13. Barking deer	Vulnerable
14. Nilgai	Vulnerable
15. Indian bison	Vulnerable

Losses of migrated Bird species in the submergence area of Polavaram area

1. Ibis leucocephalus*	Painted stork
2. Anastomus oscitans*	Open billed stork
3. Ixobrychus cinnamomeus*	Chestnut bittern
4. Sterna aurantia*	River tern
5. Fulica atra*	The coot
6. Haliastur Indus*	Brahminy kite
7. Eudynamus scolopacea*	Koel
8. Coracias benghalensis*	Blue-jay (Roller)
9. Upupa epops*	Hoopy

* Residential and locally migratory.

Loss of Herpetile species of the submergence area of Polavaram area

1. R. trigrina Daudin	Threatened
2. Lissemys punctata granosa (Schosopff)	Threatened
3. Kachuga tectum tentorica (Gray)	Threatened
4. Geochelone elegana Schospff	Threatened
5. Chamaeleon zeylanicus Laurenti	Threatened
6. Varanus bengalensis Boulenger	Threatened
7. Python molurus Linnaeus	Threatened
8. Eryx johni johni Russell	Threatened
9. Dryophis pulverulentus Dumm. & Bibr.	Threatened
10. Enhydris enhydris Schneider	Threatened
11. Bungarus caereleus Schneider	Threatened
12. Naja maja naja Linnaeus	Threatened
13. Vipera russelli Shaw	Threatened

Distribution, Habitat, Status and Abundance of Herpetile species of the Polavaram area

Family Emydidae: Lissemys punctata granosa Aquatic, Fresh water, Lotic (Running waters) (132)	Threatened
Kachuga tectum tentorica Aquatic, Fresh water, All Community type (137)	Threatened
Family Testudinae: Geochelone elegana Terrestrial, All land types, All Community type, Bushes (397e)	Threatened

Species	Status
Family Chamaeleonidae: Chamaeleon zeylanicus Terrestrial, All land types, Arboreal (393g)	Threatened
Family Varanidae: Varanus bengalensis Terrestrial, All land types, All Community type, Boreal (397b)	Threatened
Family Boidae: Python molurus Terrestrial, All land types, Boreal (393b)	Threatened
Eryx johni johni Terrestrial, All land types, All Community type, Boreal (397b)	Threatened
Family Colubridae: Atretium schistosum Aquatic, Fresh water, Lotic (Running waters) (132)	Vulnerable
Dryophis pulverulentus Terrestrial, All land types, Body, Arboreal (393g)	Threatened
Enhydris enhydris Aquatic, Fresh water, Lotic (Running waters) (132)	Threatened
Family Elapidae: Bungarus caereleus Terrestrial, All land types, All Community type, Boreal (397b)	Threatened
Naja maja naja Terrestrial, All land types, All Community type, Boreal (397b)	Threatened
Family Viperidae: Vipera russelli Terrestrial, All land types, All Community type, Boreal (397b)	Threatened

[My point - However, having given all these details of the precious loss of wildlife, the paper concludes, I think, rather oddly: "The study was suggested to develop bird, wild life sanctuaries for the mammal species which come under vulnerable endangered, threatened and rare species categories in the submergence area of Polavaram area and was suggested development of suitable habitat zones for the herpatile population which come under threatened categories. All the facilities including funds should be provided for the habitat zones. In other hand the present study also investigated that the fish resource in the submergence area of Polavaram is very rich and provides livelihood for a large number of the population in and around project region. Hence, the study suggested chalking out a plan for the development of fisheries both in the reservoir and their canals. Further, the study suggested creation of habitats for endangered and rare species of fishes in separate zones. The fisheries department is to be involved in these programmes for effective implementation of the development programmes." I couldn't help this exclamation mark !!!]

Some of People's Petitions

Some Voices of Protest from Kondamodalu Today: a Gram Sabha Resolution on Polavaram[1]

This was held on 3rd June 2006 (I reached the same day but missed the actual Gram Sabha). It was presided over by Illa Rami Reddi, President, Kondamodalu Panchayat Board, Madi Chinna Pentaiah, Vice President, Kondla Gangaraju, Devipatnam Mandal President (Mandal Parishad Territorial Constituency), Madi Mutyem, Ward Member, Chintalada Devamma, Ward Member, Palem Gopalam, Ward Member, Murali Krishna, Project Office, ITDA and Narsing Rao, RDO (Revenue Divisional Officer), Rampachodavaram.

Resolutions (Few)

"1. In spite of the Supreme Court directive, the spillway work in ongoing at Polavaram and they just stopped work for a day when the empowered committee had visited. This Gram Sabha condemns this. We demand stoppage of all works.

2. This Gram Sabha reiterates that the government must take cognizance of Dharma Rao's alternative proposal and save 276 tribal villages from displacement.

3. This Gram Sabha condemns the stoppage of all developmental activities on the pretext of the Polavaram project and demands construction of roads, specifically, Geddada to Kondamodalu, Devipatnam to Manturu and similarly to build the Tadivada (water) pumping scheme to help the farmers of Kondamodalu, to bring three phase current and dig borewells for us.

4. Only after gaining all clearances, after giving land for land and forest for forest to the tribal communities and maintaining cohesiveness of the hamlets in the R&R colonies, should the government think of the Polavaram project and only after making all these processes transparent to the people.

5. To release seeds, sprayers, etc to the farmers.

---

1. Copy of the "Kondamodalu Grama Sabha In Accordance With PESA 7 OF 98". Original in Telugu, translated here. Thumb impressions/signatures of Murla Devi Reddi, Vetla Sriramulu, Vetla Kannam Reddi, Kundla Bullabbbai Reddi. One has included only a few relevant resolutions here.

6. Keeping in mind the impending monsoons and diseases that come with it, the Gram Sabha urges the government to build a permanent building for the PHC and appoint permanent staff therein. And we demand that the government should address the unresolved issues of the school in Kondamodalu – appointment of teachers, compound wall, latrines, and introduce the 10th standard in the school.

7. To re-establish the Kondamodalu-Polavaram launch service. And to introduce special launches during floods[2] in Kondamodalu as earlier.

8. To provide drinking and irrigation water facilities and to see to it that every individual gets the complete quota of ration supplies (rice, etc) and to make new ration cards.

The Gram Sabha unanimously passes these resolutions.

***

2. Since this was a resolution handed over to the official authorities the term 'floods' (varadalu) was used here, and this is the only instance I found it being used during the course of my interactions throughout.

Right Bank Canal Affected Farmers, Let's Fight Together

Samagra Niti Hakula Parirakshana Samagra Punaravasa Sadhana Committee
(Committee for Water Rights and Rehabilitation)

Date: August 12, 2005

They have started digging canal even without basic clearances or people's consent. A 300 meter long canal is being dug, destroying 12,000 acres in West Godavari and this is not a cause for joy for farmers. On 15.05.2005 in Tadepalligudem Pumping Scheme discussion farmers said the Tadepudi and Polavaram canal are destroying farm lands. Mr. Sitapati Rao, advisor to the Government, had assured that a study would be conducted. The Chief Minister and TRS Chief, Chadrasekhara Rao had also agreed to rethink the height of the Polavaram dam. Without doing all this, contractors are going ahead with the digging. The Polavaram Right Bank Canal (*kudi kaluva*) needs to be analysed again. Tadepudi lift and Polavaram Right Bank Canal are of no use for West Godavari. The question is – who has the right to this water? From 1990s, these are for the hydel projects rather than for irrigation. As a result, Krishna barrage was destroyed by the cooling canal for the Ibrahimpatnam VTSK. Srisailam, Nagarjuna Sagar hydel electric projects dried up Krishna delta. Polavaram Right Bank Canal is meant to connect Prakasam barrage to Buckingham Canal to give water to Chennai – this is in their plan. They are already saying, if farmers grow paddy and sugarcane under the Lift Irrigation schemes, they will stop giving water. Tadepudi canal farmers grow rice, sugarcane, coconut. Tomorrow they may notify these to be stopped. It is horrible to see that the state is working to stop farmers from growing even food crops. 276 villages are to be sacrificed under Jalayagnam and the offerings will be given to cities, power projects, and later to older ayacuts. This is why all designs are secretive. Godavari water should be provided via lift to dry areas in West Godavari – Koyyalagudem, Jangareddygudem, Kamavarapukota, Chintalapudi, etc. The Kakinada gas pipeline on Godavari should be used to provide electricity free for farmers. R&R contractors and officials are threatening farmers and changing alignments as per their whims. In compensation, the canal-affected project land should be compensated in command areas upto 5 acres. And compensation as per market rate should be given for ore than 5 acre loss and besides, pump and motorshed should also be provided and land for land. We will fight facing all problems.

- West Godavari farmers to fight together against the Right Bank Canal digging
- Rights on Godavari are Farmers' alone
- Total rehabilitation is the right of the affected people
- Canal affected people are one
- We condemn wrongful arrests, harassment

Signed
Maddipatla Vira Venkata Satyanarayana
(Convenor, Chityala, Gopalapuram)
Kunjam Rama Rao
(Tellavaram), Polavaram Mandal

***

Handwritten Petitions by Devipatnam, Addateegala Mandal Girijans

Subject: Buying lands from non-tribals for the Polavaram project. The hanky-panky (*avaka-tavakalu*) of Rampachodavaram RDO and his illegal activities.

Through 49 illegal land transfers, the RDO (Revenue Divisional Officer) bought 572.81 acres of land from non-tribals landlords in Devipatnam, Addateegala, Gangavaram for Rs._______ (not clear).

1. According to our survey, these lands are under cultivation of girijans (tribals) of Indukuru, Devaram, Jeederu, Jeeyampalem and lands in other panchayats , which he bought from the non-tribals, showing records as though these lands were being cultivated by them. The ITDA is an active accomplice in this. It must be taken into account.
2. In Devipatnam mandal Pothavaram village, Survey Nos. 275, 247 (7.50+37.50 acres) land in the past was declared as ceiling land by the Government. All this land was bought from the non-tribals and the RDO has also made big money from this purchase. Part of those 45 acres of land, removing 12 acres of land, the rest is uncultivable land, rocky and filled with stones. Whereas the cultivable land in M. Ravilanka and Pothavaram villages has been in use by the girijans since several years now. In the year 1997, 19 girijans were also given D pattas on these lands. Patta survey numbers and survey numbers of non-tribal lands purchased, together with sadr lands shows that the RDO is attempting to forcibly remove the tribals from these lands. We urge for a comprehensive enquiry into the buying of the ceiling lands, survey numbers of lands bought and survey numbers of patta lands of girijans.
3. In the fair (copy) *adangal*[1] registers and other records, land shown as *banjar* and wasteland has been bought by the RDO from the non-tribals. This is a case of government land being bought with government money (eg. Indukuru village, Survey Nos 175, 176 land).
4. Girijans are cultivating lands at Kondalapalem, Marripalem, Etipally, Chinabhimpally and other villages, these lands are being bought to throw away the girijans, their produce was forcibly usurped and the police has been slapping cases against the girijans - there is evidence

1. A British Colonial legacy – a register of annual statement of occupancy and cultivation.

of the atrocities committed by the RDO. We seek an enquiry into all of these.

5. As per the provisions of the 1/70 Act, if land has to be bought from non-tribals, a public consultation is required, with appropriate notices posted at relevant Gram Panchayat offices and police stations. Any objection to the buying of these lands from non-tribals must be considered and enquired into. In violation of these provisions, the RDO directly purchased land from the non-tribals. This is illegal. If land is to be acquired for a project in Scheduled Areas, then under the Panchayati Raj Act 1998, Section 243 (F), public resolutions, passed in Gram Sabha and Panchayat consultation is necessary.

Signed (thumb impressions with names written) – Parada Yellamma, Thokala Varalakshmi, Borage Akkayamma, Sarapu Lakshmi, Midiyam Kannayi, Vadinju Nagamani, Panda Chellayamma, Chivakam Kumari, Pamula Venkatesu, Kosu Bhadram, Panda Raghupati, Seram Sitamma, Karam Subhalakshmi, Kittam Virabhadrayya, Tireti Acchiveni, Tenki Papa, Kampana Anasuyamma, Tenka Rasulu, Mindi Kondayya, Ekka Sitamma, Pattam Pentamma, Rampu Venkatesu, Komaram Bujjayya, Kandala Venkatnarayana, Ranke Nukaraju, Karam Chellanna Dora, Bellam Balu Dora, Bellam Gangaraju Dora.

***

The RDO did not implement the Central Panchayati Raj Act, or the orders of the Central Rural Development Ministry. He recently bought lands in spite of Court directives and present court cases. RDO's activities must be enquired into, to bring to light the realities and protect the land rights of the girijans. We bring to your focus the following demands:-

1. Immediately transfer the Rampachodavaram RDO; RDO's land deals to be enquired into as per records, by IAS officers and necessary action to be taken.
2. Presently cultivating girijans should not be moved away from their lands.
3. Illegal land sales to be annulled.
4. Land cultivated by girijans should not be taken for the (Polavaram) project displaced girijans under the Land for Land provision.

Dated 14.10.06
Rampachodavaram
Signed (Thumb Impression against names): Kunjam Chinabapanamma, Madakam Savitri, Ari Rajamma, Kosu Bullayamma, Kosu Lakshmi, Karu Gangamma, Kittam Chinnari, Kosu Ramalakshmi

Copies to:
Project Officer, ITDA, Rampachodavaram
Principal Secretary, Tribal Welfare Department, Hyderabad
Chief Secretary, Home, Hyderabad
Chief Secretary, Revenue, Hyderabad

Petition by Tribal Peasant-Cultivators in the Context of Illegal Buying of Lands for the R&R Colony for Polavaram Oustee Tribal Families[1]

Division Rampachodavaram

Petition Letter, typewritten. Written (in the names of) We, the villagers of Devipatnam Mandal, viz. Komaram Satyavati, Neram Sitamma, Madakam Kumari, Kosi Ramalakshmi, Karu Ganga, Pittam Virabbayi and Komaram Mulaswamy (Sarpanch, Chinabhimpally), are submitting this requisition.

Subject: Harassment by Indukur Panchayat Village Officer. Request to take action

About 70 landless poor girijans (tribal), with Survey Numbers 93, 177, 109, 110, 126/2, 175, 12, 89/1, 93, 96, 139/2, 143/2, 145/1, 146/1, 150/1-187/2, 112/2. Since 12 years we girjans are cultivating sadr land, banjar land and wastelands. In the past, these were owned by girijans. But girjans were not given pattas in their names on sadr lands. Several requisitions were made in this connection to the Project Officer (ITDA), MRO (Mandal Revenue Officer). Instead of registering these lands in our names, they registered them in the names of non-tribals Pasumarthi Mangaraju, Pulapally Krishna, Srinivasa Reddy, Badireddy Narasayya Dora, Badireddy Nakku Dora, taking bribes from them. The non-tribals have registered these lands in their names, showing these lands as being cultivated by them. When we protested that this was injustice, the MRO, Devipatnam retorted back that this is how it will be. Our panchayat secretary is not residing in our village. If we need any help we will have to go to his house in Rampachodavaram mandal, in I. Polavaram village. We plead with you to accept our requisition about our Panchayat Secretary.

Signed (Names, written out) - Komaram Satyavati, Neram Sitamma, Madakam Kumari, Kosi Ramalakshmi, Karu Ganga, Pittala Virabbayi and Komaram Mulaswamy (Sarpanch, Chinabhimpally)

1. Ms. Uma (Lakshmi) and later Ms. Sharavani (Prajashakti Publications, Hyderabad) read these petitions to me. Translations mine.

Copies Sent to
District Collector, Kakinada
Project Director, ITDA, Rampachodavaram
District Revenue Officer, East Godavari District, Kakinada
District Panchayat Officer, East Godavari District
Division Revenue Official, Rampachodavaram
MRO, Devipatnam
Mandal Development Officer, Mandal Parishad, Devipatnam
Panchayat Secretary, Indukurupeta

Various Appeals to Stop Polavaram Project (2005-06)

An Appeal for a Re-look on Indira Sagar Project Site at Polavaram[1]

To

The Central Committee
Central Water Commission
Union Ministry of Tribal Welfare and Specialists in R&R Package
Director, Ministry of Forests and Environment
Conservator of Forest
Director – Archaeological Survey of India
Secretary to Projects
New Delhi

From

1. M.S. Nagaraju
   State Committee Member
   Rythu Coolie Sangham (AP)
   C/O Kunjam Rama Rao,
   Tellavaram Post
   Polavaram Mandal
   West Godavari district

2. Chintalada Chinnabayi
   President
   Agency Girijana Sangham
   Tutigunta Post
   Polavaram Mandal
   West Godavari district

Sirs

Sub: The ongoing protest demonstration by the adivasis against the construction of the Indira Sagar Project at Polavaram as it poses a serious

1. Verbatim; relevant excerpts.

threat to their identity, culture and livelihood – Request to examine the feasibility of implementing an alternative proposal for dam construction at Dummugudem.

On behalf of our Agency Girijana Sangham (affiliated to Rythu Coolie Sangham – AP) we extremely happy over your visit to the West Godavari agency on the mission of the Polavaram project. In this connection, we would like to draw your kind attention to the raging protests by the adivasis from a cluster of habitations against the dam for quite some time. They fear the dam construction at the present site will cause a large scale displacement, submergence of villages and lands…Besides the social disturbance, the dam is also likely to adversely affect the ecological imbalance by destroying the flora and fauna.

In this background we earnestly request you to consider the alternative proposals in the place of the present site at Polavaram. In this connection, we would like to draw your attention to a proposal submitted to the government by an irrigation expert, M. Dharma Rao. Mr. Rao, a retired Chief engineer in the Irrigation department , has proposed to construct the dam at Dummugudem with the same height as designed in the proposal under execution at Polavaram at present. Saying that it is technically feasible, the expert opines that the construction of the dam at the alternative site at Dummugudem which is relatively at a higher altitude will benefit huge tracts of hitherto water-starved upland regions in Krishna, West Godavari and Khammam districts, apart form averting the damage of forest wealth, tribal displacement and submergence of lands.

Mr. Dharma Rao proposes to build another dam as an extension of his proposal at Nelakota on Sileru to supply drinking water to Vishakhapatnam and irrigation water in East Godavari, Vishakhapatnam and Srikakulam districts through the Left Canal.

Mr. Dharma Rao maintains that it is quite possible for transfer of Godavari water into Krishna through the alternative proposals.[2]

Thanking you

Yours Sincerely
(Signed)
M.S. Nagaraju
Chintalada Chinnabbayi

Dated 5-9-2005

***

2. The alternative, detailed - several pages long - proposal was attached to this letter, which I have not quoted.

Open Letter to AP Government on Polavaram Project

From E.A.S. Sarma (Former Secretary, Government of India), Vishakhapatnam

October 26, 2006

To

Shri J. Harinarayana, Chief Secretary, Government of AP

Dear Shri. Harinarayana,

Deccan Chronicle has carried a report today that the State Government has "decided to construct 43.86 km flood bank along the Godavari river to prevent the likely submergence of Bhadrachalam and girijan hamlets of Chintoor mandal in Khammam district after the Polavaram dam is constructed." Corroborating this, The Hindu has also stated that the proposed "flood bank" will have a height of 189 feet. The State Irrigation Minister has even gone to the extent of proudly describing the structure as "China Wall-like"!

I am sure that you, as a former Union Water Resources Secretary, must be wondering what all this is about! As fare as I am concerned, I am not only amused but also aghast at the cavalier manner in which such large projects are being dealt with and the motives that underlie them. In fact, this latest statement of the government questions the credibility of the Polavaram project itself!

As a citizen, I wish to raise the following questions for the consideration of the government.

i. Is the government more concerned about saving Bhádrachalam town and its temple than more than 2 lakh people who are going to be displaced by this project?
ii. The government was well aware that the highest level that Godavari had ever touched was in 1986 when eth water discharge was as high as 35 lakh cusecs. In August this year, when the discharge was 28 lakh cusecs, even without the Polavaram project in place, 322 villages were flooded. On the other hand, the government has all along been planning for the rehabilitation of people on the premise that only 276 villages would get inundated. Is there some serious flaw in the computation of the levels and area of submergence?
iii. Would it not be necessary to review the basic numbers that go into the design of the project, in view of the latest developments? What would be the area of submergence if the pond level of the project were to be 150 ft MSL?
iv. The proposed "flood bank" appears to be larger than the dam itself!

If it ever breaches, what would be the magnitude of the disaster that would follow?

v. Eminent environmentalists like Prof. Sivaji Rao and others have cautioned the government about the possible consequences of a "dam break" in the case of Polavaram. Its adverse impact would be on the downstream population ad assets. It is perhaps necessary to carry out a similar stochastic analysis of breaches in the flood bank before proceeding further with this project on which the government is spending crores of rupees that belong to the tax payer.

vi. The proposed flood bank will cost more than Rs. 300 crores, as per the news reports. Going by the lackadaisical way in which the project design and estimates have so far been handled by the government, the cost of the flood bank will be much more and it will further add to the cost of the project that is already astronomical in magnitude.

vii. While the project will render lakhs of voiceless people homeless and deprived of their Constitutional rights, one is not sure whether the already crisis-ridden finances of the State could ever absorb such lavish and senseless expenditure on the project. Would it crowd out essential sectors such as health and education from the development agenda of the State government?

I am afraid that this latest report throws suspicion and doubt on the entire project about which the government is reluctant to come clean.

I have copies of the resolutions of the Gram Sabhas in tribal villages that clearly reveal opposition to the project at the grass-root level. On the other hand, I understand the State Government has informed the Ministry of Tribal Affairs that all the Gram Sabhas have been consulted and their concurrence taken! Once again, in a democratic society like ours, these facts raise serious doubts about eth commitment of our leaders to democratic principles.

I hope that these issues will get evaluated by an independent group of experts before the State government proceeds further.

In the name of "good governance" I will appreciate if you can acknowledge the receipt of this letter. I am making copies of this letter to well-meaning individuals and the press.

Regards,
Yours Sincerely,
EAS Sarma

***

Convention on the realisation of comprehensive water rights protection comprehensive rehabilitation under Polavaram Project[3]

The convention…held on April 28, 2005 in Devarapalli village has adopted the following resolutions in the presence of many elders form peasants, girijans and workers.

1. Grama Sabhas: Many doubts are arising among the people on the right canal of Polavaram project and its survey works. The Government is not trying to convene Grama Sabhas to patiently and responsibly explain and carry on the works with people's participation and approval. In fact, the Government has earmarked April 14 – the day of Doctor Ambedkar's Jayanti – as the day for Grama Sabhas. This convention strongly condemns that the Government has not at all utilised even this occasion to get the approval of people and Grama Sabhas in regard to this project….Authorities and contractors are threatening the people that the police will be deployed and repression will be unleashed against them

Demands:

1. The Revenue and Engineering officials must hold the Grama Sabhas immediately within a specified time. By the time the Grama Sabhas are held, they must clarify in what contour a particular village and its ayacut fall and how the water will be supplied to that village and its entire ayacut.
2. The peasants are agitating that they are losing thousands of acres of land as a consequence of widening of Tadipudi and Polavaram canals. These agitations must be taken into consideration. The Government must once again examine all things like flow of water in right canal, capacity of Krishna Barrage to store it and also Godavari's water joining the Krishna river from Dummugudem…The extent of land submersion be reduced considerably…

2. Rehabilitation

…thousands of acres along the canal would face submersion because of right canal…The Government is acting against its own declared policy of rehabilitation. It is reducing rehabilitation into an act of throwing alms by calculating it in terms of money. This convention is denouncing these methods…

3. The Right over Godavari remains only with people:

…This convention resolves that the right over water especially Godavari waters, belongs to none but cultivating peasants and the people….

3. Excerpts.

Adivasi Samkshema Parishat, Andhra Pradesh[1]

To

The Hon'ble Chairman,
Central Empowered Committee
Supreme Court of India
New Delhi
Camp at Bhadrachalam **Date: 29-07-2006**

Sir

It is our privilege to submit this memorandum before your highest dignitary for your benevolent perusal and consideration.[2]

It is needless to submit here as your Hon'ble authority is well aware that lion share of our aboriginal tribal sects are dwelling in dense forest of Dandakaranya in AP adjacent to the river bank of Godavari since time immemorial and having our own traits, plights, dialectics[3], and culture with distinct customs and traditions. Mainly our isolated people who are innocents were subjugated by non tribal, more specifically in recent past, the so-called modern Governments, under the shadow of development, themselves violating all Agency laws which were intending to welfare of aboriginals…under gross violation of Agency laws urbanisation was taken place rapidly in Scheduled Area of Andhra Pradesh…In this context, at this juncture, the construction of Major Irrigation Project like Indira Sagar project popularly known as Polavaram project across the river Godavari, how will be the beneficial to the aboriginal tribes living both the sides of River Godavari banks? After construction of the said Dam whither the Tribal entity? In this connection we totally and strongly opposing the construction

1. Excerpts. Note: I am only sharing in this book a few of the thousands of petitions sent to various officials, ministers, authorities, etc. I still do not have possession of many more petitions coming forth since 2009 till date, from the ground, seeking either a rethink on the project or redressal of several grievances. My not including them here is purely a matter of personal limitation and lack of access to those petitions rather than any other reason.
2. Verbatim. Many of the people from tribal communities within political parties are semi-literate and very few Graduates. Sometimes, their petitions are written out for them by others, in similar contexts. I respect their expression, and would like to keep it verbatim.
3. They mean 'dialect'; but perhaps to say dialectics is quite appropriate as well!

of Dam as the Agency Area will be inundated and not only ancient culture of aboriginal will disappear but entity of tribe will come into end. Moreover it has no societal and legal sanctity.

We further submit that we are not demanding:-

1) to minimise the height from 132 foot to low level
2) to construct the dam across Godavari at Eturnagaram instead of at point 34 km from Rajahmundry, and
3) for best Rehabilitation package…

…Our people never viewed the land as commodity and they never utilised forest and entire natural resources of Agency area including water as commodity.

Hence we requesting our reputed committee to recommend

1) To constitute an independent committee which shall consisting with an outstanding Anthropologist, a renowned Human Rights Activist, International Racial and Ethnical Expertised intellectuals and experts of constructing in Major Irrigation projects to study:

Whither the entity of tribal who are part and parcel of nature, after construction of Dam? How so far it is beneficial to the aboriginals? Whether this type of Major Irrigation Projects are imminent essential for the developing countries? Whether it is advisable to build up a nation on tombs of socially frailed, economically poor and politically backward people and areas. Is it not submergence of cherished goals of our freedom fighters, constitution makers and people of India who adopted, enacted and enshrined the 'Equality' concept in our Constitutional Preamble?

III. Whether the recitals of V Schedule of the Constitution, in which deletion from or addition of Territory in Schedule area, have been followed the Government or not as after construction of dam some extent of Territory will disappear from the Territorial jurisdiction of Scheduled Area which has already notified following due process engrafted U/V Schedule of the Constitution. The disappearance of Territory will neither by the Act of God nor by the natural calamities, it is by act of man (person) only.

IV. The consensus of aboriginals should be obtained through Grama Sabhas and MPPs pertaining to Scheduled area. It is imperative on the part of Government but till to date Government has not obtained this consensus and not followed the Panchayat Raj (Extension to Scheduled Area) Act, 1998.

Therefore, the Hon'ble Chairman may be pleased to convey our grief and grievances to the Apex Court and recommend to stop the construction of Polavaram dam till submission of Commissions Report to be constructed by the Government at the direction of Hon'ble Supreme Court of India.

Yours faithfully
(Signed) Sonde Veeraiah
Bhadrachalam
Date: 29-07-2006

***

Polavaram Project Samagra Punaravasa Sadhana Committee – (Pamphlet) 26.06.2005[4]

The officials are filling up the Kothur *cheruvu* (rainwater pond) with soil (from work at the spillway site), but the revenue officials did not obstruct the same. The RDO (Revenue Divisional Officer) Sriramachandramurthy said 'we will continue work but will not go by the Gram Sabha's decisions. On 17th we all (tribal communities) went to meet the Collector, but at Vadapally, our tractor met with an accident, slipping off a slushy mound. 75-year-old Mada Siramayya and, Chintalada Sudhakar Reddy died in that accident while 35 were injured. Tama Bhavani also died. Siramayya was a veteran of the Telangana Armed Struggle between 1946-51. As a youth he worked with the Communists. He was a lorry cleaner and learnt driving to become a lorry driver later. These people died for the cause of the Polavaram project affected people. After April 8th 2005, G.O 68 was announced. There was lathicharge on people without caring for the decision of the Gram Sabha. As per the G.O 68, Ch.V, 11th Para, if banjar land is not sufficient, they can take private land. The 6th Chapter, 19th Para, 1st sentence points out that precedence must be given to tribal people, but does not mention giving land without fail. But only if land is found. They say there is not enough land available, and are trying to give money and send the tribals away. The 6th Chapter, 19th para , 2nd line mentions Rs. 500 as minimum wages for tribal people; Rs.40,000 (at daily wages of Rs. 80). Tribal communities collect cashew, mango, areca nuts, etc and make use of other trees and plants for medicinal use; *kalapa* (used in making broomsticks) comes from the forest, so too, soapnuts, berries such as *usiri*, tamarind, beedi leaves (*tuniki, adaku*), *kondacheepiri, musirikaya, kovela jiguru, tene* (honey), etc. In one season, one tribal family earns Rs. 15,000 from *adakulu, tuniki aku*, even if they do not do any other work. The forest is a mother. The government wants us to sell her for Rs. 40,000/-. They are saying they will give us 150 square yards in the R&R Colonies. Only 150 sq.yards! But girijans have common rights over common land. What about that? The G.O 68 mentions 'humane touch' (in dealing with tribal communities). They destroyed Venkatareddigudem graveyard, by digging up the same with proclainers. Is this their 'humane touch"? The Collector and RDO Sriramachandramurthy announced in newspapers that they will use police force and dig up the lands. The Revenue officials are saying that setting up a committee of four or five people is enough (to discuss and implement R&R). They are violating panchayati raj guarantees. West Godavari officials are acting as contractors' personal secretaries. Without surveys, they are trying to vacate villages.

4. Written in Telugu, read out by Sharavani (Prajashakti), translations, unless otherwise mentioned, mine. This is from a Committee for Comprehensive Rehabilitation of Polovaram project.

We demand building of barrages of smaller heights, to reduce submergence.

To protect land from submergence
The Rights on Godavari are with the farmers
Do not give them over to private companies.
Do not continue with the project without completing rehabilitation

Signed, Chitta Ranga Rao, President
Kunjam Rama Rao, Secretary

***

Memorandum Submitted to the Central Empowered Committee of Hon'ble Supreme Court of India By Girijana Sangham (R.No. 242/2002)

Hon'ble Sir/Madam

You are well aware of displacement of more than one lakh tribals and submersion of forest due to the construction of Indira Sagar (Polavaram) project in the three districts of AP. We humbly bring some important aspects with regard to environment, rehabilitation and resettlement of STs and other related constitutional , statutory and other violations being done by the government.

1. Out of the 276 villages that would submerge, 274 are in Schedule V area. More than one lakh are Scheduled Tribes out of 1, 80,000 people that would displace. The Rehabilitation and Resettlement attracts Schedule V of Constitution of India, Article 338 (9) of the same and Panchayati Raj (PESA) Act of GOI, Panchayati Raj Act of AP, R&R Policy of GOI and GOAP and also the National Tribal Policy (draft) recently released by GOI.

People of Scheduled area shall be rehabilitated and resettled in scheduled area itself. But the first R&R colony, that is being built near the dam site itself is out side the Scheduled Area. Hence the poor tribes will be losing constitutional rights and provisions because of rehabilitation. If this is the case in west Godavari, where only 29 villages are under submersion, one can imagine what will happen for the people of Khammam district where 205 villages will be submerged.

The National Tribal Policy draft in para 8.6 (released by GOI on 6th July 2006) states that there shall be a threshold of displacement viz the maximum number of persons that can be displaced in one project. Project involving displacement of more than a fixed number, say 50,000 would not be considered, if the majority are STs, or would be subjected to more stringent appraisal norms. The proposed Indira Sagar displaces 1, 80,000 of which more than one lakh are Tribes. So, construction of Indira Sagar would be a greater violation to the National Tribal Policy (draft). The R&R policy of AP didn't make land to land for STs. Though it says about constituting

Project level/District level R&R Committees , Monitoring committees, Redressal mechanism, etc, nothing was observed in this regard though more than one year has elapsed since the policy came into. It is astonishing to note that only about Rs. 16 (sixteen) crores only were spent for R&R aspects out of the total spending of more than 1000 crores so far according to newspaper reports. Such is the misery of R&R. The previous experience of R&R in other minor irrigation projects like Surampalem are of most dissatisfaction to the Tribes.

2. The project submerges 3700 ha of reserve forest including 190 ha of wildlife sanctuary. Now the people of the villages in and around the forest areas are dependent on minor forest produce for day to day needs and the forest as a whole for the entire life. But if the forest disappears, the people will be losing their livelihood. The flora and fauna in the forest can not survive and that will be a national loss…The Tribes and other forest dwelling people can't live without having forest as part of their environment. One can't build up that sort of environment in three years or five years. So, the people that will be displaced will be fish out of water and such situation must be avoided.

3. The Government of AP has been violating the Constitution of India, Statutes of the land and misleading Government of India. It started construction of the project without having even site clearance. The public hearing was conducted hastily and undemocratically in October last. It gave self contradictory statements with respect to R&R to tribes in EIA report. Most of the Gram Sabhas opposed the construction of the proposed project causing huge submersion. But they were not taken into consideration, violating PESA Act and AP Panchayati Raj Act. Work in non forest area also shall not be commenced without having Forest clearance according to Environment (Protection) Act. But the government knowing well of this commenced work and spent about thousand crores….Only after the advice form the Hon'ble CEC, the government suspended construction of the project.

In this background we humbly request the Hon'ble committee that the state government may be directed to go for alternative to the proposed project, which causes no submersion of Notified Area and to avoid Tribes' misery. Actually eminent engineers like Sir Dharma Rao, submitted alternatives to the proposed Indirasagar, causing no submergence. We once again urge upon the Committee to protect the interests of the Scheduled Tribes.

Thanking you,
Yours Sincerely,
(Signed) K. Lakshmana Rao (Secretary), R. Sriram Naik (Gen Secretary)
Bhadrachalam
30.7.2006

***

Letter to CEC Chairperson, P.Vijay Krishnan, from Convenor, ASDS, Gandhi Babu[5]

Subject: Polavaram Projects Impact Environment

Dated 29.07.2006

1. It will cause permanent damage to the delta, according to the former engineer, K.Ramakrishnaya who had looked at the water resources and utilisation in AP in 2004. and expressed objections to Polavaram The eastern canal of the (Dowlaishwaram Cotton anicut) has already caused damages, 250 tmc from Polavaram (after the project is built) in the kharif season will mean worse damage. Environmentalists are not looking at this aspect, but only talking of R&R.Rs. 10,000 crores are being spent which will lead to further damage of the delta.
2. Wetlands will also be affected. A study of the Singareni collieries by the Godavari shows that Bhaskar colony there has graphite ores in excess of 250 tonne. The Polavaram dam submergence will lead to graphite dust killing organisms and aquatic life in Godavari where it mixes in water.
3. Health impacts are many. Water-borne diseases will increase - filaria, diarrhoea, liver diseases will increase.
4. Wildlife- Species such as the crocodile (*mosali*), tiger, leopard, wild buffalo, etc will be affected.
5. An Andhra University study of Godavari river between 1977-2002 warned that…between 1976-2001 the sea bed was 40.83 kilometers. Alluvial soil (*ondru*) was, 2029.67 km. Mangroves reduced from 200 km o 140 km. Polavaram dam will cause problems to the people in the command area. We request the government through your kind offices, to look at alternatives.

Gandhibabu, ASDS, V.R. Puram, Khammam District.

***

5. Excerpts of the same. In Telugu. Read out by Sharavani (Prajashakti Book House), translated by me.

Glossary

Conversions

1 Lakh	100,000.00	100 Thousands
10 Lakhs	1,000,000.00	1 Million
1 Crore	10,000,000.00	10 Million
10 Crores	100,000,000.00	100 Million
100 Crores	1,000,000,000.00	1 Billion

1 Acre – 4,046.9 square meter

Acronyms

TMC – Thousand Million Cubic Feet *TMC* ft = 28.3 million cubic meters. 1 cubic meter = 1000 litres of water.

FRL – Full Reservoir Level [It is the level corresponding to the storage – 'inactive' and 'active' storages and also the flood storage, if provided for. This is the highest reservoir level that can be maintained without spillway discharge or without passing water downstream through sluice ways.]

Dead Storage Level (DSL)

[Below the level, there are no outlets to drain the water in the reservoir by gravity.]

ST – Scheduled Tribe
SC – Scheduled Caste
BC – Backward Caste
OC – Other Castes
MRO – Mandal Revenue Officer (of the administrative unit called Mandal; known as Block in other parts of the country, comprising a number of villages and panchayats)
MDO – Mandal Development Officer
SDC – Sub-Divisional Collector
RDO – Revenue Divisional Officer
P.O – Project Officer
ITDA – Integrated Tribal Development Agency
S.I. – Sub-Inspector
VO – Village Officer
MoE&F – Ministry of Environment & Forests

Select Bibliography (Oral and Written)[1]

Primary Sources

People from East, West Godavari and Khammam districts (now in two states)

For history of colonial forestry, reserved forests, tribal resistance: Kondareddis - Illa Rami Reddi, Kondla Gangaraju; Sitamahalakshmi; Saroja, Ranga Reddi of Pochavaram

For leading me to tracing a certain history of tribal resistance and the Rampa-Rekapally connection: Koyas – Posi Babu, Kotaragummu Kondareddis, Mutchika Suramma (late), Mutchika Ramalakshmi,

For my understanding on tribal-tribal conflict, land alienation, state repression and the other side of R&R: Kammaras of Pedabhimpally, Chinabhimpally, Indukur – Rajanna Dora, Koyesi Ramalakshmi, Komaram Mulaswamy…

For new insights on the Forest Rights and National Park: Jaya and other Naikpods of Mulakalapally

For leading me to understanding of river, fish cycle, fishing zones, etc: Fishermen of Warangal, Vadapally, Manturu, Kachluru, Tuthigunta, Singanapally, Kapileswarapuram – Rama Rao, John Babu, Suri Babu, Malladi Posi, Gangadharam, Balapparaj, others

For my understanding on the dalit question in the submergence zone: Malas and Madigas of Pedapolipaka, Pydipaka, Rudramakota, Valairpadu – Sarojinamma, Errapotu Chinipiri, Andru Pedaraju, Jadla Mary, and Babji of APVVU –

For understanding of crossing overs: Sivagiri Punnam Mutyam

For a new angle to the submergence zone; the wetlands post-Godavari's coming, and marginal livelihoods: Chitturi Subba Rao

Colonial Documents

Arthur Cotton, "Two letters on I. Public works: Being A Rejected Letter to The "TIMES," in Answer to Sir James Stephens, And a Letter To The India Office, In Answer To Lord George Hamilton" Bristol Selected Pamphlets, (1878),

1. I wish to acknowledge the role of these people in leading me to getting a fresh perspective, or refining my perspectives, by referring back to history, understanding the present better, etc. Every conversation with them, was an eye opener. I mention a few whose role in those specific contexts and issues (that I have mentioned here) was very important to my writings throughout the journey, as and when they happened (the journeys and the writings). However, on the whole, the people of Godavari have been my primary and most respected source of knowledge, but I am unable to mention all the names here.

Published by: University of Bristol Library Stable
Colonel Arthur Cotton, *Profits upon British Capital expended on Public Works in India as shown by the Results of Godavery Delta Works of Irrigation and Navigation*, Richardson Brothers, London, 1856
F.R. Hemingway, *Madras District Gazetteer: Godavari*, Superintendent, Government Press, Madras, 1915
Henry Morris, *Sir Arthur Cotton: Philanthropist Engineer*, The Christian Literature Society, London, Madras, Colombo, 1907
L.A. Cammiade, 'A Primitive Oil-Extractor from the Godavari District', *Man*, Vol. 32 (Apr., 1932), Royal Anthropological Institute of Great Britain and Ireland; p. 84, Stable URL: http://www.jstor.org/stable/2789318
Major General Sir Arthur Cotton, "Irrigation and Navigation in Connection with the Finances of India", Address delivered at the Calcutta Chamber of Commerce, May 7th 1863. Reprint from Calcutta Englishman, London, 1863
Report on the Land Revenue Settlement of the Upper Godavery District, Central Provinces, effected by Chief Commissioner's Office Press, 1869
Selections from the Proceedings of the Godavery District Association in 1912 and 1913, From the India Office Records, British Library, London.

Select Books, Articles, etc

A Century of Irrigation: Godavari-Krishna Deltas, 1859-1959, The Institution of Engineers (India), Andhra Pradesh, Hyderabad, 25th August 1959
Amit Prakash, Decolonisation and Tribal Policy in Jharkhand: Continuities with Colonial Discourse, *Social Scientist*, Vol. 27, No. 7/8 (Jul. - Aug., 1999)
Antonio Gramsci, *Selections from the Prison Notebooks*, edited and translated by Quintin Hoare and Geoffrey Nowell Smith, International Publishers, New York, 1971, 11th Print, 1992
Benjamin Weil, 'Conservation, Exploitation, and Cultural Change in the Indian Forest Service, 1875-1927', *Environmental History*, Vol. 11, No. 2 (Apr., 2006), pp. 319-43, Forest History Society and American Society for Environmental History, accessed through JSTOR, (http://www.jstor.org/stable/3986234)
Biksham Gujja, et al, eds, *Perspectives on Polavaram: a Major Irrigation Project on Godavari*, Academic Foundation, Delhi, 2006
Bimal Kanti Paul, "Perception of and Agricultural Adjustment to Floods in Jamuna Floodplain, Bangladesh", *Human Ecology*, Vol. 12, No. 1 (Mar., 1984), pp. 3-19
Carol Boyack Upadhya, "The Farmer-Capitalists of Coastal Andhra Pradesh", *EPW*, Vol. 23, No. 27 (Jul. 2, 1988), pp. 1376-1382
Cheryl Colopy, *Dirty, Sacred Rivers: Confronting South Asia's Water Crisis*, Oxford University Press, New York, 2012
Christophe and Elizabeth von Fürer-Haimendörf, *The Reddis of Bison Hills: A Study in Acculturation*, Macmillan and Co Ltd, London, 1945
Christophe von Fürer-Haimendörf, ed, *Tribes of India: The Struggle for Survival*, University of California Press, California, Los Angeles, 1982
Dahdouh-Guebas, S Collin, T Ravishankar, et.al, Analysing ethnobotanical and fishery-related importance of mangroves of the East-Godavari Delta (Andhra Pradesh, India) for conservation and management purposes, *Journal of Ethnobiology and Ethnomedicine* 2006, 2:24, Published: 08 May 2006, http://www.ethnobiomed.com/content/2/1/24 (Open Access)

Daniel Thorner, "Coastal Andhra: Towards an Affluent Society", *EPW*, Vol. 2, No. 3/5, Annual Number (February, 1967) - pp.241-252

David Ludden, 'Spectres of Agrarian Territory in Southern India', in Sanjay Subramanyam, ed, *Land, Politics and Trade in South Asia*, OUP, Delhi, 2004

———, "Development Regimes in South Asia: History and the Governance Conundrum", *EPW*, Vol. 40, No. 37 (Sep. 10-16, 2005), pp. 4042-4051

Emanuel Wallerstein, "The Ideological Tensions of Capitalism: Universalism versus Racism and Sexism", in Etienne Balibar, Wallerstein, *Race, Nation, Class: Ambiguous Identities*, Verso, London, 1991

G.N. Rao, *Constraints on Agricultural Growth in a Subsistence Economy – a Study of Godavari District 1860 – 1890*, CDS, Working Paper, No. 136, 1981

Historical and Polemical Documents of the Communist Movement of India, Vol. II (1964-1972), published by the Tarimela Nagi Reddy Memorial Trust, Vijayawada, 2008

H.M. Kasim, "Mangrove Ecosystem and its Relation to Fisheries", Discussion Papers, National Workshop on Restoration and Conservation of Mangroves through Participatory Mangrove Management, Rajahmundry, 12th February 2002

"India: Land Policies for Growth and Poverty Reduction", Report No. 38298-IN, July 9, 2007 (Agriculture and Rural Development Sector Unit, India Country Management Unit, South Asia Region)

Jayprakash Rao, "Konda Reddis in Transition", in Haimendörf, *Tribes of India: The Struggle for Survival*, University of California Press, California, Los Angeles, 1982

K. Balagopal, "Land Unrest in Andhra Pradesh: Impact of Grants to Industries", *EPW*, September 29, 2007 (pp.3906-3911)

K. Balagopal, "Land Unrest in Andhra Pradesh-III", *EPW*, October 6, 2007 (pp. 4029-4034)

Klara Feldes, Modernity in India: a Case Study of the Polavaram Dam Project, Unpublished PhD Thesis submitted to Humboldt-Universität zu Berlin, 2013

Kuntala Lahiri-Dutt, "Nadi O Nari: Representing the River and Women of the Rural Communities in the Bengal Delta", in Kuntala Lahiri-Dutt, ed, *Fluid Bonds: Views on Gender and Water*, Stree, Kolkata, 2006

M. Venkatarangaiya, ed, *The Freedom Struggle in Andhra Pradesh (Andhra), Vol. II (1906-1920 AD)*, The Andhra Pradesh State Committee Appointed for the Compilation of a History of the Freedom Struggle in Andhra Pradesh, 1969

M.L.K. Murthy, "Costly Mistake!", in Biksham Gujja, et.al, *Perspectives on Polavaram: a Major Irrigation Project on Godavari*, Academic Foundation, Delhi, 2006

Mahesh Rangarajan, *Environmental Issues in India: a Reader*, (Pearson-Longman, Delhi), 2007

P. Jaswantha Rao, "Tribals' Struggle against Land Alienation", *EPW*, Vol. 33, No. 3 (Jan. 17-23, 1998), pp. 81-83

Patrick McCully, *Silenced Rivers: the Ecology and Politics of Dams*, Orient Longman, Hyderabad, 1998

P. Sundarayya, *Telangana People's Struggle and its Lessons*, Foundation Books, Hyderabad (reprint, 2006)

Ramachandra Guha, 'Forestry in British and Post-British India: A Historical Analysis', in *Economic and Political Weekly*, Vol. 18, No. 44 (Oct. 29, 1983), pp. 1883-1885

Ramaswamy Iyer, ed., *Water and the Laws in India*, (Sage, Delhi), 2009

———, *Towards Water Wisdom: Limits, Rights, Justice, Harmony* (Sage, Delhi), 2007

———, River Linking Project: A Disquieting Judgment, *Economic & Political Weekly*, April 7, Vol. XLVII, No. 14, 2012

Robin Tennant-Wood, "Silent Partners: the Third Relationship between Women and Dammed Rivers in the Snowy Region of Amsterdam", in Kuntala Lahiri-Dutt, ed, *Fluid Bonds: Views on Gender and Water*, In Conjunction with National Institute for Environment (NIE) and Amsterdam National University, Canberra, Stree, Kolkata, 2006

Rukmini Rao, Tony Stewart, *India's Dam Shame: Why Polavaram Dam Must not be Built*, Gramya Resource Centre for Women, Secunderabad, 2006

S. Prabhakar, 'Availability and Utilisation of Godavari Waters for the Upland Areas of Andhra Pradesh', Proceedings of a National Seminar on Regional Identity and Articulation, Department of Geography, Osmania University, March 18-19, 2005

Shereen Ratnagar, "Our Tribal Past", *Social Scientist*, Vol. 31, No. 1/2 (Jan-Feb, 2003)

Sunil Vaidyanathan, Shayoni Mitra, *Rivers of India*, Niyogi Books, Delhi, 2011

T. Ravishankar, R. Ramasubramanian, D. Sridhar, N. Srinivas Rao, M. Maqbool and D. Ramakrishna, "Community Participation in Joint Mangrove Forest Restoration and Management", in S.K. Patnaik, H.N. Thatoi, eds, *Mangrove Conservation and Restoration: Proceeds of the National Workshop on Mangrove Conservation*, Bhubaneswar, 2001

Tim Robbins, *Connemara: Listening to the Wind*, Penguin, Ireland, 2006

William G. Robins, "Narrative Form and Great River Myths: the Power of Columbia River Stories", *Environmental History Review*, Vol. 17, No. 2 (Summer 1993)

Index

Names (Place, People, Associations, Political Parties, Species)

Themes/Issues/Concepts

*****